S0-DQX-396

Magnetic Recording

Magnetic Recording

Charles E. Lowman

Manager, Instrumentation Technical Writing, Ampex Corporation
Senior Instructor, Ampex Training Department

McGRAW-HILL BOOK COMPANY

New York St. Louis San Francisco Düsseldorf Johannesburg
Kuala Lumpur London Mexico Montreal New Delhi
Panama Rio de Janeiro Singapore Sydney Toronto

Library of Congress Cataloging in Publication Data

Lowman, Charles E
Magnetic recording.

Includes bibliographies.
1. Magnetic recorders and recording. I. Title.
TK7881.6.L69 621.389'32 72-4057
ISBN 0-07-038845-8

4 5 6 7 8 9 0 HDMM 7 6 5 4 3 2

The editors for this book were Tyler G. Hicks and Stanley E. Redka, the designer was Naomi Auerbach, and its production was supervised by George E. Oechsner. It was set in Caledonia by The Maple Press Company.

It was printed by Halliday Lithograph Corporation and bound by The Maple Press Company.

To Louvain

Contents

Preface *xiii*

Chapter 1. Why Magnetic Recorders **1**

Introduction *1*
Some Ways in Which Magnetic Recorders Are Used *2*
Aerospace Acquisition of Data *2*
Time-base Expansion and Contraction *3*
Frequency-division Multiplexing *3*
FM/FM Systems *4*
Magnetic Recording Devices in Science and Medicine *4*
Electrocardiograph *5*
Electroencephalograph *6*
Radioactive Tracers and Transducers *6*
Underwater Seismology *7*
The Tape Recorder as a Filing Clerk *7*
The ERMA System *7*

Chapter 2. The Development and Advantages of Magnetic Recording **9**

The Development of Magnetic Recording *9*
Inherent Advantages of Magnetic Recording *14*
References *15*

Chapter 3. Introduction to the Basic Elements of the Magnetic Recorder 16

General *16*
Functions of the Basic Elements *17*
The Magnetic Head *17*
Magnetic Tape *19*
The Tape Transport *20*
The Record Amplifier *21*
The Reproduce Amplifier *23*
REFERENCES *25*

Chapter 4. The Theory of Magnetism 26

General *26*
Ferromagnetism *26*
The Process of Magnetization *28*
REFERENCES *30*

Chapter 5. Magnetic Heads 31

General *31*
Construction of a Magnetic Head *31*
Longitudinal (Analog) Head *31*
Wideband Rotary Heads *39*
REFERENCES *44*

Chapter 6. Magnetic Tape 45

Introduction *45*
Design Considerations *45*
Manufacture of Magnetic Tape *49*
Oxide Conversion *49*
Mix Preparation *50*
Base Film *52*
The Coating Process *52*
Particle Orientation *54*
The Drying Process *55*
Tape-surface Polishing *55*
Slitting *56*
Testing *56*
Magnetic Tape Applications and the Parameters Required to Meet Them *56*
Audio Tape *56*
Longitudinal-instrumentation Tape *57*
Computer Tape *59*
Rotating-head Video and Instrumentation Tape *60*
REFERENCES *61*

Chapter 7. The Recording and Reproducing Process 63

How Recording Takes Place *63*
How Reproduction Takes Place *65*

Details of the Recording Process 67
The Hysteresis Loop 67
The Need for Bias 69
Recording Losses 75
Demagnetization 75
Bias Erasure 75
Eddy Currents 76
Details of the Reproduce Process 76
Reproduce Losses 77
Gap Losses 77
Head-azimuth Loss 78
Separation Loss 80
Eddy-current Loss 81
Surface Loss—Magnetic Medium 81
Long-wavelength, Low-frequency Losses 82
The Need for Equalization 85
Amplitude Equalization 86
Phase Equalization 89
REFERENCES 90

Chapter 8. The Transport **92**

Selecting a Magnetic Recording System 92
The Ideal Transport 93
Transport Components 94
The Baseplate Assembly 94
Longitudinal Drive Systems 94
Drive Motors—The Printed-circuit Motor 99
Capstan Servo Systems Used with Longitudinal-tape Transports 100
Tape Drive and Servo Systems for Rotary-head Transports 107
Capstan Drive Systems 107
Rotary-head Drive Systems 107
Capstan-drive Servo Systems 110
Rotary-head-drum Servo Systems 111
Helical-scan Drive and Servo Systems 112
Why Single and Dual Heads? 112
Compatibility between Helical-scan Recorders 115
Capstan Drive Systems for Single-head Helical Scan 117
Capstan Drive Systems for Dual-head Helical Scan 117
Servo Systems Used in Helical-scan Recorders 118
Disc Recording and Reproducing Systems 121
In-contact Head-to-disc System 121
Television Disc Systems 124
Instrumentation Disc-recorder Servo Control 125
Spaced-head Systems 126
Spaced-head Servos 127
Reel- or Tape-storage Systems 128
Functions 128
Reel Servo and Tensioning Systems 128
Tape Storage and Servoing Using Vacuum and Photoelectric Sensing 131
Reel Servo and Tensioning Systems Using Mechanical Control 134

Sensing the End of the Tape *136*
Broken-tape Sensing *137*
REFERENCES *139*

Chapter 9. Direct Record and Reproduce Signal Electronics 141

General *141*
Low-band Direct Record/Reproduce Electronics—to 100 kHz *141*
Typical Voice-log System *142*
Typical Audio and Cue-track Systems *143*
Low-band Reproduce Amplifiers *144*
Intermediate-band Direct Record/Reproduce Electronics—to 600 kHz *145*
Single-speed Reproduce Electronics and the Process of Equalization *147*
Amplitude Equalization *147*
Phase Equalization *148*
Reproduce Circuitry after Equalization *151*
Multispeed Reproduce Electronics *151*
1.5-wideband Direct Record/Reproduce Electronics—to 1.5 MHz *151*
2.0-wideband Direct Record/Reproduce Electronics—to 2 MHz *153*
Multiband Wideband Direct Systems *154*
2.0-wideband Direct Systems—Single-band *155*
Record System *155*
Reproduce System *157*
Cosine Equalizer—Aperature Corrector *157*
Limitations and Advantages of the Direct Record/Reproduce System *159*
REFERENCES *160*

Chapter 10. Frequency-modulation Record and Reproduce Electronics .. 161

General *161*
Intermediate-band FM Record/Reproduce Electronics *162*
Modulator Types *163*
Demodulator Types *166*
Wideband Group I FM Record/Reproduce Electronics *168*
Multispeed Modulator *169*
Multispeed Demodulator *172*
Wideband Group II Record/Reproduce Electronics *174*
Multispeed, Multiband Modulators *175*
Multispeed, Multiband Demodulators—Recirculating-charge-dispenser Type *177*
Advantages, Limitations, and Major Applications for the FM Recording *180*
REFERENCES *181*

Chapter 11. The Television Recorder 182

Introduction to the Television Recorder *182*
The Quadruplex Video Recorder *183*
The Quadruplex Head *183*
Female Guide *184*
Control-track Head *187*
Audio and Cue Erase Heads *187*
Tape-speed Control *188*

Timing Rings and Tachometer Systems *189*
The Three Servo Systems of Video Recorders *190*
Capstan and Head-drum Systems *190*
Capstan Servo—Record Mode *195*
Control-track Signal in Record *195*
Head-tip Penetration—Female-guide Horizontal-sync-pulse Relationship *197*
Venetian-blind Effect *197*
Scalloping *198*
Quadrature Error *199*
Record Process *200*
Modulators *201*
Record Process Block Diagram *202*
EE Signal *203*
Record-system Accessories *204*
Playback Process *204*
Preamplifier *205*
Channel Switching and Playback Equalization *206*
Limiting before Demodulation *208*
Demodulator *208*
Auto-chroma and Dropout Accessories *208*
Auto-chroma *208*
Dropout Sensor *209*
Time-base Error *211*
Correction with Amtec *211*
Correction with Colortec *212*
Color Processor—Proc Amp *213*
Capstan Servo—Playback Mode *214*
Operation of Intersync—Preset and Normal Modes *216*
Operation of Intersync—Vertical, Horizontal, and Automatic *218*
REFERENCES *222*

Chapter 12. Cassette and Cartridge Systems **223**

Introduction to the Cassette *223*
Introduction to the Cartridge *224*
Uses of the Cassette Recorder *225*
The Audio Cassette *226*
The Digital Cassette/Cartridge *230*
Cassette Tape Drive Systems *231*
The Video Cassette/Cartridge Recorder for Television Broadcasting *233*
The Video Cassette/Cartridge Recorder for Home and Office *238*
The Videocassette System *238*
The Instavideo System *240*
REFERENCES *242*

Appendix A. Glossary of Audio and Instrumentation Terms **243**

Appendix B. Glossary of Television Terms **263**

Index *275*

Preface

As the magnetic recording industry continues to expand, the demand for trained technicians with a workable knowledge of the basic principles involved with magnetic recording has grown also. The author, formerly the senior instructor for the instrumentation Training Department of the Ampex Corporation, was hard-pressed to find an adequate reference book that contained the basic principles of magnetic recording. To meet this need, he coauthored a company-sponsored publication titled "General Magnetic Recording Theory." It was an instant success, and copies of this publication are used all over the world as a reference on tape recording and for training tape-recorder technicians.

Unfortunately, that publication has several limitations with respect to current tape recorders. It was written over ten years ago, and much of the information is not up-to-date. In addition, it does not contain information on cassette/cartridge, rotary-head, and helical-scan recording, and does not delve into the servo-control and signal record/playback circuitry. To correct these limitations, a new book has been written. This book includes a discussion of the most up-to-date circuitry, the newest techniques, and the latest innovations in the magnetic recording field. It also enlarges on the basic descriptions contained in the original coauthored publication.

It presents the basic concepts of magnetic recording in easy-to-understand language, with a minimum of mathematics and many simplified sketches. Although this book presents no new theories on magnetic recording, it does provide a convenient reference on the basic concepts of magnetic recording and related equipment. It is designed for use by the layman, student technician, or engineer that is entering the field of audio recording, broadcast and closed-circuit television, instrumentation recording, and computer data storage. It also covers areas of interest associated with magnetic recording, such as tape manufacturing and design and construction of magnetic heads. Definitions of terms used for audio, instrumentation, and television recording are included as appendixes.

Grateful acknowledgment is made for the generous assistance of Mathew McGillicuddy, Loren Clark, Wayne Eaton, Thomas Oliver, James Alford, Robert Kane, and James Lawson, who made available technical data or read sections of the manuscript and offered invaluable suggestions.

Thanks are also due to the following manufacturers and to the entire magnetic recording industry for assistance in providing information that helped to make this book complete: American Telephone and Telegraph Co., Long Line Dept.; Ampex Corporation; Data Disc, Inc.; Minnesota Mining and Manufacturing Co.; Philips Corporation of Holland; Printed Motors, Inc.; Radio Corporation of America; and Sony Corporation of America.

Finally, general acknowledgment is made to all the technicians and engineers whose published achievements and practical developments have made the magnetic recording industry what it is today.

Charles E. Lowman

Magnetic Recording

CHAPTER ONE

Why Magnetic Recorders

INTRODUCTION

Nearly everyone is familiar with the uses of a magnetic tape recorder for audio purposes, since the recording of speech and music for entertainment purposes is a common fact. Undoubtedly, the progress made in this field has been dramatic. But even more spectacular, and much less understood and appreciated, are the advances that have been made in magnetic recording for instrumentation, digital, and video purposes.

These types of magnetic recording concern themselves with measurements, data acquisition and storage, data translation, and data conversion and analysis. They are an important part of recording, editing, storage, and public presentation of entertainment material over the television networks. They play an important part in modern closed-circuit television, computers, and industrial machine control. They are an integral part of our research programs in medicine, space, telemetry, and undersea exploration. They play a vital role in mathematics, and in our culture in general. With the aid of the magnetic recorder, the transportation industry has given us safer, more comfortable, and easier-handling automobiles, buses, trains, and airplanes. Millions of subscribers to numerous weekly and monthly magazines and journals get their copies on time because a magnetic recorder was used to print the address labels. The purchase of an airline ticket and the rapid confirmation of space on connecting flights are served by magnetic recording.

SOME WAYS IN WHICH MAGNETIC RECORDERS ARE USED

Aerospace Acquisition of Data

Often the components of a data acquisition system are separated by large distances. This could be the separation between a rocket and its receiver station. It could be the distance from the data center to the data analysis center. The transmission of information can be done by simply carrying a reel of magnetic tape from one point to another or via telephone line, microwave links, or a telemetry system. In Fig. 1-1, transducers are inserted throughout the test vehicle to measure the stresses and strains, temperature, pressures, and fuel-flow information, etc. These transducers are normally wired to an encoder, in which their outputs are combined. If the test vehicle is nonrecoverable, the data will be transmitted to various airborne or ground receiving stations. However, as backup, it is usual to include a magnetic recorder in the vehicle. This will permit the ground operator to extract set

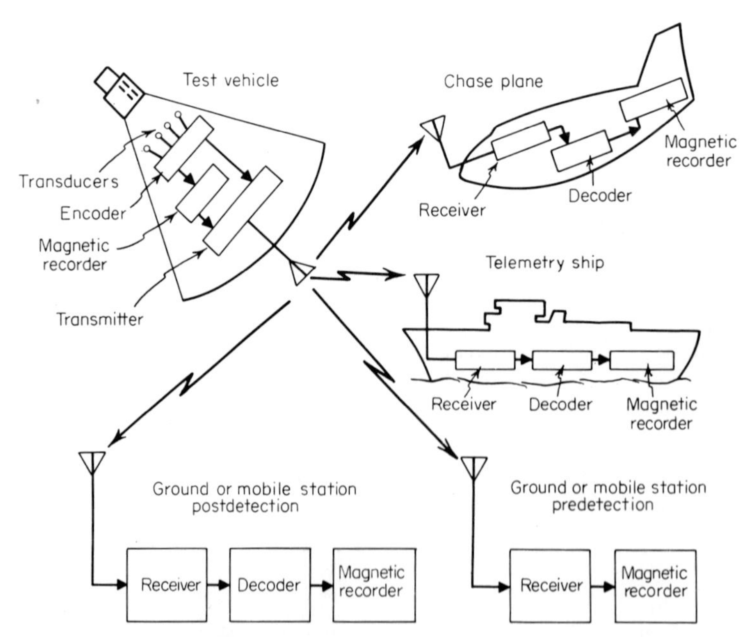

Fig. 1-1 Aerospace acquisition of data via telemetry.

amounts of information from the vehicle upon command. In this fashion, repeat information may be obtained in place of weak or bad telemetry transmissions that are momentary in nature.

Time-base Expansion and Contraction

On occasion, the bandwidth of the data is very much wider than the transmission line. For example, a telephone line can handle information up to about 3,000 Hz, whereas the data could be as much as 100,000 Hz. To solve this particular problem, a magnetic tape recorder could be used. The data are recorded at a high speed, then reproduced at a slow speed into the telephone lines. This speed reduction reduces the frequency of the data from the magnetic tape. On the other end of the telephone link, a slow-speed magnetic tape recorder would be used to record the incoming data from the line. When the data are played back at the analysis center, a higher rate of tape speed is used to make the time base the same as it was when the data were originally recorded. Thus a magnetic tape recorder may be used to overcome the limitations of a low-frequency transmission line. Conversely, if the data were relatively low frequency and the transmission link were capable of handling information of much higher frequency (as through a microwave link), a very small proportion of the microwave-link capability would be used if the data were transmitted in its original form. In this case the original data would be recorded on a magnetic tape recorder at low speeds and played through the microwave link at much higher speeds. Many hours of recording could be transmitted in minutes across the microwave link, resulting in a reduction of cost and full utilization of the capabilities of both the magnetic tape recorder and the microwave link.

Frequency-division Multiplexing

When thousands of measurements are to be made, the object of instrumentation recording is to encode, package, and mix all these thousands of measurements in such a way that a magnetic recorder can be used most effectively. There are various ways of doing this, one of which is frequency-division multiplexing. Referring to Fig. 1-2, it can be seen that if the output of 18 transducers were fed into 18 separate voltage-controlled oscillators (VCO), each of the VCOs would act as a carrier for its transducer input. The carrier frequencies are normally chosen to conform with IRIG Standards (Inter-Range Instrumentation Group). The output of each VCO can be treated as a separate entity, since it occupies a discrete band of frequencies, much as each one of the radio stations in any one area occupies its own band of frequencies. The output of all the VCOs could then be combined in a mixer stage

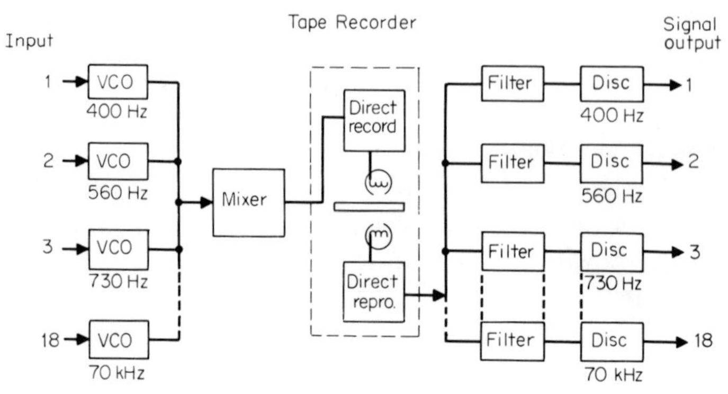

Fig. 1-2 Frequency-division multiplexing.

and fed to a tape-recorder channel. In this way 18 separate signals from 18 transducer sources can be recorded on a single track of a magnetic tape recorder. Upon reproduction, the multiplex signals are fed to a number of filters, each of which accepts the band of frequencies corresponding to the individual VCO output. From the filters, the signals are fed to individual FM discriminators, where the data are extracted and sent to the readout devices.

FM/FM Systems

When a magnetic tape recorder has a wide bandwidth, by using a system of translators, it is possible to multiplex many IRIG groups of 18 carriers, each group occupying an 80-kHz bandwidth. Using a 2-MHz recorder, 20 groups of 18 signals may be recorded on each of the tracks of the recorder. If the system contains 14 tracks, a total of 5,040 individual signals can be recorded at one time, provided that the frequency requirements of the output of each transducer are low (Fig. 1-3). Other multiplex systems are used when the frequency requirements of the transducers are wide.[1]

MAGNETIC RECORDING DEVICES IN SCIENCE AND MEDICINE

There are many applications for magnetic recording devices in science and medicine, although a particular device may take different forms, such as memory cores, discs, drums, loops, and reels.

[1] Refer to IRIG Standards, Document 106–71, and H. L. Stiltz, "Aerospace Telemetry," Prentice-Hall, Inc., Englewood Cliffs, N.J., for more details on optional multiplex systems.

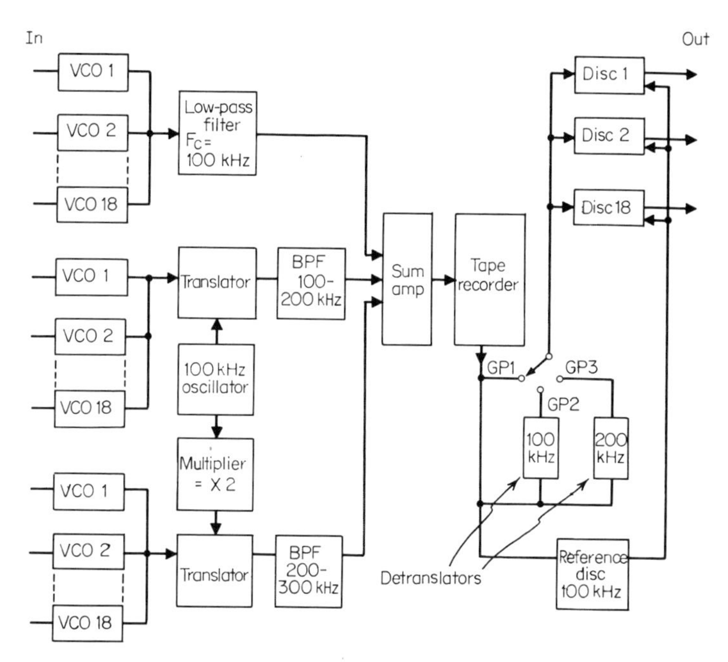

Fig. 1-3 Typical translated FM/FM system.

In the Apollo program, close watch is kept on the body functions of the astronauts through the combination of medicine and electronics. This combination is called medical electronics. It might be defined as the application of electronic techniques to medicine, including diagnosis, laboratory work, and therapy.

Electronics may be used to help find out what is going on inside our bodies, since the body itself acts as a generator of signals. The brain, for example, generates electrical signals, as does the heart and other muscles. The heart and the lung generate audible signals. The kidneys and the thyroid do not produce either sound or electrical potentials, but they can be tricked into making signals that can be monitored. As the mechanical action of the heart produces sounds, so also does it set up electrical-potential differences that can be measured each time it contracts and relaxes. Whenever electrical potentials or audible sounds are produced, they can be recorded on a magnetic recorder.

Electrocardiograph

The electrical waveform produced by the heart action is familiarly known as the electrocardiograph, or EKG. EKG is widely used to study heart action. Figure 1-4 illustrates the setup generally used for these measurements.

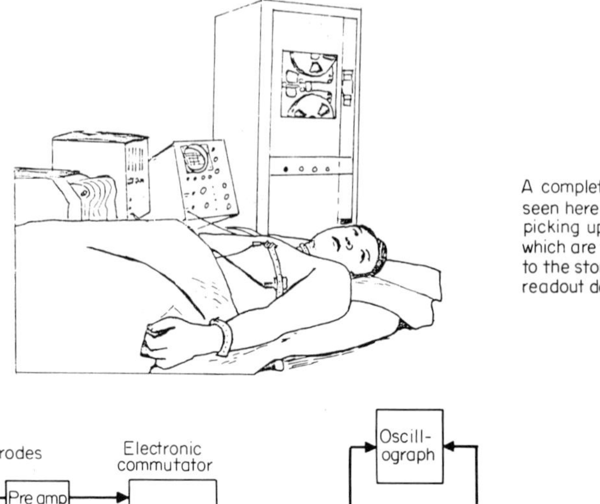

A complete EKG system can be seen here. Note the electrodes picking up the heart potentials which are then amplified and fed to the storage system and visual readout devices

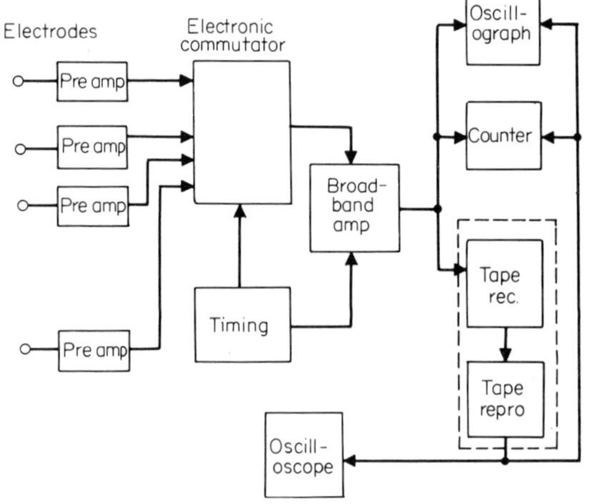

Potentials generated during contraction and relaxation of heart muscle.

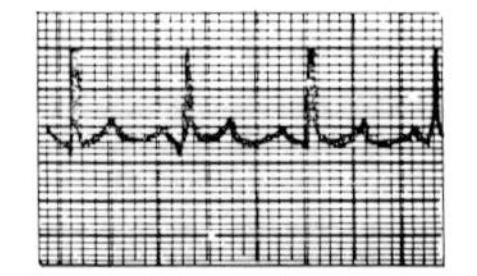

Fig. 1-4 Electrocardiography.

Electroencephalograph

The brain, like the heart, gives out electrical signals. These signals can be monitored with an electroencephalograph, or EEG. The EEG records from pairs of electrodes placed over the scalp. The exact waveshape that will be produced will depend on the location of the pairs of electrodes and the activity and the condition of the brain. When information of this type is stored on a magnetic tape recorder, the doctor can play back the information for leisurely study. He can speed it up or slow it down. He can play it as many times as he wishes so that nothing is missed.

Radioactive Tracers and Transducers

When an organ in a body does not itself give off electrical energy, radioactive tracers are often used. The rate at which the tracer is absorbed by the organ is used to determine the condition of that organ, as in thyroid examinations. A method such as this is often used to detect cancer. Likewise, patients are often given small electronic transducers to swallow. These tiny combination transducers and radio transmitters are used to report the condition of the patient's stomach and intestines, etc. Many persons are living today because of this new and exciting field of medical electronics.

UNDERWATER SEISMOLOGY

Another application of magnetic recording is underwater seismology. In this field scientists are enabled to calculate, to a very accurate degree, the depth of the ocean floor and the types and thicknesses of the sedimentary formations to be found there.

Explosive charges are set off at predetermined depths from a moving ship. Other ships stationed at discrete intervals pick up the sounds of the underwater explosion with hydrophones. The timing of the sounds reflected from the ocean floor and the various rock layers is compared with that of the original explosion. From this relationship calculations can be made regarding depth, type of material, and thickness of the various formations.

THE TAPE RECORDER AS A FILING CLERK

We have all been overwhelmed from time to time by the shear mass of paperwork that has been fed to us. Insurance companies have this problem on a large scale. They are hard-pressed to take care of the many thousands of transactions that occur each day. A misplaced data card could create havoc. Using a magnetic tape recorder as a filing clerk, the records can be changed quickly. They are always correct, and they reflect the business just transacted. Records on tape save time, because the master file can be brought up to date in a single day. This new filing system also saves space, because the recorders take up less room than paper files or punch cards.

THE ERMA SYSTEM

Banks also have turned to the tape recorder and the computer to keep up with the increase in today's business transactions. One of these banking systems, the Bank of America's ERMA (electronic method of accounting) system, is shown in Fig. 1-5. ERMA takes thirty-two millionths of a second for each step in processing a checking account. It can do all sorts of jobs, such as sorting, reading, posting, and record

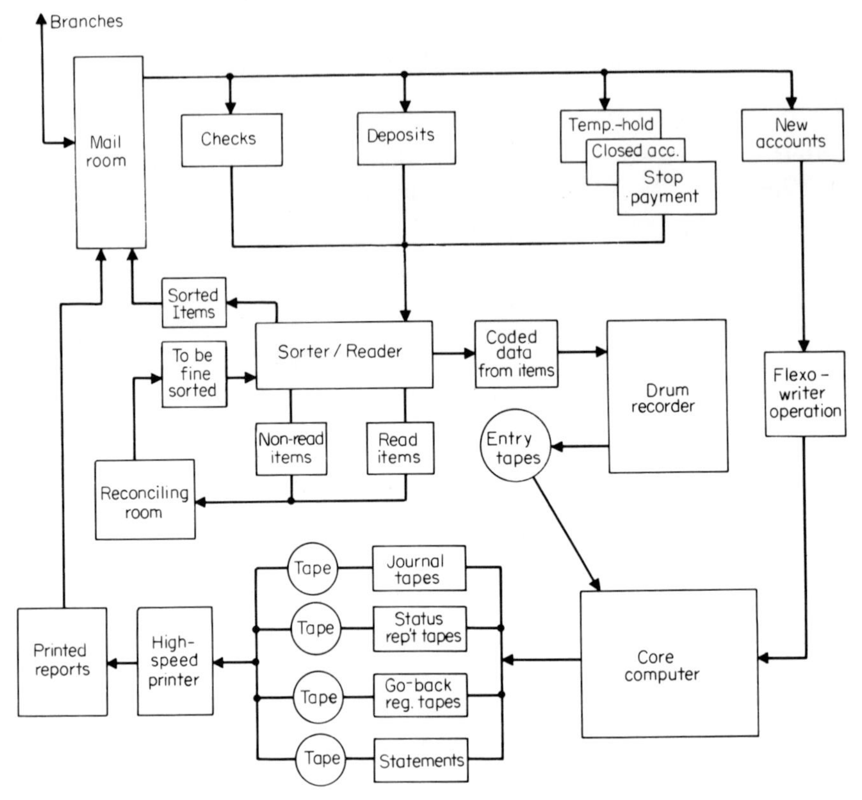

Fig. 1-5 ERMA system.

keeping of 110,000 accounts in slightly over three hours. Compared with a human clerk's 245, ERMA can post 33,000 accounts per hour.

Magnetic recording is an exciting, ever-expanding field. Its growth is continuing at an almost unbelievable rate. As such, it demands of the industry a rate of development that ten years ago was believed impossible.

This book describes the development of magnetic recording, its inherent capabilities, and the principles of operation of magnetic recorders.

CHAPTER TWO

The Development and Advantages of Magnetic Recording

THE DEVELOPMENT OF MAGNETIC RECORDING

Magnetic recording did not just happen, but rather it was the marriage of many sciences such as acoustics, magnetics, electromagnetism, and electronics, in countless years of research and study. Its development is closely related to man's attempt to capture and preserve the sound of the human voice.

As early as the third century B.C., Heron of Alexandria developed a device that could imitate the cries of animals. Much later, in A.D. 1791, DeKempelein of Vienna published a description of a "talking machine that could be made to speak short sentences." Neither of these devices recorded sounds, but rather were used to imitate them.

There is no record of anyone recording the human voice until a young French typographer, Leon Scott, deposited a paper with the Academy of Sciences, in 1857, in which he described his invention, the phonautograph. His device consisted of a barrel-shaped plastic speaking horn. The upper end was left open, while the lower end was fitted with a 4-in-diameter brass tube, across which was stretched a flexible membrane. A stiff pig's bristle was fastened to the outside of the membrane to act as a stylus or pen. A smoked-paper cylinder was rotated beneath

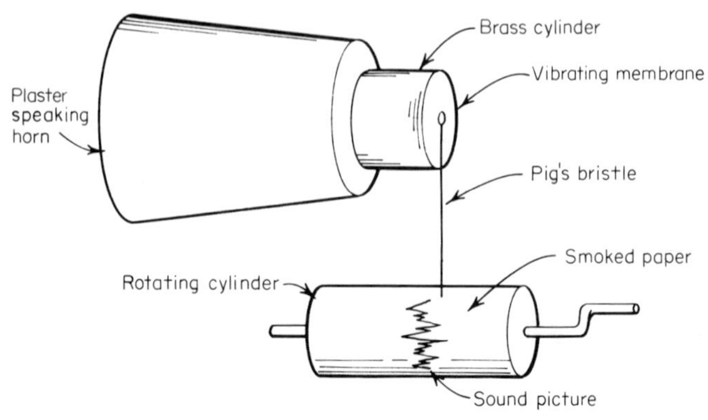

Fig. 2-1 Principles of Scott's phonautograph.

the pig's bristle. When sounds were directed into the horn, the membrane and bristle moved back and forth and traced the waveforms as a wavy line on the smoked-paper cylinder. There was no way in which the wavy lines could be reproduced as sound, but for the first time man was able to see "sound pictures of his voice."

Refer to Fig. 2-1 for an illustration of the principles of Scott's phonautograph and the first side-to-side, or lateral, recording of voice.

From this simple start came the development of the phonograph (Thomas Edison, 1877). In essence, Edison's invention consisted of a membrane to which was attached a steel stylus and a cylinder covered with tinfoil. At first the membrane did the work of the human ear, vibrating up and down in response to sound waves. As it vibrated it transmitted these movements via the stylus to the foil-covered cylinder. Edison rotated the cylinder, as well as moving it laterally beneath the stylus. As the membrane moved up and down, the stylus made a series of spiral "hill-and-dale" grooves in the foil surface. When the stylus was made to travel over the spiral grooves a second time, it made the membrane vibrate up and down in response to the depressions in the grooves. Now the membrane acted like the human vocal cords. By vibrating up and down in obedience to the motion of the stylus it re-created the original sounds.

It was difficult to remove the tinfoil from the cylinders or to replace it without distorting the material and injuring the indentations. Thus it was necessary to have a separate cylinder, screw, and crank for every new record. Later, Edison improved his recording process by using wax in place of tinfoil cylinders. The grooves were cut with an agate or sapphire point. During reproducing, the up-and-down movement of the sapphire was passed to the membrane via a series of weights and levers.

Some ten years later, another inventor, Emile Berliner, developed the flat recording disc with a lateral-cut groove. This technique gave birth to the gramophone, or record player, as it is called today. A "morning-glory" horn was added to improve the sound amplification, doing away with the need for cumbersome headphones. These improvements, plus Berliner's process for the mass production of disc records through the shellac method, ushered in a new era of home entertainment.

Even with all their improvements, the phonograph and gramophone had definite limitations. These could be traced to the lack of power of the human voice or musical instrument to drive the recording stylus. Often the changing vibrations of the voice or musical instruments were so feeble that many tone values were lost. Even under the best conditions the records often sounded thin and reedy. It was necessary, when recording, for the artist and musicians to gather as close as possible around the recording horns. The bands were kept small, and played as loud as they could, while the singer sang at the top of his voice. Expression as we know it today could hardly exist under such conditions.

All this changed when electronic amplifiers became available as a result of the discovery of the vacuum tube. Now sound waves could be amplified thousands of times. The making of records was a costly process, however. A single mistake during cutting, and the record had to be started all over again. Editing of a record was a slow, tedious job and rarely completely satisfactory.

In 1948, with the introduction of the first commercial magnetic tape recorder, the making of a record became very much simpler. Now it was possible to make the master recording on a plastic-backed magnetic tape that could be cut, spliced, dubbed, and re-recorded without affecting the quality of the final result. For making high-quality disc records, it could be used thousands of times without serious deterioration. For cheaper records, the magnetic tape master was used to produce a disc from which thousands of stamped records were made. Thus, in this age of modern miracles, the magnetic tape recorder has become inseparably interlinked with the science of acoustics.

In its early stages of development, magnetic recording was closely tied to telegraphy and telephony. It started with Oberlin Smith's article in *The Electrical World*, in 1888, wherein he suggested the possibility of the use of permanent magnetic impressions for recording sound. Smith visualized a cotton thread impregnated with iron particles that could be magnetized in accordance with an undulating current created by a microphone. Although Smith theorized the feasibility of magnetic recording, he never put his theories to practical use. That was to wait for the Dane, Valdemar Poulsen (1869–1942), some ten years later.

There is no record that Poulsen saw or heard of Smith's theories; therefore full credit for the invention of the magnetic recorder is given to Poulsen alone. Poulsen was the first to record and reproduce sound through the orientation of magnetic domains. He filed his first patent in Denmark on Dec. 1, 1898, under the title "Methods of Recording and Reproducing Sound or Signals." On Nov. 13, 1900, he was also granted a patent in the United States for a device titled "For Effecting the Storing Up of Speech or Signals by Magnetically Influencing Magnetized Bodies." Poulsen's invention was called the telegraphone. When he entered it in the Paris Exposition of 1900, it created a terrific stir among the scientific and technical bodies. In fact, it won the grand prize of the Exposition.

A sketch of Poulsen's drawings, which he submitted with his patent application, is shown in Fig. 2-2. In this figure the steel wire (*g*) was wound in a spiral groove of a large brass cylinder (*d*). The wire rested against two pole pieces of an electromagnet (*p*). The cylinder was kept stationary, while the electromagnet, or recording head, was rotated. As the recording head moved along the wire, it magnetized the wire by an amount corresponding to the voice current. On reproduction, the electromagnet was used again, but now as a reproduce head. The output of the head was fed to a telephone receiver. The varying levels of magnetization of the wire caused a varying amount of current to flow in the electromagnet (reproduce head) and telephone receiver, producing sound.

Poulsen's early recorders had extremely low signal levels and high distortion. It was not until 1909 that he discovered that the addition of a small amount of properly polarized direct current fed to the recording head improved the reproduction of the signal. The adding of dc bias to an incoming signal is the oldest electronic method of improving recording linearity. Although this technique gives a relatively low signal-to-noise ratio, the simplicity of the circuit makes it useful in some special cases even today. A great number of portable magnetic tape recorders, particularly those selling for under $40, use dc bias because the simplicity of construction of the bias circuit reduces cost, weight, and size.

One of the primary objections to Poulsen's early recorder was the lack of amplification. This lack was solved with the invention of the vacuum tube and the vacuum-tube amplifier. Unfortunately, the amplifier not only amplified voice signals, but also background noise. Therefore a search was carried on for some means to improve Poulsen's low signal-to-noise ratio. The U.S. Navy, interested in using magnetic recorders for high-speed transmission of telegraph signals, sponsored investigations into various methods of improvement.

Under such sponsorship, William L. Carson and T. W. Carpenter made the next great discovery in magnetic recording. In August 1927

No. 661,619. Patented Nov. 13, 1900.

V. POULSEN.

METHOD OF RECORDING AND REPRODUCING SOUNDS OR SIGNALS.

(Application filed July 8, 1899.)

(No Model.) 3 Sheets—Sheet 1

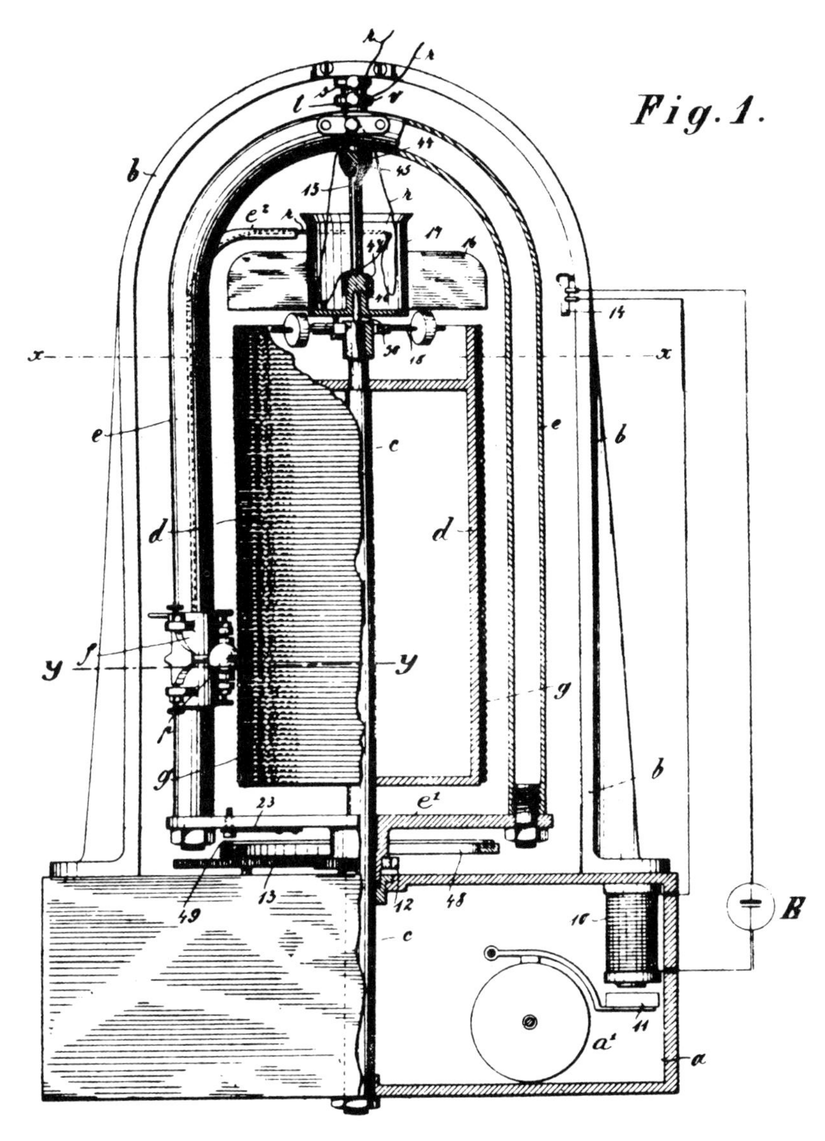

Witnesses:

Waldo M. Chapin

Inventor:

Valdemar Poulsen

by

Atty.

Fig. 2-2 Valdemar Poulsen's sketch of the telegraphone.

they were issued U.S. Patent 1,640,881, in which they described their newly discovered ac-biasing method. The ac-biasing technique effected such a basic improvement in the quality of magnetic recording that it is today in universal use in high-quality magnetic recorders.

Today, the magnetic tape recorder is to be found in computers; in entertainment, both audio and visual (music and television); and in an ever-expanding field called instrumentation (the recording of measurements from precision instruments used in aerospace, transportation, rocketry, undersea exploration, and laboratory and medical research, to name a few).

INHERENT ADVANTAGES OF MAGNETIC RECORDING

In the recording/reproducing business, strip-chart and hot-stylus recorders, recording oscillographs, phonograph records, photographic films, galvanometers, and many other types of recording devices have been used to preserve the output from a transducer. No single one of these devices has all the advantages of the magnetic tape or disc recorder as listed below:

1. Wide frequency range. Magnetic tape or discs permit the recording of information from direct current up into the multimegahertz range.
2. Very wide dynamic range. Magnetic tape or discs permit a range of recording in excess of 40 dB, giving accurate and linear recording from full scale down to 0.3 percent full scale.
3. Low inherent distortion. When an overload occurs with magnetic tape, it occurs gracefully, as contrasted with many electronic devices.
4. Immediate playback. No time is lost in processing magnetic tape, cores, or discs. Recordings are available for immediate playback.
5. Multichannel recording. Thousands of bits of information may be recorded simultaneously on a single width of tape using various multiplexing techniques. Additionally, accurate time and phase relationships can be maintained between these simultaneous signal channels.
6. Very high density storage. Several million bits of data can be contained on a single reel of tape or a single magnetic disc.
7. Signal is in its electrical form. The data are preserved in electrical form, so that the event can be re-created at any time. Such a format lends itself to automatic reduction of data.
8. Economic advantage. Magnetic tape, cores, or discs can be erased and reused many times.
9. Repeated playback. Magnetic tape or discs can be played back thousands of times to provide complete recovery of data (e.g., delayed television presentations of taped programs to suit area scheduling).

10. Time-base altering. Magnetic recordings permit the recording of data at one speed and its reproduction at another, thus providing the ability to alter the time base, which no other medium can do. This altering of the time base permits events to be re-created either faster or slower than they actually occur, with the resulting multiplication or division of all the frequencies involved.

REFERENCES

1. Smith, O.: "Some Possible Form of Phonograph," *Elec. World,* vol. 12, pp. 116–117, September, 1888.
2. Poulsen, V.: The Telegraphone: A Magnetic Speech Recorder, *Electrician,* vol. 46, pp. 208–210, November, 1900.
3. Lindsay, H., and M. Stoaroff: Magnetic Tape Recorder of Broadcast Quality, *Audio Eng.,* vol. 32, pp. 13–16, October, 1948.
4. Begun, S. J.: "Magnetic Recording," Rinehart & Company, Inc., New York, 1949.
5. Ginsburg, C. P.: Interchangeability of Video Recorders, *J. SMPTE,* November, 1958.
6. Anderson, Charles E.: Signal Translation through Videotape Recorder, *J. SMPTE,* vol. 67, no. 11, pp. 721–725, 1958.
7. Lowman, C. E.: The Magnetic Tape Recorder/Reproducer and Concept of Systems Used for Recording FM Analog Test Data," vol. 1 of "Fundamentals of Aerospace Instrumentation," Instrument Society of America, Pittsburgh, 1968.

CHAPTER THREE

Introduction to the Basic Elements of the Magnetic Recorder

GENERAL

All magnetic recorder/reproducer systems consist of five essential elements:

1. Magnetic heads
2. Magnetic tape
3. Tape transport
4. Record amplifier
5. Reproduce amplifier

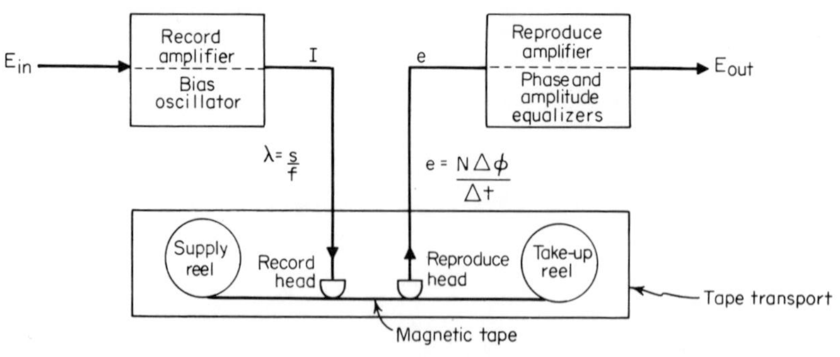

Fig. 3-1 The basic elements of a magnetic tape recorder/reproducer.

Sometimes only a portion of the total system is used, i.e., a record-only or reproduce-only system. But often a magnetic recorder system is made more complex by the addition of accessory equipment such as voice monitors, display cathode-ray or meter monitors, remote controls, etc. Figure 3-1 shows the typical magnetic recorder system and the relationship of one basic element to the other.

THE FUNCTIONS OF THE BASIC ELEMENTS

The Magnetic Head

Magnetic heads may be divided into three categories:

1. The record head
2. The reproduce head
3. The erase head

The Record Head Basically, the function of the record head is to change the current I produced by the record amplifier into magnetic flux ϕ. As record current varies, so also will the magnetic flux. The change of ϕ will be affected by both the amplitude and the direction changes of the record current. The changing current through the head produces a proportional changing magnetic force H in the head.

$$H = I \sin \omega t \tag{3-1}$$

where H = magnetic force, oersteds
I = maximum current
$\omega = 2\pi f$
t = time, sec

The magnetic force H in turn produces magnetic flux ϕ in the tape. Unfortunately, there are some losses to be considered in the conversion process. These are (1) the hysteresis effect of the head and tape materials (lagging or persistence of action in magnetic substances), (2) eddy-current losses in the record head, and (3) partial demagnetization. Thus there will always be less magnetic flux left in the tape, ϕ_r, than was originally produced by the magnetic force H. This remaining magnetic flux is called remanent flux ϕ_r. Thus, substituting in Eq. (3-1),

$$\phi_r = KI \sin \omega t \tag{3-2}$$

where ϕ_r = remanence flux
K = conversion constant required due to losses
$\omega = 2\pi f$
t = time, sec

The size (wavelength) of the signal recorded on magnetic tape will be a function of tape speed and frequency. Thus

$$\lambda = \frac{s}{f} \tag{3-3}$$

where λ = wavelength, in.
s = tape speed, ips
f = frequency, Hz

The Reproduce Head The basic function of the reproduce head is to change the magnetic field pattern found in the magnetic tape into a voltage e. The reproduce head will act as a miniature generator following Faraday's law, i.e.,

$$e = N\frac{d\phi}{dt} \tag{3-4}$$

where e = instantaneous voltage
N = number of windings around reproduce head core
d = symbol for change
ϕ = magnetic flux
t = time, sec

The voltage created in the reproduce head follows a 6 dB/octave curve (within certain limits). By this it is meant that as the frequency doubles, so also will the value of voltage output from the head. Thus, if a constant level of input were fed to the record amplifier at increasing frequencies, the output of the reproduce head would not be constant, but rather would increase at a 6 dB/octave rate (Fig. 3-2).

The deviation from the straight line of the 6 dB/octave curve is caused

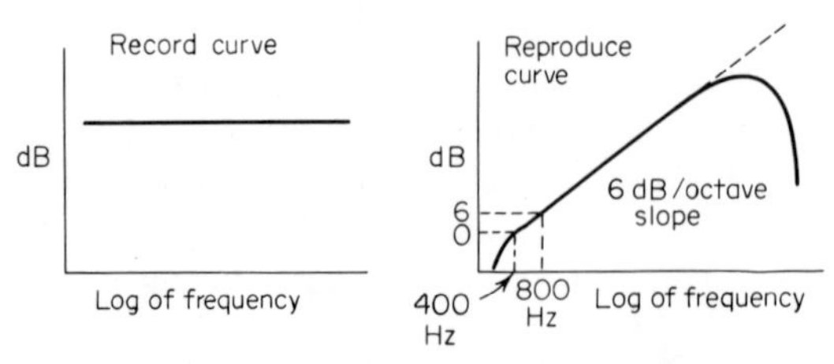

Fig. 3-2 Record and reproduce curves showing resultant output of a reproduce head to a constant-level record signal over a wide frequency range.

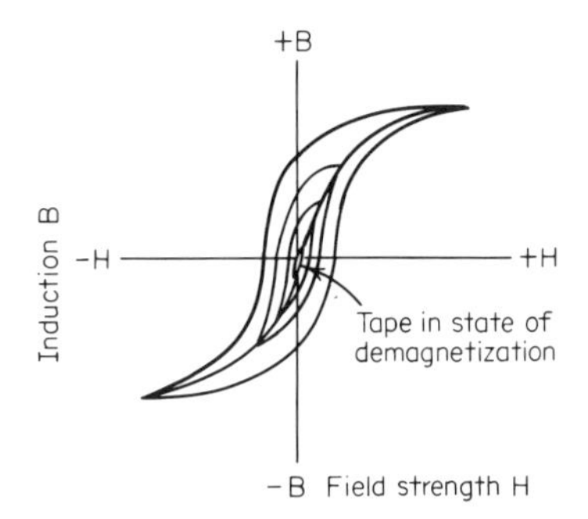

Fig. 3-3 The demagnetization curve.

by losses associated with head construction, magnetic tape characteristics, the speed of tape movement, and many other factors. Detailed explanation of these and other losses will be found in Chap. 7. It is sufficient to say at this time that most of the correction needed to straighten the reproduce curve to one comparable with the record curve will be done by circuits called equalizers, which are mounted in the reproduce amplifiers.

The Erase Head The basic function of an erase head is to demagnetize the magnetic signal on tape. A very heavy high-frequency alternating current (100 mA or more) is fed through the head. This current drives the tape into saturation first in one direction and then in the other. As the tape is pulled away from the head, a slowly diminishing cyclic field is presented to the tape. The diminishing cyclic field leaves the magnetic vectors in the tape in a completely random state. In such a random condition no single magnetic vector, or group of magnetic vectors, has more strength than the others, and the tape is demagnetized, or degaussed. See Fig. 3-3 for the demagnetization curve.

Magnetic Tape

The accepted medium for magnetic recording is a tape consisting of a backing coated with particles of the gamma form of ferric oxide ($Fe_2O_3\gamma$). The backing material, either acetate or polyester film, provides the base to which the magnetic particles are fixed with a synthetic-resin binder. Each particle (approximately 25 μin. in length) forms an elemental magnet. The north and south poles of these magnets lie along some axis of the iron oxide crystal lattice, the particular axis depending upon the crystalline structure of the oxide used. There will be specific types of oxides used for specific bandwidths of data signals and types of recording techniques, i.e., low-, medium-, and high-frequency response and digital, audio, video, and longitudinal-instrumentation recording techniques.

When magnetic flux ϕ from the record head penetrates into the magnetic tape coating, it reorients the vectors of the magnetic domains (an elemental magnetic volume). With a low value of magnetic flux, a small number of magnetic vectors are aligned with the flux field. When a large magnetizing flux is used, a large number of vectors are aligned. If the tape is moved linearly across the record-head gap, the vector pattern will be altered to produce, at any point, a net amount of magnetic vector alignment. This alignment has a magnitude and direction that is a function of the magnetizing field intensity (flux at record-head gap) that existed at the instant the tape left the gap.

The Tape Transport

The tape transport must move tape across the record and reproduce heads at a constant linear velocity, with the least amount of disturbance to tape motion. It must also provide some means of tape storage, either in a form of loop, bin, reel, or magazine. If speed or tension of the tape varies, dynamic time-base errors will be introduced into the system. These will adversely affect the amplitude, phase, and frequency response of the magnetic recorder. (Magnetic discs are discussed in Chap. 8.)

The effort in designing a tape-transport mechanism must be directed toward instantaneous and long-term speed control of the recording medium (tape). Instantaneous tape-speed errors (commonly known as flutter and wow) and long-term speed errors (drift) are introduced by inaccuracies of the drive system. They are usually expressed as percentage variation from absolute selected speed. Logical development of each major assembly and careful selection of components, combined with numerous tests and a highly efficient quality-control system, will result in good transports with reliable performance. Although tape transports may vary from each other in appearance, size, and operation, the design principles are basically the same. Even the best tape transports fall short of ideal specifications. Further improvements in any respect are, however, becoming progressively more difficult and costly to achieve.

Important Design Criteria Generally speaking, the five following points are the most important criteria when designing or purchasing a tape-transport mechanism.

1. It must have reliability and dependability, achieved for the most part by keeping the configuration as simple as possible.
2. It must have high performance, that is to say, the ability to transport tape at the desired speed with no introduction of variables in the signal to be recorded or reproduced and no disturbance of tape motion.
3. It must be standardized and compatible with other equipment. It must have standard speeds, tape widths, reel sizes, track spacing, etc.

4. It must have low downtime and maintenance cost. It must be easily maintained, and parts must be easily accessible. It must have a long mean-time-between-failure rate.

5. It must have flexibility for modification and freedom from obsolescence. This is generally achieved with modular construction and completely independent major subassemblies so that modification and updating can be easily accomplished.

The Record Amplifier

The record amplifier must change the incoming data signal E_{in} into a form that is suitable for recording on tape. That is, the amplifier must change the E_{in} into current I which, when fed to the magnetic heads, will be converted into magnetic flux ϕ. The magnetic flux will be used to magnetize the tape so that a given amplitude, a given polarity, and a given point on the tape represent the data E_{in} at a given instant of time.

Direct Recording The direct record amplifier will produce current that is analogous to the frequency and amplitude of the incoming ac data signal, the exception to this being the audio record amplifier. With audio, preemphasis is added to increase the low- and high-frequency current so that when the tape is reproduced, the sound is more compatible to the human ear, which is a nonlinear device. Bias must be added to the record signal to place it in the linear portion of the magnetic tape response curve. Both ac and dc bias are used with audio recorders. But almost without exception, ac bias only is used with instrumentation and video systems. Direct recording uses the maximum bandwidth capability of the recorder and tape, but is limited in low-frequency response, due to the reproduction losses that cannot be compensated for (Faraday's law, etc.—greater detail will be found in Chap. 7).

Frequency-modulation Recording (FM) The function of the FM record amplifier (FM modulator) is to convert the input data into a series of frequencies (carriers and sidebands). A particular frequency is selected as the center (carrier) frequency corresponding to a zero data input signal. A +dc signal will deviate the carrier frequency a given percentage in one direction, while a —dc signal will deviate the carrier frequency a given percentage in the opposite direction. An ac signal would deviate the carrier alternately on both sides of the center frequency, at a rate equal to the frequency of the input signal. The amount of deviation is controlled by the amplitude of the input signal, while the rate of deviation is controlled by the frequency of the deviating signal.

The primary consideration for the choice of center-carrier frequency is such that the range from maximum to minimum deviation will fall within the bandwidth limitations of the recorder. After conversion (E_{in} to f), the FM signal is passed through a power amplifier to the heads and recorded on tape (Fig. 3-4).

It is customary when using wideband FM systems (frequency response of dc to 500 kHz and a center-carrier frequency of 900 kHz)

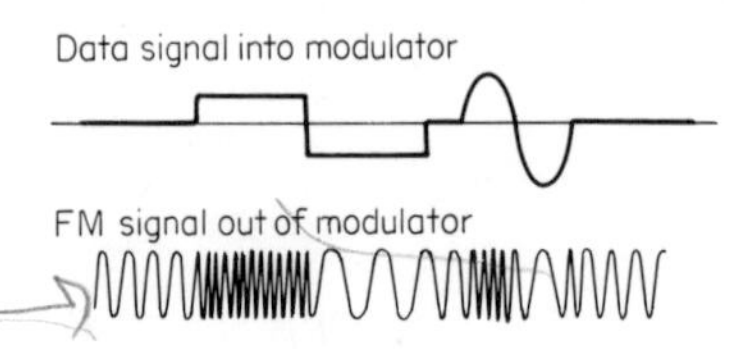

Fig. 3-4 Data-to-FM signal conversion.

to add ac bias to the record system. This improves the linearity of the system on reproduction (lower distortion). Video (rotating-head) systems use center-carrier frequencies of 7.5 to 9 MHz and data bandwidths of dc to 6 MHz. Alternating-current bias is not used with video systems; other means are used to compensate for any distortion.

Pulse-type Record Amplifiers In recording processes that use coding techniques, the function of the record amplifier is somewhat different from that of direct and FM. In these systems the pulse-type amplifiers produce pulses of current or frequencies that are used to designate one of two binary states (1 or 0) or the beginning and end of a pulse of definite length.

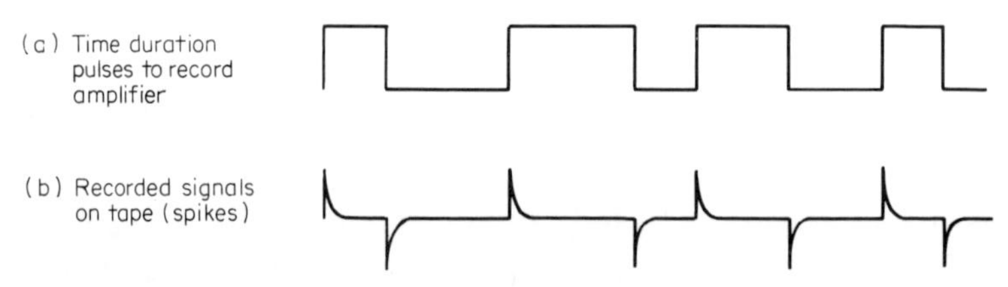

Fig. 3-5 Conversion of data signals to PDM or TSM spikes.

Some of the pulse-type record amplifiers are:

1. Frequency-shift modulation (FSM)
2. Pulse-duration modulation (PDM)
3. Time-sharing modulation (TSM)
4. Non-return-to-zero (NRZ) (digital)
5. Return-to-zero (RZ) (digital)
6. Pulse-code modulation (PCM)

See Fig. 3-5 for an illustration of the output from PDM and TSM amplifiers.

The Reproduce Amplifier

The reproduce amplifier should provide an output that is the same as or similar to that fed to the record amplifier, that is, $E_{out} = E_{in}$.

Direct Reproduce Amplifier Since the output of the reproduce head follows Faraday's law, $e = N d\phi / dt$ within specified limits and is nonlinear in other regions (Fig. 3-2), the reproduce amplifier must compensate for this type of irregular output. In direct systems where the frequency response is relatively narrow (up to 100 kHz), only an amplitude equalizer is included as part of the circuitry. As can be seen in Fig. 3-6, the equalizer characteristic curve is such that amplitude variations from the reproduce head are compensated for, resulting in a flat reproduce-amplifier output over a specified range of frequencies. In audio reproducing, the curve is not maintained flat, but rather is designed to vary from a straight line in order to be pleasing to the response curve of the human ear.

When the direct system is one where the frequency response is extended beyond 100 kHz, it is normal that a phase equalizer be included in the reproduce system, in addition to the amplitude equalizer. The need for this arises, primarily, because phase shifting of the data signal is introduced by the components of the broadband amplitude equalizer. Additionally, if the data are nonsinusoidal in nature (pulse or squarewave), the reactance of the reproduce head, which follows the formula $X_L = 2\pi f L$, tends to distort the waveshape, lengthening rise and fall times and creating peaks and dips in tops and bottoms. The phase equalizer essentially compensates for the induced delay (phase shift) and reconstructs the reproduced signal into its original form, $E_{out} = E_{in}$.

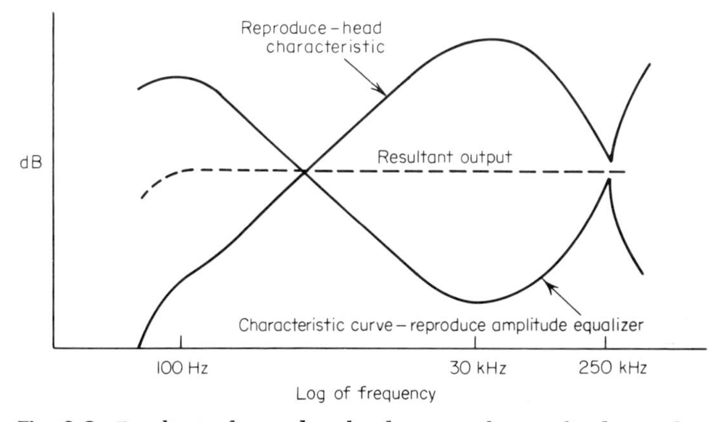

Fig. 3-6 Resultant of reproduce-head output plus amplitude equalization.

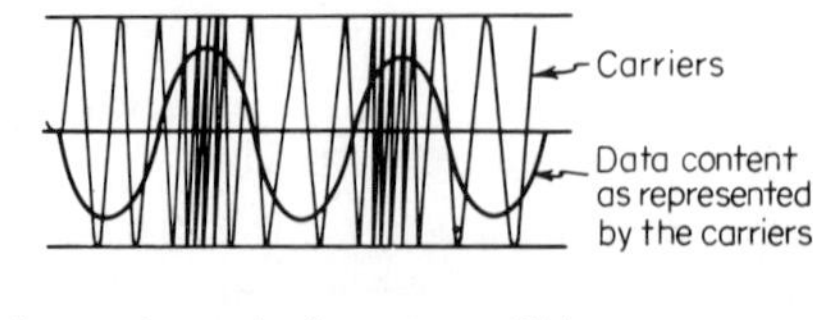

f_0 = center carrier frequency = 0 V

f− = CCF − 30% = −1.414 V

f+ = CCF + 30% = +1.414 V

Fig. 3-7 FM carrier frequencies and the resultant data signal on demodulation.

Frequency-modulation Reproduce The signal that is recorded on tape in an FM system is a frequency representation of a dc voltage input, or a number of frequencies (carrier and sidebands) representing an ac voltage input. The job of the FM reproduce amplifier is to change these frequencies back into their respective dc or ac values (Fig. 3-7).

Pulse-type Reproduce Amplifiers Where pulse-type recording was used, the reproduce amplifier will be required to reconstitute the time-duration pulses from a series of pulses that are the differentiated forms

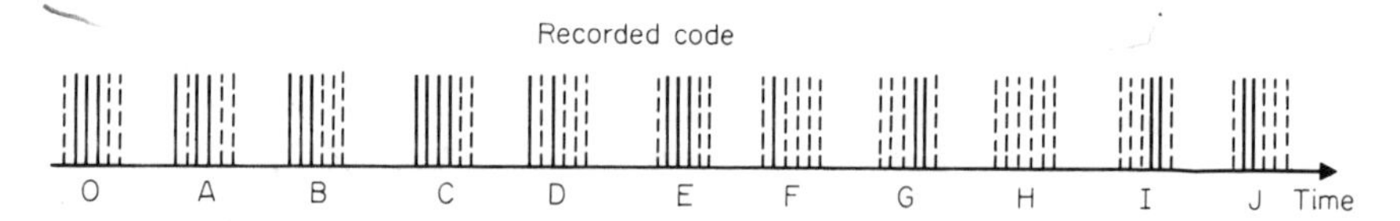

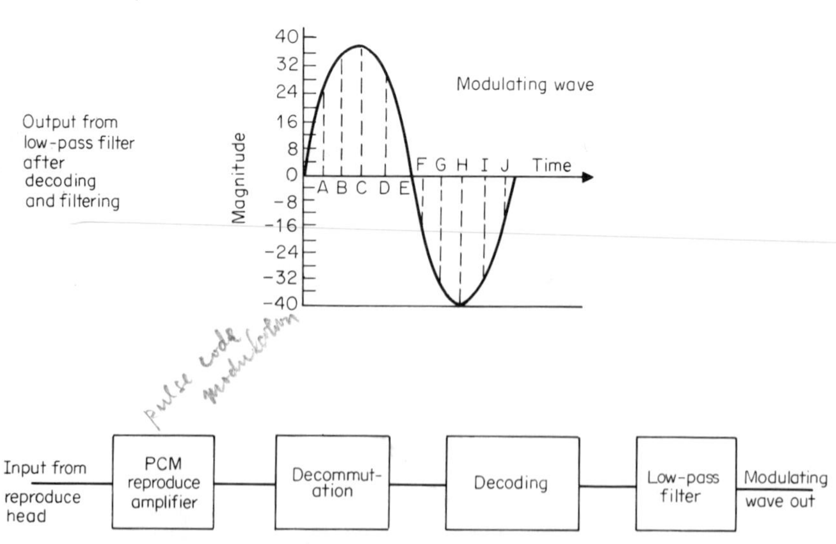

Fig. 3-8 Transforming a sequence of PCM code into sine-wave output.

of the original recorded spike (Fig. 3-5). A triggered multivibrator is most often used for this type of conversion. In the case of FSM, the reproduce amplifier must sense which of the two representative frequencies was recorded on tape (one frequency used to represent a one and another frequency used to represent a zero). High-quality filters followed by shaping amplifiers are used for this purpose. With PCM, since, during the recording process, samples of the data are quantized into discrete steps and fed as a code group of 1s and 0s to the record amplifier, the reproduce amplifier need only reproduce the recorded digital bits and feed them to the output stages. There the code groups will be used to re-create the original sampled data. Refer to Fig. 3-8 for the PCM reproduce sequence.

REFERENCES

1. Weber, P. J.: "The Tape Recorder as an Instrumentation Device," Ampex Corporation, Redwood City, Calif., 1967.
2. Haynes, N. M.: "Elements of Magnetic Tape Recording," Prentice-Hall, Inc., Englewood Cliffs, N.J., 1957.
3. Spratt, H. G. M.: "Magnetic Tape Recording," The Macmillan Company, New York, 1958.
4. Athey, S. W.: "Magnetic Tape Recording," National Aeronautics and Space Administration, Washington, D.C., 1966.
5. Stewart, W. E.: "Magnetic Recording Techniques," McGraw-Hill Book Company, New York, 1958.
6. Davies, G. L.: "Magnetic Tape Instrumentation," McGraw-Hill Book Company, New York, 1961.
7. Stiltz, H. L.: "Aerospace Telemetry," Prentice-Hall, Inc., Englewood Cliffs, N.J., 1961.
8. Lowman, C. E., and G. J. Angerbauer: "General Magnetic Recording Theory," Ampex Corporation, Redwood City, Calif., 1963.
9. Kietz, E.: Transient-free and Time-stable Signal Reproduction from Rotating Head Recorders, *Natl. Space Elec. Symp. Paper* 4.3, IEEE Professional Technical Group on Space Electronics and Telemetry, Miami Beach, Fla., 1963.

CHAPTER FOUR

The Theory of Magnetism

GENERAL

To understand the basic principles of ferromagnetism, we must take into account the modern theories of spinning electrons and the group behavior of molecular masses called domains.

FERROMAGNETISM

In ferromagnetic materials, it has been theorized that their atoms have permanent magnetic moments. These magnetic moments come from the electrons rather than the nucleus and are, primarily, the result of electron spin rather than orbital motion. It has also been theorized that there is a kind of atomic force that keeps the magnetic moments of many atoms parallel to each other, so that there is a strong magnetic force built up in each part of the ferromagnetic material. Pierre Weiss (1865–1940) explained this action in 1907 with his *theory of domains.*[1] In his theory he assumed that ferromagnetic materials were made up of large numbers of elemental volumes (which were called *domains*), all of which were magnetized to saturation in some direction. If the

[1] Pierre Weiss, *J. Phys.*, vol. 4, no. 6, p. 661, 1907.

magnetic moments of the separate domain volumes are oriented in different directions, the material was unmagnetized, since the various forces cancel each other; whereas if there is a degree of alignment of the magnetic moments of each domain, the material was magnetized.

The discovery of the electron spin was closely associated with the quantum theory, the basic idea being that the electron can be likened to the earth and its action in the solar system. Like the earth in its diurnal rotation about its own axis as it rotates around the sun, so the electron spins on its own axis while rotating about the nucleus of the atom. The effects of the electron spin were first presented by two Dutch physicists, George E. Uhlenbeck and Samuel A. Goudsmit, in 1925. Their findings explained Wolfgang Pauli's proposal that there should be a fourth quantum number, or index, m_s, which can take on the values of $\pm\frac{1}{2}$. Based on the four quantum numbers, Pauli introduced his exclusion principle (for which he was later awarded the Nobel Prize), which stated that no two electrons can have all quantum numbers n, 1, m_1, m_s the same. The quantum numbers were defined as follows:

n = principal quantum number
1 = azimuthal quantum number
m_1 = magnetic, or equatorial, quantum number
m_s = spin orientation

The significance of m_s was a complete mystery when Pauli first published his paper in 1925. The theory of Uhlenbeck and Goudsmit was to clear up this mystery and show that $m_s = \pm\frac{1}{2}$ determines the two possible orientations of the spin angular momentum of the electron relative to the axis of quantization.

In the iron-group elements the fundamental magnetic moment is derived from the electron spin, and not its orbital motion. In the rare earths and their compounds, however, the orbital moment contributes a large part of the magnetic moment. The electrons that are responsible for the magnetic properties of iron, cobalt, and nickel and their alloys are in the third shell of their atomic structures, whereas in the rare earths they are in the fourth shell.

In Fig. 4-1 it can be seen that there are a number of shells of orbiting electrons in the iron atom. In the outermost shell ($4s$) the two electrons are free electrons and are responsible for electrical conduction. In shells $1s$, $2s$, $2p$, $3s$, and $3p$ there are equal numbers of electrons with + (clockwise) spins as there are − (counterclockwise) spins. In these shells the magnetic moments neutralize each other. In the $3d$ shell, however, there are five + spins and only one − spin. Therefore there is an unbalance of spins, and the shell can be said to contain uncom-

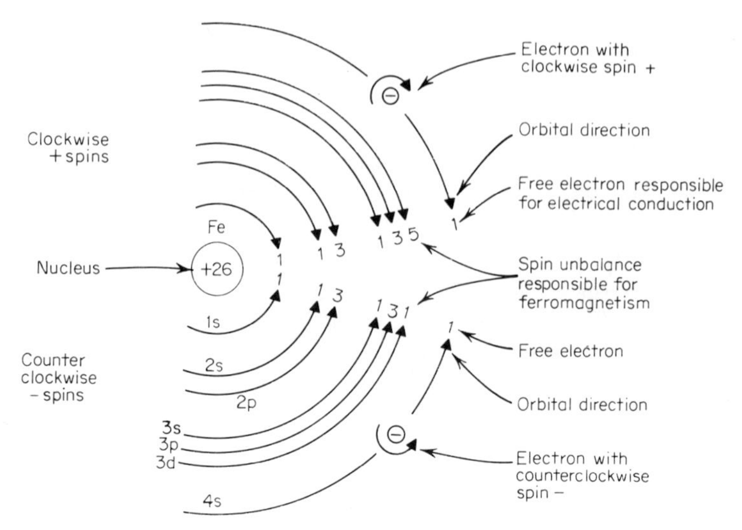

Fig. 4-1 Atomic structure of the iron atom.

pensated spinning electrons. Under these conditions, the atom as a whole will behave as a small permanent magnet.

From Fig. 4-1 we should be able to say that iron has a magnetic value of 4. This is not true, however, since the electrons tend to shift from one rotational level (shell) to another. With such movement constantly going on, it is not possible to quote the spin balance or unbalance at any specific time, but rather it must be stated as an average: for iron 2.22, cobalt 1.70, and nickel 0.61.

To iron, cobalt, and nickel can be added chromium, manganese, and such rare earths as gadolinium that also exhibit magnetic properties, either in their straight metallic form or in their alloys and compounds.

THE PROCESS OF MAGNETIZATION

During the process of magnetization of a ferromagnetic material, two types of changes occur:

1. A change in the volume of some of the magnetic domains at the expense of the others
2. Changes in the direction of magnetization of the domains

It is generally conceded in the magnetic recording industry that a domain consists of 10^{18} molecules of iron oxide ($Fe_2O_3\gamma$). It is defined as the smallest volume of magnetic material that can be considered to be a usable magnet. The domains are envisioned as being bounded by a partition, or wall, called a Bloch wall (after F. Bloch, who first studied the nature of the change in spin direction between magnetized domains). The position of the walls is associated with points of strain.

As illustrated in Fig. 4.2*a*, when magnetic material, which consists of many domains, is subjected to a very weak magnetizing force, the Bloch walls move outward in those domains whose direction of magnetization is roughly approximate to the magnetizing force. This occurs at the expense of the other domains with more diverse directions of magnetization. Since these domains are now slightly larger in size, they become predominant, and the material is slightly magnetized in the direction of these domain vectors. It should be noted that the magnetizing force did not go beyond the instep of the magnetization curve. Removal of the magnetizing force causes the Bloch walls to return to their original positions. Thus the action involved with weak magnetizing forces is reversible: remove the force, and the magnetization automatically disappears.

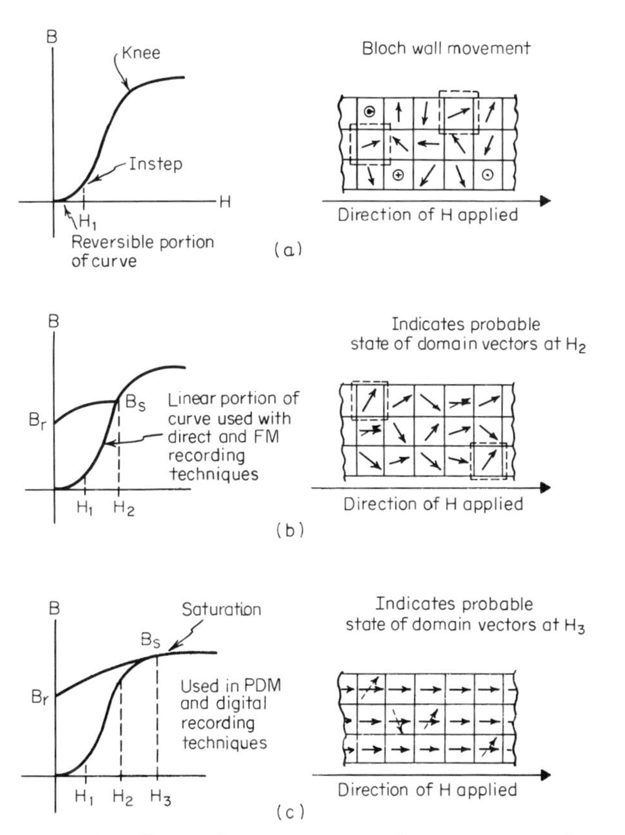

Fig. 4-2 Effects of a magnetizing force on a number of domains.

As illustrated in Fig. 4.2*b*, when stronger forces are used to magnetize a number of domains, the resulting induced magnetism increases at a relatively linear rate, as shown by the curve between the instep and knee. This causes changes in the orientation of the magnetic vectors from one direction to another. It should be noted in Fig. 4.2*b* that two of the domains are still undergoing volume changes. Also, two more have been stressed beyond their normal stable states. As a result, when the magnetizing force is removed, these four domains will tend to return to their more stable states. Thus there will be a small change in the induced magnetism of the material. This is shown in Fig. 4.2*b* as B_r.

When a very strong magnetizing force is used, that is to say, a force that goes beyond the knee of the curve, H_3, the material is saturated. In this state all or nearly all the domain vectors have been reoriented into a direction very close to that of the magnetizing force. Quite naturally, some of the vectors are in the strained state and will revert back to a lesser stressed condition as soon as the magnetizing force is removed. Thus, similar to the B_r value of Fig. 4.2*b*, the remanence magnetism B_r will have a lesser value than B_s.

REFERENCES

1. Stewart, W. E.: "Magnetic Recording Techniques," McGraw-Hill Book Company, New York, 1958.
2. Bozorth, R. N.: "Ferromagnetism," D. Van Nostrand Company, Inc., Princeton, N.J., 1951.
3. Haynes, N. M.: "Elements of Magnetic Tape Recording," Prentice-Hall, Inc., Englewood Cliffs, N.J., 1957.
4. Bardell, P. R.: "Magnetic Materials in the Electrical Industry," Philosophical Library, Inc., New York, 1955.
5. Greiner, R. A.: "Semiconductor Devices and Applications," McGraw-Hill Book Company, New York, 1961.
6. Galt, J. E.: Motion of Individual Domain Walls in a Nickle-Iron Ferrite, *Bell Syst. Tech. J.*, vol. 33, no. 5, p. 1023, September, 1954.
7. Bloch, F.: Nuclear Magnetism, *Am. Sci.*, January 1955, p. 48.
8. Weber, P. J.: "The Tape Recorder as an Instrumentation Device," Ampex Corporation, Redwood City, Calif., 1967.
9. Lowman, C. E., and G. J. Angerbauer: "General Magnetic Recording Theory," Ampex Corporation, Redwood City, Calif., 1963.
10. Kittel, C.: Magnetism, in L. N. Ridenour (ed.), "Modern Physics for the Engineer," McGraw-Hill Book Company, New York, 1954.

CHAPTER FIVE

Magnetic Heads

GENERAL

Having had a brief introduction to the basic elements of the magnetic tape recorder/reproducer, we are now ready to study the individual components in detail. There is no better place to start than with the component that might be called the "heart" of the magnetic recording system, the head.

CONSTRUCTION OF A MAGNETIC HEAD

Longitudinal (Analog) Head

The magnetic head has gone through a number of changes as the frequency range has been broadened and the types of recording expanded. Although some of the differences in the heads result from the wide range of frequencies in common use, there are also differences between those heads used for longitudinal as against rotating-head type of recording. Each head is designed for a specific task, a particular type of recording or a band of frequencies. All, however, employ similar principles in their construction.

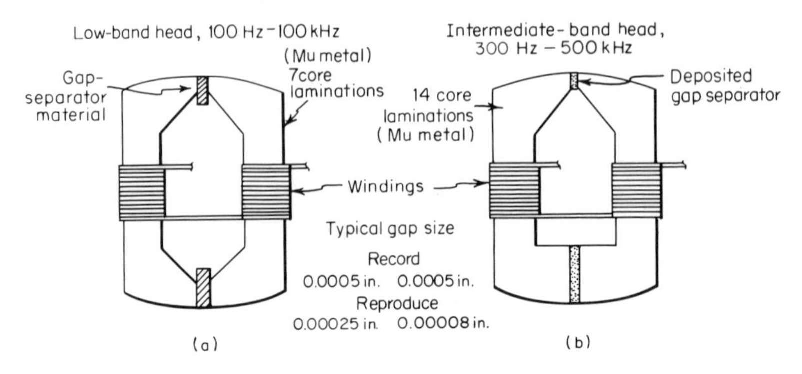

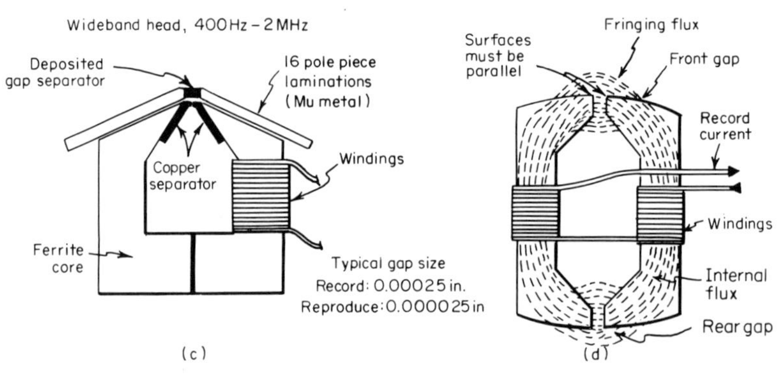

Fig. 5-1 Head-core design.

Basically, the head consists of two cores, or cores and their associated pole pieces (Fig. 5-1). In Fig. 5-1*a* and *b* it can be seen that there are windings around both of the cores. Each core is separated from the other, front and rear, by some form of gap-separator material. Figure 5-1*d* illustrates the effect of core separation when a current is fed to the windings of the record head. The magnetic flux developed in the core, fringing away at the gaps, will be used to magnetize the tape. The fringing flux at the rear gap is a loss factor that is taken care of by increased current through the head. It is, of course, pertinent for the reader to ask the question, "If the fringing flux at the rear gap is a loss factor, then why have a rear gap?" Such a question should be answered simply as follows: In the early stages of head design and manufacturing, there was no method available to cut a slot in a Mumetal ring of the size required (0.0005 to 0.00025 in.) while keeping the sides of the slots parallel. For this reason the core pieces were stamped out. Unfortunately, the stamping left the edges rough and lacking in precision, so that it was necessary to grind or lap the surfaces smooth. The surfaces of the front gap must be kept parallel to maintain the frequency response of the head constant. It was therefore necessary to grind the front- and rear-gap surfaces at different angles if the rear gap was to be closed. This type of grinding technique is incredibly difficult and costly. Thus the head manufacturers put up with the rear-gap losses and compensated for them in other ways.

Figure 5-1*c* illustrates a wideband head that would be used to record or reproduce a band of frequencies from 400 Hz to 2 MHz, at a tape speed of 120 ips. Only one of the cores is wound, and there is no rear gap. This core is not laminated and is not made of Mumetal. It is a formed, solid ferrite core with laminated pole pieces.

Figure 5-2 shows the laminations of the typical low- and intermediate-band heads. As the frequency range of the heads is broadened, the number of laminations is increased. They are used to reduce eddy currents.

Figure 5-3 shows the construction details of a low-band head stack. The nonmagnetic gap-separating material could be copper or gold, silver or platinum, depending upon design and use. In intermediate-band and wideband heads, however, the gap size is so small that the use of pieces of nonmagnetic insert material is impractical. A technique of deposition is used instead. In this process the core pieces are placed in a vacuum chamber, and silicon monoxide or some similar material is deposited over the core piece to the desired depth. Then, when the cores are butted together, the resultant separation between the cores, due to the deposition on both cores, becomes the gap spacing. As the frequency-response requirements go up, the gap is made smaller. Refer to Table 5-1 for gap-size details.

Referring once again to Fig. 5-2, it will be noted that several dimensions are indicated, i.e., gap depth, gap width, and track width. The depth of the gap determines to some extent the ultimate head life. This depth will be gradually reduced as the tape wears away the head surface. It is common to find gap depths of 0.016 to 0.021 in.

The width of the gap is dependent upon the intended use of the head. In a record head it must be wide enough to permit the magnetic flux to fringe (leak) far enough away from the gap to provide deep penetration into the oxide coating of the magnetic tape. On the other hand, the gap must be small enough so that sharp changes or gradients of the flux may be generated, to permit small changes in data level to be recognized. Thus it is that the record-gap width is a compromise between a wide gap, for strong recorded signals, and a narrow one, for definition of small increments of change.

The output of a reproduce head is maximum when one-half a wavelength of the recorded data is equal to the reproduce-head gap width ($\lambda/2 = l$). The output will decrease at a -6 dB/octave rate on either

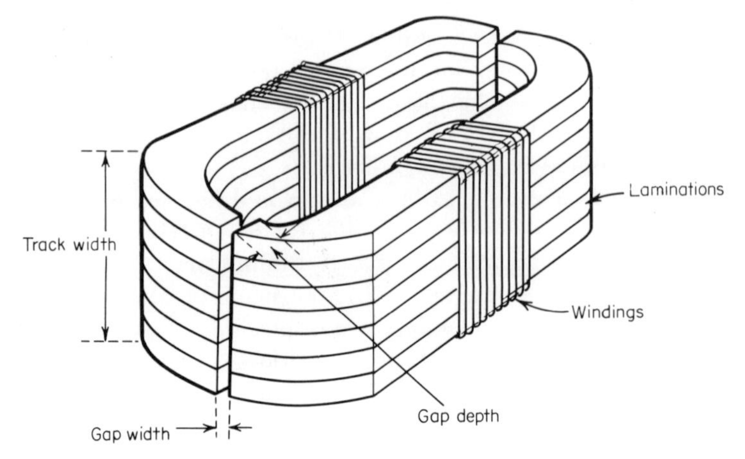

Fig. 5-2 Typical instrumentation head, low- and intermediate-band.

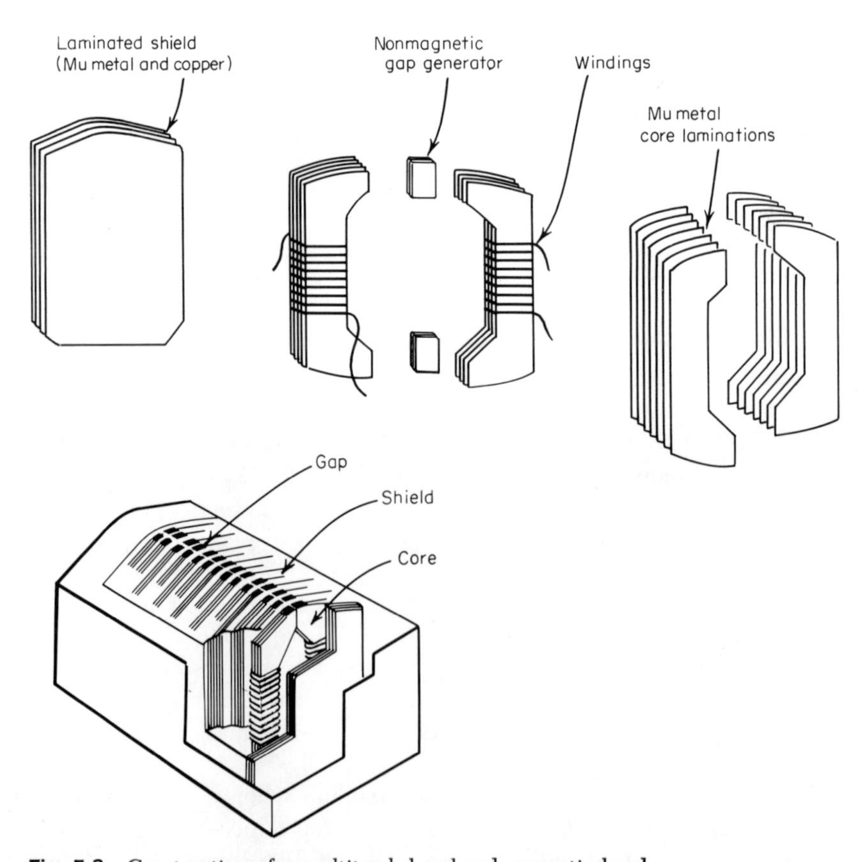

Fig. 5-3 Construction of a multitrack low-band magnetic head.

TABLE 5-1 Typical and Special Longitudinal-instrumentation-head Characteristics (Analog Heads)

	Record	Reproduce
Wire size	42	46
Number of turns	60	300
Standard track width—½- or 1-in. tape	0.050 ± 0.005 in.	0.050 ± 0.005 in.
Front-gap width:		
100-kHz response, tape speed 60 ips	0.0005 in.	0.00025 in.
600-kHz response, tape speed 120 ips	0.0005 in.	0.00008 in.
2-MHz response, tape speed 120 ips	0.00025 in.	0.000025 in.
Rear-gap width:		
100-kHz response, tape speed 60 ips	0.0005 in.	0.00025 in.
600-kHz response, tape speed 120 ips	0.0005 in.	0.00008 in.
2-MHz response, tape speed 120 ips	None	None
Peak record current	18–49 mA	
Peak reproduce voltage	....	50 μV min
Track spacing on tape—7 or 14 tracks	0.070 in.	0.070 in.
Track spacing within stack—7 or 14 tracks	0.140 in.	0.140 in.
Gap-to-gap spacing between head stacks	1.5 ± 0.001 in.	1.5 ± 0.001 in.
Gap azimuth	90° ± 1′	90° ± 1′
Head tilt	90° ± 3′	90° ± 3′
Gap scatter	100 μin.	100 μin.
Point of recording	Trailing edge	
Point of reproducing	....	Across center of gap
Special heads		
Track width—16 tracks, 1-in. tape	0.025 ± 0.002 in.	0.025 ± 0.002 in.
Track width—31⁄32 tracks, 1-in. tape	0.020 ± 0.001 in.	0.020 ± 0.001 in.
Voice-monitor track width—½-in. tape	0.005 ± 0.001 in.	0.005 ± 0.001 in.
Voice-monitor track width—1-in. tape	0.010 ± 0.002 in.	0.010 ± 0.002 in.
Track spacing on tape—16 tracks on 1 in.	0.060 in.	0.060 in.
Track spacing on tape—31⁄32 tracks on 1 in.	0.030 in.	0.030 in.
Track spacing within stack—16 tracks on 1 in.	0.120 in.	0.120 in.
Track spacing within stack—31⁄32 tracks on 1 in.	0.060 in.	0.060 in.

side of this point. Going up in frequency, a point will be reached where the full wavelength of the data frequency is equal to the gap size. At this wavelength the output of the reproduce head will be zero. Similarly, going down in frequency will produce a decreasing amount of output at a −6 dB/octave rate. When the frequency becomes too low, the reproduced data signal is so weak that it cannot be distinguished from the system noise. The span of frequencies between an acceptable low-frequency and high-frequency output level is called the dynamic range of the recording/reproducing system. To increase the dynamic range and frequency response at the high end, the reproduce-head gap must be made smaller, or the tape speed increased. Unfortunately, a penalty must be paid when the gap size is reduced. As the tape is pulled across the head, the edges of the head gap tend to wear and build a bridge of core metal across the gap (called gap smear). Naturally, once the gap is closed, no further flux penetrates the core and cuts the windings, with the result that there is no output voltage. The smaller the gap, the sooner the smear affects the output. The wear of the head is accelerated with an increase in tape speed. Thus the reproduce-gap width must be a compromise between the desired upper-frequency response, the dynamic range, and the head life.

Track width determines the width of the signal recorded on tape. This is normally 0.050 in. for standard instrumentation heads.

Figure 5-4 shows some of the head-stack parameters for instrumentation heads. Figure 5-4*a* indicates that the heads are not in line, but rather are mounted in such a fashion that the heads in the odd stack

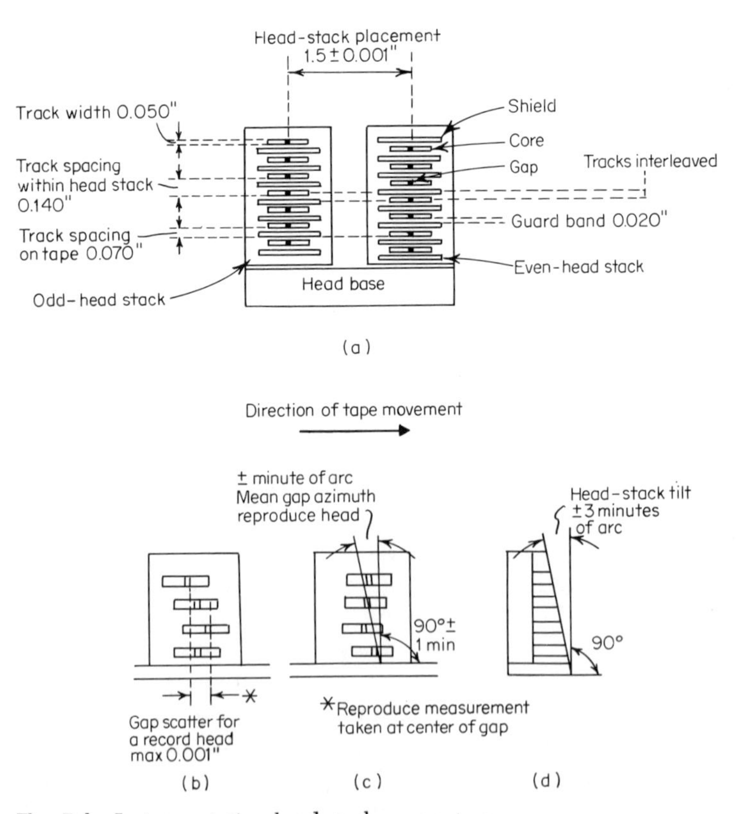

Fig. 5-4 Instrumentation head-stack parameters.

are opposite the shields of the even stack, and vice versa. This type of head construction is called interleaving. It permits the heads to be spaced farther apart in the individual stacks. In this way the interaction (crosstalk) is reduced. All standard instrumentation longitudinal heads contain 50-mil tracks spaced 140 mils apart. Thus the track spacing on tape will be 70 mils. Some of the other head-stack parameters are gap scatter, gap azimuth, and head-stack tilt. These are illustrated in Fig. 5-4*b*, *c*, and *d*, respectively. Gap scatter is defined as the distance between two parallel lines that enclose the trailing edge of the record-head gaps or the center lines of the reproduce-head gaps, between which all the gap edges or center lines fall. Effectively, it can be seen by referring to Fig. 5-4*b* that if a number of data signals were being recorded simultaneously, some signals would be physically positioned on the tape ahead of others. Quite naturally, this would give rise to timing errors in the data signals on reproduction, since the time correlation would be incorrect.

In Fig. 5-4*c*, if all the center lines of the reproduce heads are not perpendicular to the movement of tape, signals which were recorded simultaneously on magnetic tape would be reproduced with a time difference. Wideband heads are fitted with azimuth-adjustment screws that are used to compensate for this type of error. The azimuth adjustment cants the head to the right or left to compensate for the error in a single head or to effect a compromise for all the heads in the stack. Figure 5-4*d* shows the amount of forward or backward tilt permitted in any instrumentation head. This parameter is controlled during the manufacturing process.

It can be seen from Fig. 5-4 that the parameters of head construction are extremely critical. As a result, the reader is cautioned that instrumentation head stacks *should not* be removed from their precision baseplates. Their replacement requires many thousands of dollars of precision optical equipment and the skills of highly trained technicians. Normally, these skills and equipment are available only at the factory. Generally, audio heads may be excluded from this precaution. Most audio systems make provisions for individual-head-stack replacement and alignment in the field.

Figure 5-5 illustrates the two instrumentation-head numbering systems in general use. Until about 1962, heads using the old standard numbering system were the most popular. Today, however, heads using the IRIG (Inter-Range Instrumentation Group) Standard numbering system are generally used. The employment of the two standards has created a problem. When an operator who made a recording using Old Standard heads attempts to reproduce the data on IRIG heads, he finds that time correlation is lost. The data are located on different-numbered

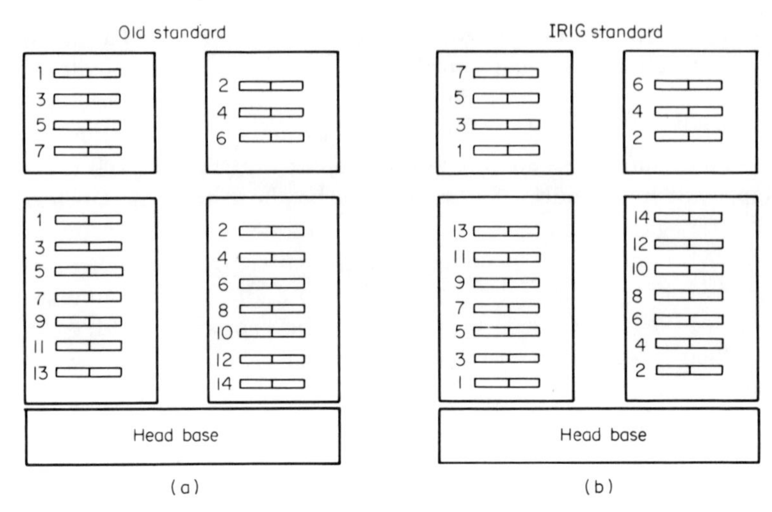

Fig. 5-5 Head-numbering systems.

tracks (i.e., if a signal were recorded on Old Standard head tracks 3 and 4, it would be found on tracks 11 and 12 on an IRIG head, and the time sequence of the two recordings would be completely reversed).

Figure 5-6*a* and *b* illustrates the action in the head and tape that occurs during the recording and reproduction of a data signal. It should be noted that recording takes place at the trailing edge of the gap, whereas reproduction takes place across the gap. The strength of the recording signal (magnetic flux ϕ) at any instant is a function of the amount of current being passed through the head coils, $\phi = KI \sin \omega t$;

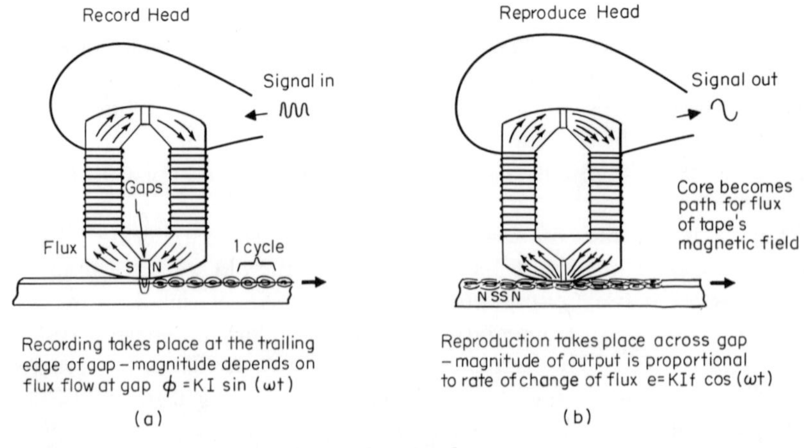

Fig. 5-6 Action at record and reproduce heads.

whereas the size of the reproduced signal e is a function of the rate of change of the magnetic flux across the head gap, $e = KIf \cos \omega t$.

Wideband Rotary Heads

The bandwidth of any magnetic recorder is essentially limited by the size of the magnetic wavelength that can be recorded and reproduced on magnetic tape. To that end, the size of the magnetic oxide particle that makes up the tape coating, the reproduce gap width, and the head-to-tape contact speed must be taken into consideration. The magnetic tape particle size averages 20 to 25 μin. Therefore, assuming that a minimum of four magnetic particles are required to record a full wavelength, the shortest wavelength that could be recorded and reproduced would be 4×20, or 80 μin. In order to be able to record a band of frequencies between 1 Hz and 6 MHz on a standard longitudinal instrumentation recorder, an FM recording technique would have to be used. This system would require a center-carrier frequency in the vicinity of 9 MHz with upper sidebands in the 15-MHz range. To record and reproduce such frequencies, the tape would have to move at better than 960 ips. Maintaining a constant tape speed, tension, and head-to-tape contact, as well as accurate tape guiding at this speed, was, and still is, nearly impossible using "state-of-the-art" longitudinal recording techniques. As a result, a new approach to the problem was developed. In 1956 the Ampex Corporation introduced the first rotary-head recorder, the VR-1000 videotape television recorder. Since that time, similar rotary techniques have been applied to the instrumentation field. It is now possible to record from 1 Hz to 6 MHz on a transient-free instrumentation rotary-head transport.

The rotary head differs considerably from the conventional stationary longitudinal head. Instead of a single head per channel, four magnetic heads are mounted on a rotating drum for single-channel recording, and eight heads are mounted on a single head drum (Fig. 5-7) for dual-channel operation. Two drums with four heads each have also been used for dual-channel operation. As the head drum is rotated and the magnetic tape moved longitudinally, each of the heads will contact the tape in turn. Figure 5-8 shows single-channel operation. Here it may be seen that the recorded signal is a series of bursts of data that overlap each other. Each video pulse (burst) contains redundant data at the beginning and end of the pulse. On reproduction, the total signal will have to be reconstituted. To increase the number of channels (dual), the longitudinal tape speed must be doubled to allow space for the second-channel heads to record on tape. To provide the necessary frequency response for wideband video recording, the head-to-tape contact speed must be high. The head-drum diameter

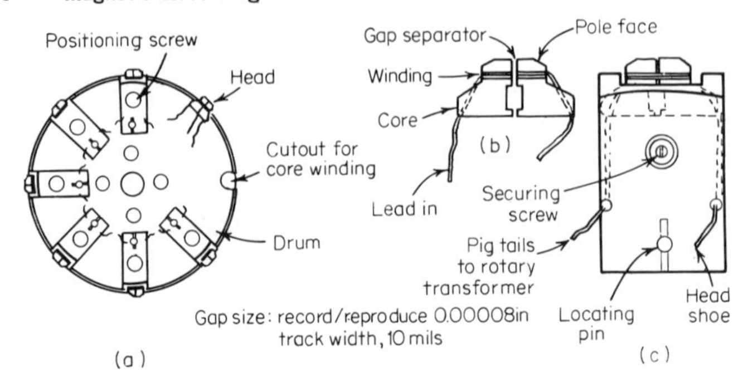

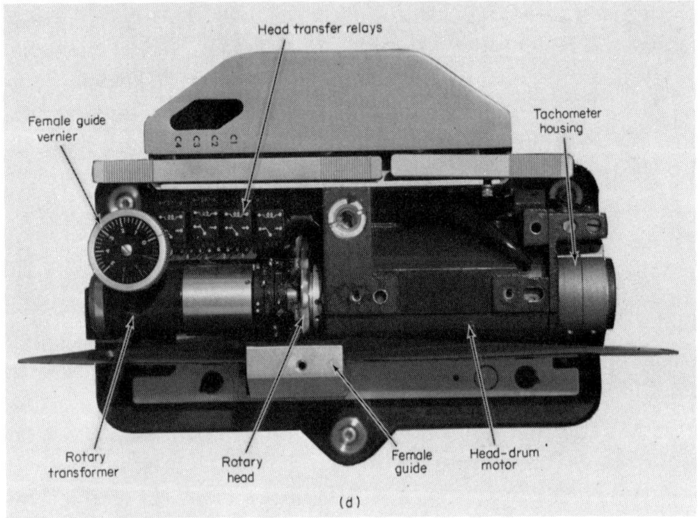

Fig. 5-7 Six-MHz rotary-head assembly—two-channel.

is approximately 2 in., and the rotational speed is 244 rps for a 6-MHz system.[1] Thus the actual head-to-tape contact speed is 1,570 ips. At this speed the wavelength recorded is comfortably larger than the average tape-particle size, so that the system is not tape-limited.

To utilize the rotating-head principle, some form of continuous connection between the signal source and the heads had to be developed. In the earlier versions of the rotating head, a set of graphite brushes was used in a slip-ring assembly to provide low-capacitance connections. The modern version of the rotary-head recorder uses a series of rotary transformers to provide this function. Quite naturally, since a single head is not in constant contact with the tape, some form of switching

[1] Modern video systems for television production use drum speeds of 240 rps. See Chap. 11 for details.

had to be used that would feed the data to and from the particular head that is in contact. Since the head drums are round, it is necessary to contour the magnetic tape for the best possible head-to-tape contact. A concave tape guide was developed (called a female guide) that curved the 2-in.-wide tape around the rotary-head drum (Fig. 5-9). To ensure that the tape assumes positions which conform to the shape of the female guide, a vacuum is applied to the guide side of the tape.

As can be seen in Fig. 5-9, the head tips actually stretch the tape as they traverse it. This stretching is utilized to compress or extend the traverse time base during the reproduction process. Manual adjustment of the amount of tip penetration into the tape (tape stretch) serves as a course correction for changes in tape dimension due to temperature changes and head-tip size (due to wear). This adjustment

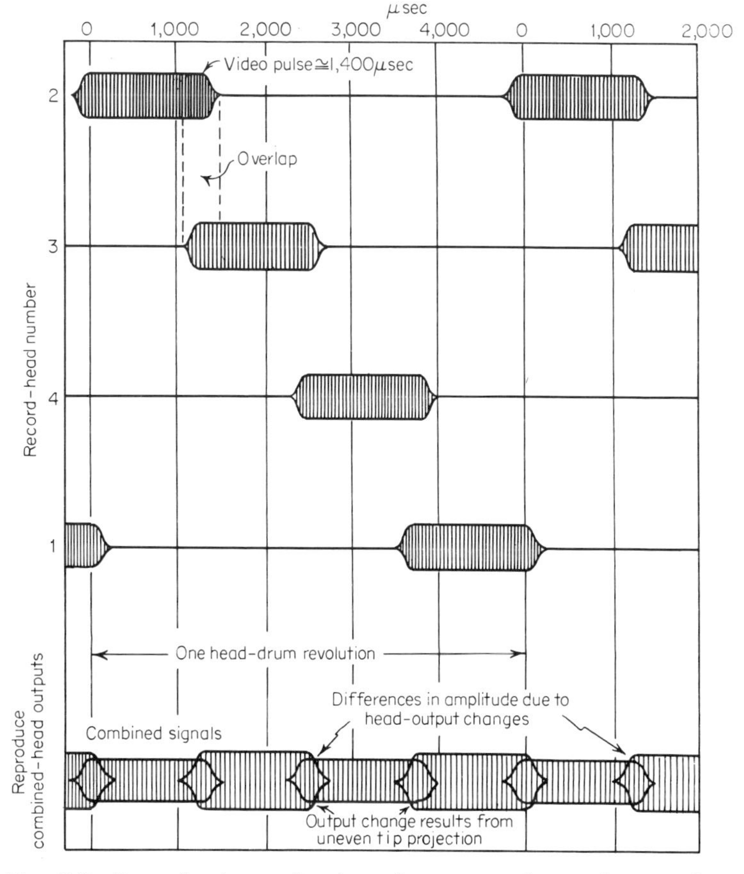

Fig. 5-8 Rotary-head recording/reproducing—signal recombination before correction.

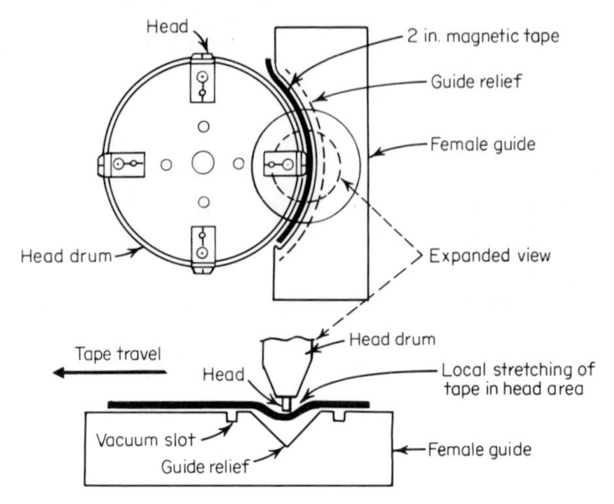

Fig. 5-9 Stretching of tape over rotary-head tips.

of the female guide effectively maintains the time base established during recording.

The high head-to-tape contact speed creates a great deal of tip wear due to friction. The first heads had a very short life, and as a result much research went into finding a better pole-piece material, one that would give longer wear for the heads and yet not decrease the life of the tape appreciably. The final choice was Alfesil. By the use of this material the life expectancy of the heads has been almost trebled.

In addition to maintaining the head-tip penetration relatively constant, the head-drum speed is also held constant. Generally, this is done by referencing the head-drum drive system to a crystal-oscillator frequency standard.[1] This oscillator has a stability of at least 3 parts in 10^5. In the record mode, the longitudinal speed of the tape is controlled by the speed of a synchronous capstan-drive motor. This motor is driven from an inverter controlled by the frequency standard. During the recording cycle, a frequency derived from the frequency standard is recorded on the tape by a fixed head. On reproduction, this control-track signal is compared with the frequency standard. If there is a phase or frequency difference, it will be recognized as an error, and the speed of the capstan changed appropriately. To control the head-drum speed, the modern 6-MHz head assembly uses tachometer outputs as the reference. The tachometers, two in number, are mounted on the end of the head-drum drive shaft, at the end opposite to the head drums. During the record mode the tachometer pulses are compared

[1] Video systems used with television production are somewhat different. The reader is referred to Chap. 11 for details.

in frequency and phase with a crystal reference which alters the phase of the 244-Hz three-phase power that drives the head-drum motor when any errors are detected. Also, during the record mode, a control-track signal is placed on the tape via the control-track head (one pulse per head revolution). In the reproduce mode, it is retrieved and compared with the head tachometer. Any difference in frequency or phase is detected as an error in the capstan motor speed. A servo system corrects for such errors.

Video systems also use stationary heads for voice annotation, cuing, and monitoring. The details of such heads are shown in Fig. 5-10,

Fig. 5-10 Stationary-head assembly—video recorder.

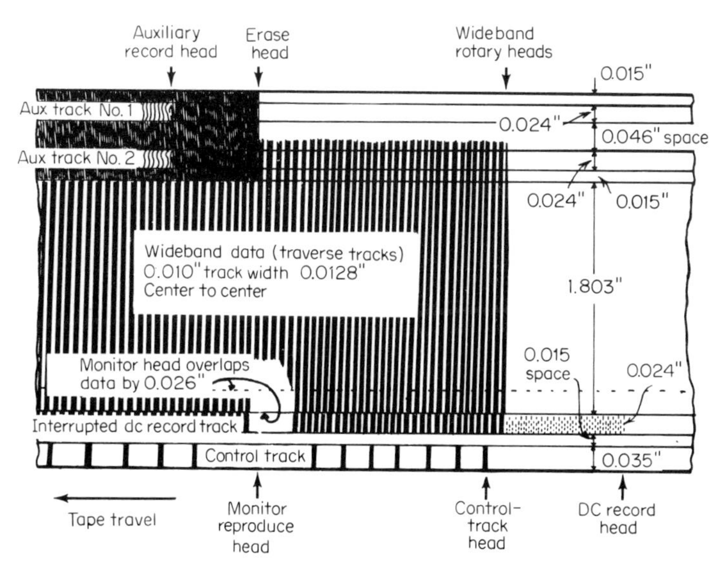

Fig. 5-11 Arrangement of recorded tracks on rotating-head-instrumentation magnetic tape.

and the arrangement of the tracks on tape in Fig. 5-11. The operation and construction of the stationary heads are the same as already described at the beginning of this chapter, the erase head being the only exception. This head is provided to ensure noise-free tape surface on which to record the auxiliary channels. A large amount of 100-kHz current is passed through a 130-mil-wide erase head, which is located before the auxiliary data heads. With this large current the video signal is demagnetized and the tape is left clean for the auxiliary tracks. Two front gaps are used in this type of head to give better degaussing action.

To provide confidence that recording is taking place, a monitor track (interrupted dc track—see Fig. 5-11) can be used to indicate that the equipment is working. The monitor record head is mounted on a fixed guide located ahead of the rotary-head assembly.

REFERENCES

1. Stewart, W. E.: "Magnetic Recording Techniques," McGraw-Hill Book Company, New York, 1958.
2. Weber, P. J.: "The Tape Recorder as an Instrumentation Device," Ampex Corporation, Redwood City, Calif., 1967.
3. Davies, G. L.: "Magnetic Tape Instrumentation," McGraw-Hill Book Company, New York, 1961.
4. Duinker, S.: Durable High-resolution Ferrite Transducer Heads Employing Bonding Glass Spacers, *Philips Res. Rep.*, vol. 15, no. 4, pp. 342–367, August, 1960.
5. Eldridge, D. F., and A. Baaba: The Effects of Track Width in Magnetic Recording, *IRE Intl. Conv. Rec.*, vol. 8, pt. 9, pp. 145–155, 1960.
6. Lowman, C. E., and G. J. Angerbauer: "General Magnetic Recording Theory," Ampex Corporation, Redwood City, Calif., 1963.
7. Kornie, O.: Structure and Performance of Magnetic Transducer Heads, *J. Audio Eng. Soc.*, vol. 1, no. 3, p. 225, July, 1953.
8. Pritchard, J. P.: Instrumentation Wideband Magnetic Recording, *Elec. Eng.*, vol. 32, no. 394, pp. 762–766, December, 1960.
9. Ginsburg, C. P.: Interchangeability of Video Recorders, *J. SMPTE*, vol. 67, no. 11, pp. 739–743, November, 1958.
10. Anderson, C. E.: Signal Translation through Videotape Recorder, *J. SMPTE*, vol. 67, no. 11, pp. 721–725, November, 1958.

CHAPTER SIX

Magnetic Tape

INTRODUCTION

As the magnetic heads have been called the "heart" of the magnetic recording system, so might the magnetic tape be called the "blood." Like the heads, the tape has often been taken for granted. Frequently, little care is exercised in its selection, handling, and storage. Just as a recording system is designed to perform a specific function, to cover a set range of frequencies or to use a particular type of modulation technique, so must the tape be designed to meet the wide and various requirements of such magnetic recording and reproducing systems.

DESIGN CONSIDERATIONS

The manufacturing processes of magnetic tape are full of complications and contradictions. Many of the design considerations appear to be in direct conflict with each other. For example, where magnetic tape should be strong (and strength is generally associated with thickness and stiffness), it should also be pliable. Pliability is associated with thinness and limpness. In order for magnetic tape to have high recording and reproducing resolution, the magnetic particles should be in intimate contact with the head. For good head wear, however, the individual magnetic particles (which are extremely abrasive) should not contact the head. In order to have a high signal-to-noise ratio, the tape should have as many magnetic particles per unit volume as possible (high density); whereas, to have good pliable, long-wearing qualities, the tape should have few magnetic particles per unit volume. In fact, when you come to think of it, the manufacturing of magnetic tape is close to impossible. This, perhaps, would explain the reason why so many of the techniques and materials used by the various magnetic tape manufacturers are classed as proprietary.

The construction of magnetic tape must take into consideration the magnetic coating and the nonmagnetic backing layers. Each of these layers should be optimized for its specific task. The coating should have optimum magnetic properties, a flat surface, and a constant thickness. It should have minimum roughness and be mechanically strong enough to retain the magnetic particles in position in spite of the frictional stresses introduced when the tape is pulled across the heads. The backing should be thin enough to permit a large amount of tape to be wound on a reel, yet it must be thick enough to prevent excessive amounts of print-through (magnetic printing effect between two tape layers when wound on a reel). Additionally, the backing should have long-term humidity and temperature dimensional stability. It should have good flexibility and uniform thickness.

The smoothness of the head and tape surfaces affects the speed at which the magnetic particles are moved past the head. Roughness of either surface causes irregular amounts of separation between the tape and head (tape flap). This results in apparent irregular tape speed which causes noise to be introduced into the system. Within the tape coating, nonuniform dispersion of the magnetic particles will also introduce noise, because the signal-to-noise ratio of magnetic tape is roughly proportional to the square root of the number of particles. Thus it might be stated that the magnetic tape must be smooth (uniform) inside and out to provide optimum recording/reproducing conditions.

Many factors influence the choice of the thickness of the magnetic coating of magnetic tape. Quite obviously, the total tape thickness will influence the number of feet of tape that can be wound on a reel. However, this criterion is based primarily on the thickness of the backing, since 70 to 80 percent of the total tape thickness is due to the backing material. Of greater importance is the frequency response of the tape. Essentially, the surface of the tape is the prime, or optimum-level, point for the ac-bias fields (see Chap. 3 for a discussion on the need for ac bias). It is at this surface level that the magnetic field for short wavelengths (high frequencies) are recorded. The longer wavelengths use more and more of the sublayers. As the tape-coating thickness is increased, the various layers of magnetic particles are less and less subjected to optimum bias currents (bias fields), and recording sensitivity decreases. Since the tape-coating thickness must be chosen, in most cases, to suit a range of frequencies, the thickness must be a compromise. It must suit the longest and the shortest wavelengths that are to be recorded. Generally, the coating thickness is chosen so that the long wavelengths are limited by the thickness, and the short wavelengths limited by overbiasing. With our modern requirement for recording 60-μin. wavelengths, most tape manufacturers have resorted to thin-coat tape. The thin-coating thickness is 0.18 to 0.2 mil as compared with thick coating of 0.41 to 0.46 mil. A definite improvement in high-frequency performance has been obtained using thin coating. However, there is a limit to the reduction of coating thickness, since the long-wavelength surface field reduces with the thickness reduction, and unfortunately, the tape noise does not also reduce. This is another way of saying that the fundamental signal-to-noise ratio of magnetic tape depends on the number of samples (magnetic particles) involved. In general, the more particles in the magnetic medium, the better the signal-to-noise ratio. Thus, when the thickness of the tape coating is reduced, fewer magnetic particles are available, and all other thing being equal, the signal-to-noise ratio reduces. In summary, we could say that to have high-frequency tape we need a thin coat, to have good low-frequency tape we need a thick coat, and to have a good signal-to-noise ratio we must have a dense coating.

To achieve a peak recording amplitude, which improves the signal-to-noise ratio of magnetic tape, the tape-coating particles are oriented in the preferred direction of magnetic recording, i.e., longitudinally along the tape for audio, computer, and longitudinal instrumentation and transversely (across the tape) for rotating-head instrumentation and video. After the tape backing has been coated, and while the coating layer is still wet, the magnetic tape is passed over a strong magnetizing field. This field produces an orientation of the magnetic particles in the tape. The effect of this action is to increase the peak amplitudes of the signal along the preferred recording direction by approximately 6 dB. This is a signal-to-noise improvement of ± 3 dB over unoriented tape.

In spite of the complex tape-design criteria, with its attendant difficulties and compromises, the magnetic tape manufacturers have made very significant improvements in the wide variety of tapes they produce. They have been able to keep pace with the increased requirements for quality and quantity of the audio, computer, instrumentation, and video users by making large-scale improvements in their manufacturing techniques and plants.

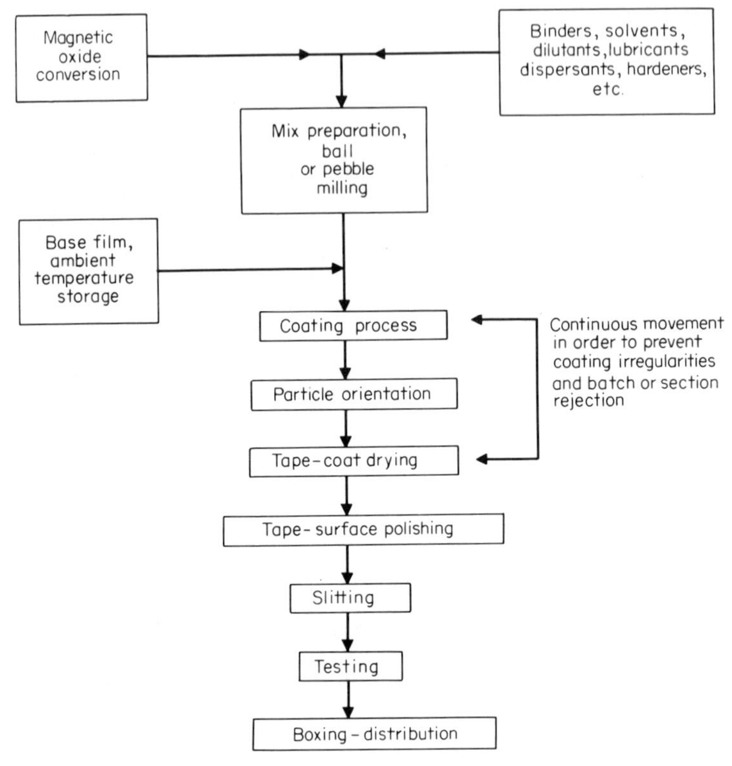

Fig. 6-1 Basic manufacturing processes of magnetic tape.

They have even developed different materials from which to manufacture tape. One of these is chromium dioxide. This material was first produced from chromium trioxide by Dr. Paul Arthur, Jr., of Du Pont's Central Research Department. United States Patent 2,956,955 was issued to Dr. Arthur and assigned to Du Pont based on Dr. Arthur's discovery. With this material Du Pont has developed a chromium dioxide magnetic tape called Crolyn.[1]

Chromium dioxide is synthesized in the form of acicular single-domain particles that can be varied in length over a wide range (4 to 400 μin.). The particles have an aspect ratio of 10:1. Because of the uniform crystal structure, chromium dioxide makes extremely smooth tape. It also has a low-noise and a low-print-through factor. Crolyn magnetic tapes have been made with coating thicknesses of from 80 to 250 μin.

Because the particle size can be kept to a narrow range and there is excellent dispersion of the particles of chromium dioxide in Crolyn magnetic tape, a higher-frequency response at low tape speeds and a

[1] A registered trademark of Du Pont.

better signal-to-noise are achieved. All this is not without cost, however. The recording hardware had to be modified to accept the smoother tape. Record current and bias current had to be greatly increased. Generally, this meant that recorders not originally designed to operate with Crolyn had to undergo some rather extensive modifications to permit using Crolyn magnetic tape. Crolyn costs more than conventional magnetic tape. It is for these reasons that Crolyn has not achieved major popularity in the recording industry.

MANUFACTURE OF MAGNETIC TAPE

As may be seen in Fig. 6-1, the actual making of magnetic tape requires several preliminary steps before actual coating can take place. Once the coating process has been started, it must be a continuous one. Even splices, if required, are made while the tape is moving. The basic requirements of each step in the manufacturing process are described in the following paragraphs.

Oxide Conversion

Figure 6-2 shows the processes by which the raw nonmagnetic lemon-yellow alpha ferric oxide ($Fe_2O_3 \cdot H_2O$) is converted to the rust-brown-colored acicular-shaped gamma form of ferric oxide ($Fe_2O_3\gamma$). During

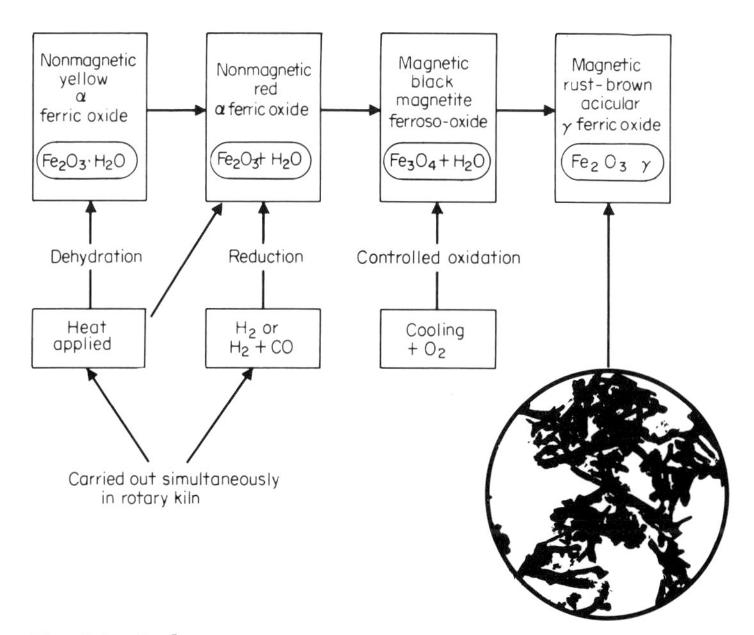

Fig. 6-2 Oxide conversion.

this conversion, carefully controlled dehydration and oxidation processes result in regulation of the $Fe_2O_3\gamma$ particle sizes. Those batches which result in particle sizes that average lengths of approximately 30 μin. are used for low-frequency, long-wavelength tape, whereas those batches that have average particle sizes of 20 μin. are used for high-frequency, high-resolution tape. Lately, experimental tape has been made using particles as short as 7 μin. Such tapes have very high frequency response and high signal-to-noise. However, constancy from batch to batch remains a problem for large-scale production of such tape. The magnetic particles have a length-to-width ratio of approximately 8:3.

Today's high-frequency tapes use $Fe_2O_3\gamma$ particles with a coercivity factor of 270 oersteds and a retentivity value of 1,200 gauss. Coercivity H_c may be defined as the magnitude of the magnetizing force required to reduce the remanence magnetism to zero. Retentivity B_r is defined as the remaining, or remanence, magnetism, i.e., the amount of magnetization or flux density remaining in a magnetic material after the magnetizing force has been returned to zero.

Mix Preparation

The dispersion of the magnetic particles and the preparation of the coating mixture for making magnetic tape employs many of the technologies developed for the manufacture of paint. The processes are complicated, however, by the fact that each of the small magnetic particles acts as a miniature bar magnet. Thus they tend to cling together and form agglomerates. During the preparation of the coating mixture, it is necessary to separate these small masses into their individual particles and hold them apart. Additionally, gamma ferric oxide particles are highly abrasive, by the very nature of their crystalline structure, shape, and size. Thus each individual particle must be completely coated with binder formulation which will serve many purposes, as follows:

1. It will encapsulate each of the small magnetic particles with a relatively soft layer that will be used to keep each of the particles separated from the other.
2. It will form a protective layer around each magnetic particle to act as a cushion between the oxide particles on the tape surface and the head. This reduces the head wear.
3. The binder will have adhesion (hold the magnetic particles firmly to the backing material).
4. The binder will have cohesion (allow no oxide or binder shed that would clog the heads).

Tape wears out because the binder system degenerates and loses its ability to hold the particles to the backing material (loses its adhesion qualities). Head wear increases as the particles loosen, because they will then tend to expose surfaces to the heads that are deficient in protective binder material. The steps involved in processing the coating material are shown in Fig. 6-3. The ball or pebble mill is used to force the agglomerated particles apart, coat each with binder material, and disperse them uniformly throughout the mixture. Steel ball bearings are used in the ball mill, and specially selected pebbles are imported from France to use in the pebble mill. The pebbles are chosen for their smooth surfaces and uniformity of size and hardness. The pebbles do not contaminate the oxide material.

The first step in the process is to wet down the particles and separate the agglomerate masses. To do this, part of the solvents, dilutants, and dispersants and all the ferric oxide material are placed in a ball or pebble mill. The rotational movement of the mill cascades the balls or pebbles against each other and against the mixture. This action separates the oxide masses into their individual particles and holds them in dispersion. At this point the balance of the solvents, dilutants, and dispersants plus the binders, stabilizers, conductive agents, plasticizers, and lubricants are added. The whole mixture is then milled for 17 to 100 additional hours, or until tests prove that the individual magnetic oxide particles are coated with binder material and uniformly dispersed throughout the mix. Uneven dispersion will result in uneven tape coating. This in turn will produce pronounced level variations in the record

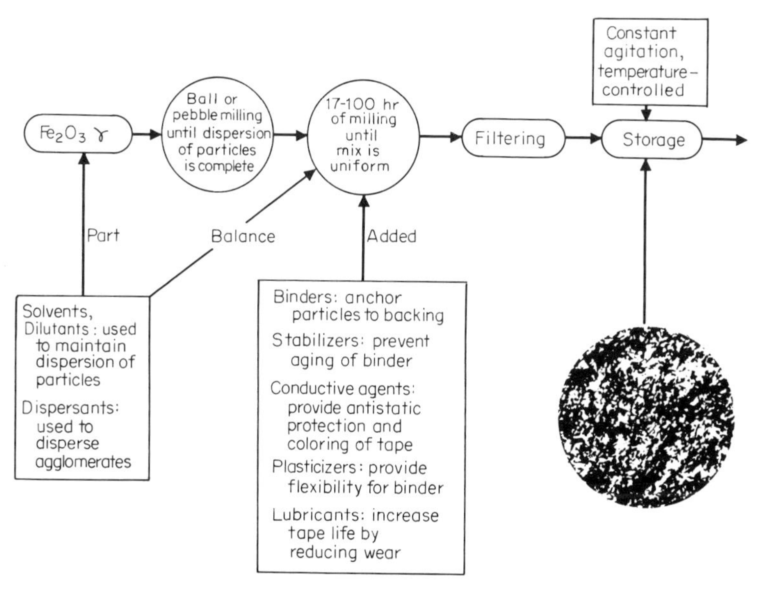

Fig. 6-3 Mix preparation.

and reproduce signals, particularly at the higher frequencies. Additionally, the overall sensitivity of the tape will be reduced and the noise level increased.

Base Film

Modern high-frequency, high-speed magnetic tape recorders have placed critical demands on the base film used as a backing material for magnetic tape. Fast start-stop times, such as experienced with digital recorders, require a tough base that will withstand sudden applied loads. Long-term storage requires that the tape does not become embrittled or deformed over relatively wide ranges in temperature, and is not attacked by fungus or mildew. High frequencies require the tape to be pliable, so that good head-to-tape contact is made. Thus we must examine the base film used for magnetic tape for qualities of dimensional stability, strength, toughness, pliability, resistance to attack by mildew and fungus, and the cost of all these critical demands to the end user.

Although many types of base materials have been used over the years, such as paper, cellophane, thin metal strips, cellulose diacetate, cellulose triacetate, polyvinyl chloride, and polyethylene terephthalate (polyester), only the cellulose acetates and the polyesters have become commercially important in the United States. Polyvinyl chloride is used to a large extent for making audio tape in Europe.

Table 6-1 illustrates the essential differences between the acetates and the polyesters when based on a film thickness of 1.5 mils and a tape width of 0.25 in. There are advantages and disadvantages in both types of material; thus the manufacturer of magnetic tape must weigh one criterion against another before choosing the backing film for a particular magnetic tape.

It is obvious from Table 6-1 that polyester film is superior to the acetates in all but two areas, uniformity of thickness and cost. Therefore, for those markets where cost is of less importance than quality (i.e., professional audio, instrumentation, digital, and video), polyester film is used almost exclusively.

The Coating Process

Once the coating process is started, the tape movement must be continuous or the coating, particle orientation, and tape-drying processes will be incomplete and irregular and the tape will be damaged beyond recovery. Additionally, in some tape-manufacturing plants the tape-polishing process is also part of the in-line processes. The raw backing film, once started at one end of the line, would have become polished tape at the other.

TABLE 6-1 Basic Differences between Acetate and Polyester Backing Materials

	Acetate	Polyester
Manufacturing techniques	Compounding of cellulose acetate and plasticizers	Synthesized from petroleum derivatives
Cost	Less expensive	More expensive
Thermal dimensional stability	Coefficient of thermal expansion per 1° change of temperature = 3.0×10^{-5} in./(in.)(°F)	Coefficient of thermal expansion per 1° change of temperature = 1.5×10^{-5} in./(in.)(°F)
Humidity dimensional stability	Coefficient of expansion per 1% change of relative humidity = 15×10^{-5} in./(in.)(% RH)	Coefficient of expansion per 1% change of relative humidity = 1.1×10^{-5} in./(in.)(% RH)
Effect of exposure to high temperatures on size	Tends to contract because shrinkage due to dehydration exceeds the expansion due to temperature increase	Relatively unaffected up to 80°C, beyond which it then behaves like an elastomer
Effect of exposure to high temperatures on toughness	Loses plasticizer and becomes very brittle	Remains flexible
Tensile strength	5.6 lb (2.545 kg)	11 lb (5.0 kg)
Tear strength	4 g	25 g
Mildew and fungus resistance	Low	High
Uniformity of thickness	Excellent	Good
Width tolerance	0.248 + 0.000 − 0.004 in.	0.248 + 0.000 − 0.004 in.
Length tolerance	−0 + 30 ft per reel	−0 + 30 ft per reel

Many types of coating methods have been used successfully for making both thick-coat (0.41 to 0.46 mil) and thin-coat (0.18 to 0.2 mil) tape. Some of these are:

1. Reverse roll. The coating thickness is in direct relation to the distance between the surfaces of the rollers.
2. Gravure coating. The depth of the engraving will control the coating thickness, and the viscosity of the mix will control the dispersion of the small beads of coating material. See Fig. 6-4.
3. Knife-blade coating. The coating is applied directly to the film web, and a fixed blade is used to scrape off the excess. The distance between the blade edge and the base-film surface determines the thickness of coating. See Fig. 6-5.
4. Extrusion coating. This method uses a large-diameter, highly polished roller. Carefully controlled amounts of coating mix are extruded

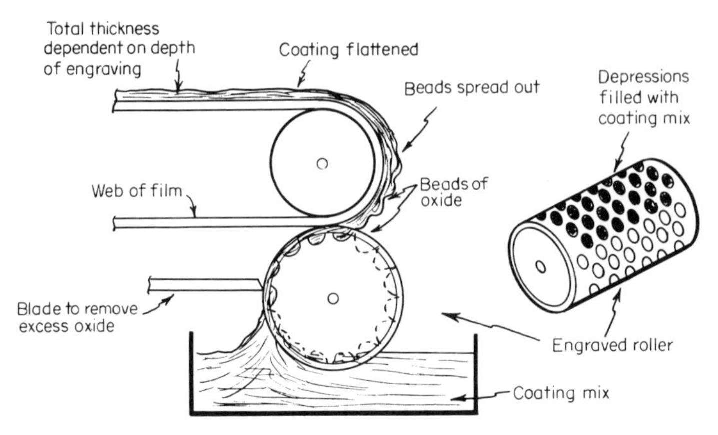

Fig. 6-4 Gravure coating process.

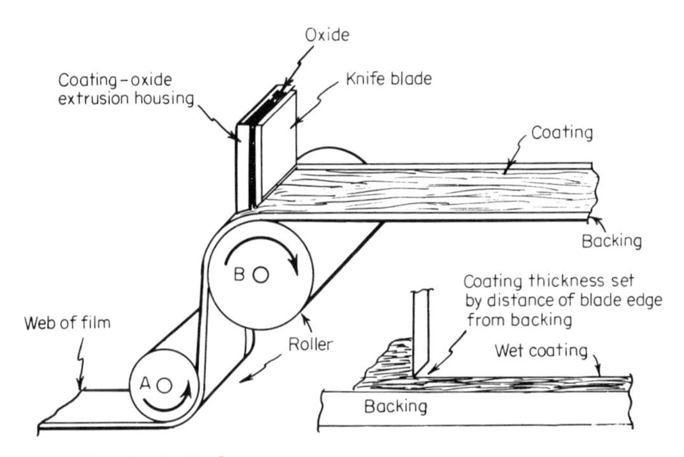

Fig. 6-5 Knife-blade coating process.

onto the surface of the roller at the beginning of its revolution. Later in the revolution, the coating mix is transferred by pressure to the film web.

5. Spray coating. This method uses pneumatic or electrostatic forces to spray the coating mix onto the surface of the film web.

Particle Orientation

After the coating process, and while the coating is still viscous enough to permit rearrangement of the oxide particles, the web is passed through a magnetic field. This will orient the particles in the preferred direction of recording, i.e., longitudinally for fixed heads and those used for helical-scan video and transversely for rotating heads, both instrumentation and video. Particle orientation improves the signal-to-noise ratio by ±3 dB.

The Drying Process

The tape coating is usually dried in a bank of large ovens. Both microwave and infrared drying techniques have been used, in addition to the hot-air oven. Each bank of the oven uses a different temperature and direction of air flow. Tape with a wet-coat thickness of 1 to 1.5 mils will be reduced to 0.2 mil in the dry state.

Tape-surface Polishing

Since the tape-coating material ($Fe_2O_3\gamma$) is in the form of long needle-shaped particles, some of them are bound to project from the surface of the tape. Quite naturally, this would cause unwanted head wear. Therefore the tape surface must be treated in some way, with the ultimate objective of having a smooth, highly polished surface of usable magnetic oxide. Many methods have been used to achieve this objective. The surface has been buffed with an Arkansas stone and brushed with a high-speed rotary brush made of horsehair or nylon. The tape has also had a razor blade or surgical knife-edge applied to scrape off the projecting coating particles. Additionally, the tape has been passed back over itself between pressure rollers which allowed the oxide coatings to polish each other. By these methods large quantities of oxide dust are formed, and there is a tendency to tear the oxide particles from the tape surface, causing voids. There may also be fractured particles on the tape surface which produce noise. See Fig. 6-6.

Another tape-polishing method is called Ferrosheen.[1] This process uses a technique that does not tear out or fracture the surface particles. It does, in fact, literally "iron the surface flat." By pressing the

[1] Registered trademark of the Ampex Corporation.

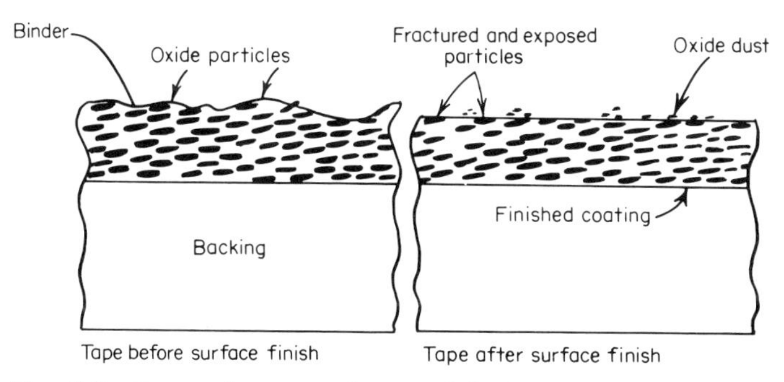

Fig. 6-6 Tape-surface imperfections left by some surfacing techniques.

particles back into the main body of the tape coating, instead of pulling them out, it produces an extremely smooth surface finish of from 6 to 10 μin. Essentially, this is a process of softening the binder material with heat and ironing the surface of the tape with a mirror-finished roller.

No one of these processes has been universally accepted. Often a combination of two or more of the techniques is used, particularly for tape used for high-frequency instrumentation and video.

Slitting

The wide webs of polished magnetic tape must be slit into usable widths to fit the various tape recorders, i.e., 150 mils wide for the small audio tape cassettes and up to 2 in. wide for rotating-head video and instrumentation recorders. The slitting is usually done on a high-speed precision slitting machine that uses a set of rotating-disc cutters (rotary sheers). The razor-blade type of cutter has also been used. It is general practice in the industry to cut the tape under size; i.e., for ¼-in. tape, cut to 0.246 ±0.002 in., for ½-in. tape, cut to 0.498 ±0002 in., and for 1-in. tape, cut to 0.998 ±0.002 in. This means that the width of the tape will never exceed the size of the tape guides of the tape recorder.

Testing

Just as there is a wide variation in the requirements of audio, video, digital, and instrumentation recorders, so also is there a wide variation in the specifications of the tape used for each type of recording. For this reason the amount and type of testing performed on the various families of tape are diverse. It is true that many of the qualities are common to all tapes. The final testing of tape prior to boxing is really only the last step in a series of tests and inspections that were carried out during the manufacturing processes. This time, however, as a finished product, the tape must pass a series of electrical, magnetic, and mechanical specifications that are laid down for the particular type of use, i.e., audio, digital, instrumentation, or video.

MAGNETIC TAPE APPLICATIONS AND THE PARAMETERS REQUIRED TO MEET THEM

Audio Tape

Audio tape is divided into three classes:

1. Mastering tape. Used as the master tape from which records and prerecorded tapes are made.

2. Professional tape. That used by professional high-fidelity sound-recording studios (i.e., radio broadcasting, etc.).
3. Consumer tape. That used by industrial, educational, and home-recording users.

The first two types of tape are normally supplied on reels, the latter on reels and cartridge, or the cassette form.

Mastering tape is normally used at 15 ips and over a frequency range of 20 Hz to 20 kHz. The coating thickness is unusually heavy (approximately 0.55 mil) to provide an extremely good signal-to-noise ratio and low-frequency (long-wavelength) response. It must have low distortion and a very wide dynamic range (better than 63 dB) and provide a negligible amount of output change even after 1,000 record/reproduce cycles.

Professional tape must perform on professional-type magnetic recording equipment at normal tape speeds of 15 and 7½ ips and have a frequency response of 20 Hz to 20 kHz. It should have a signal-to-noise of at least —59 dB, print-through of —43 dB maximum, and low distortion (0.5 percent at standard record levels).

Consumer tape, which forms by far the largest market, is used at normal tape speeds of 7½, 3¾, and 1⅞ ips. The signal-to-noise ratio should be at least —55 dB, print-through —40 dB maximum, and distortion less than 1 percent at standard record levels. The dynamic range should be 56 dB or better.

Typical magnetic properties of audio tape are:

Intrinsic coercivity H_{ci}, 260–290 oersteds
Retentivity B_r, 800–1,050 gauss
Squareness factor ϕ_r/ϕ_m, 0.70–0.75
Erasure with 1,000-oersted field, better than 60 dB

Longitudinal-instrumentation Tape

Magnetic tape used for longitudinal instrumentation may be classified as one of four major frequency-range groups, and one or more oxide types, wear, or coating-thickness subgroups. These groups and subgroups are as follows:

1. Frequency ranges
   a. 100 Hz to 100 kHz at 60 ips
   b. 300 Hz to 500 kHz at 120 ips
   c. 400 Hz to 1.2 MHz at 120 ips
   d. 500 Hz to 2 MHz at 120 ips

2. Oxide categories
   a. A oxide. Used where short wavelengths (high frequencies) are not a critical factor.
   b. B oxide. Used where high resolution and short wavelengths are required. This category has become the standard oxide for mid-frequency-range instrumentation tape.
   c. E oxide. This oxide is qualified under federal specification WT 0070/5 for wideband applications (500 Hz to 2 MHz).
3. Wear characteristics
   a. Regular wear. Standard of wear is "no appreciable change in tape characteristics after 1,000 record/reproduce cycles."
   b. Long wear. Provides 10 to 20 times the wear of regular-wear tape.
4. Coating thickness
   a. Thick—0.40 mil nominal
   b. Thin—0.20 mil nominal

As may be seen, there could be many combinations of the tape characteristics listed above. Each would fulfill a particular instrumentation need.

As the frequency-response requirements rose, the need for better tape became apparent. With higher frequencies, smaller head gaps were used, the angle of wrap around the head was changed (see Chap. 5), and smoother tape was needed if excessive head wear or gap smear was to be avoided (Fig. 6-7). Additionally, the tape must have fewer dropouts, lower print-through, and less skew. Higher-frequency response also required very close head-to-tape contact to prevent separation loss, which follows the formula $55d/\lambda$ dB (where d = distance of separation in inches and λ = wavelength in inches), this being a second reason for smooth tape. The higher tape speeds caused higher friction and its resulting heat. This caused binder breakdown of the older-type tapes and resulted in head clogging by the buildup of loose binder on the trailing edge of the gap. Coincident with head clog, there was also an oxide buildup on the leading edges of the gap. The new high-frequency, high-resolution instrumentation tapes are a far cry from their predecessors.

Typical magnetic properties of instrumentation tape are:

Intrinsic coercivity H_{ci}, 255–275 oersteds
Retentivity B_r, 1,000–1,200 gauss
Squareness ratio ϕ_r/ϕ_m, 0.75–0.79
Coating roughness, 5–15 μin. peak to valley
Dropouts per 100-ft tape, 3–4 as defined by federal specification WT 0070

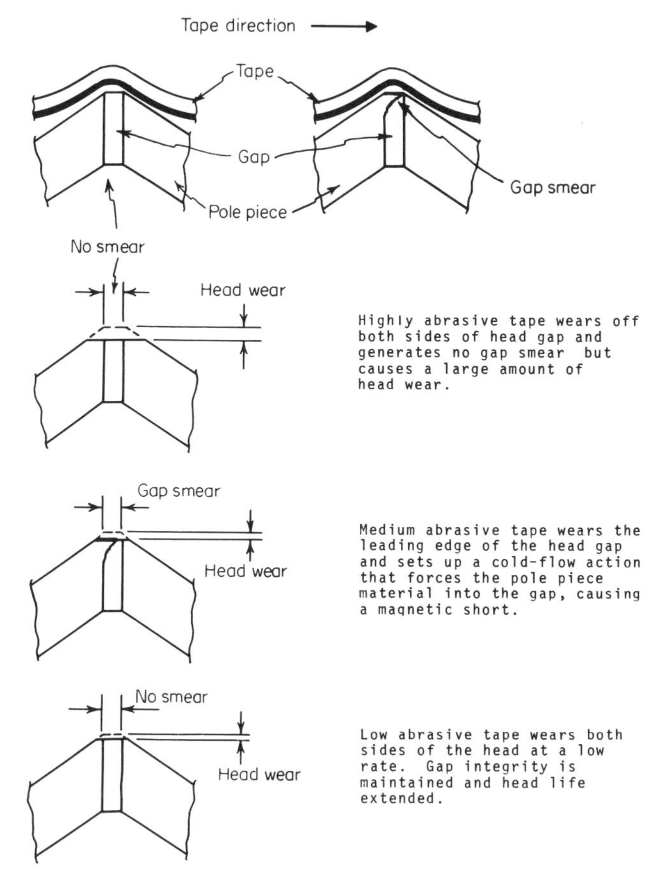

Fig. 6-7 Head-wear gap-smear problems.

Computer Tape

Computer tape is designed for use with the digital-tape transport and the associated digital computer. As such, since the computer format consists of bits of information (digital format 1 and 0), the loss of a single bit will jeopardize the associated data. The most critical requirement of computer tape is its freedom from error even after extended use.

Computer tape must be clean and free of irregular widths that cause skew, and it must have a uniform coating thickness. Also, by virtue of the "search, read, and write" techniques used with digital recorders and the rapid acceleration and deceleration that results, the tape pack tends to cinch (slippage of part of the pack on the reel with relation to the adjacent section), and this will fold over and damage the tape. To prevent cinch, antistatic and lubricating compounds are added to the tape mix. A second reason for the use of antistatic compound is that any static charge attracts dust and lint and causes dropouts.

Because of the stringent requirements of the computer industry, most computer tapes are or can be bought with the manufacturer's certification of being "error-free" (dropout-free) at a particular-bit packing density, i.e., zero errors at 200,556, or 800 bits/in.

Typical magnetic properties of computer tape are:

Intrinsic coercivity H_{ci}, 250–270 oersteds
Retentivity B_r, 990–1,290 gauss
Squareness ratio ϕ_r/ϕ_m, 0.70–0.78
Magnetization saturation, 90% value below 900 oersteds

Rotating-head Video and Instrumentation Tape

This tape may be divided into four major categories:

1. Transverse video tape
2. Rotating-head instrumentation tape
3. Helical-scan video tape
4. Longitudinal, stationary-head video tape

Each of these categories may be further subdivided into smaller groups having specific frequency ranges, widths (that is, 2, 1, ¼ in., etc.), and wear properties.

The principal characteristics of each main category are as follows:

1. *Transverse video tape.* 2 in. wide, transverse-oriented; base film (backing) 1 mil polyester; coating thickness 0.40 mil nominal; signal-to-noise ratio, longitudinal, —46 dB, signal-to-noise ratio, transverse, 53 dB; wear life better than 900 passes with 1.5-mil tip penetration (head engagement) and 250 passes with 2.5-mil tip penetration.

2. *Rotating-head instrumentation tape.* 2 in. wide, transversely oriented; base film 1 mil polyester; coating thickness 0.40 mil nominal; signal-to-noise, longitudinal, —46 dB, signal-to-noise, transverse, —53 dB; wear life better than 200 passes with a 2.5-mil tip penetration.

3. *Helical-scan video.* 2 or 1 in. wide, longitudinally oriented; base film 1 mil polyester; coating thickness 0.40 mil nominal; video signal-to-noise 42 dB minimum, audio signal-to-noise —46 dB minimum; stop motion operation 5 min minimum.

4. *Longitudinal, stationary-head video.* ¼ in. wide, longitudinally oriented, all other properties same as for wideband-instrumentation tape.

Because of the requirements of helical-scan and rotating-head video and instrumentation recorders, there is a comparatively high head-to-tape contact speed, i.e., for 1-in. helical-scan recorders 1,000 ips, 2-in. helical-scan 640 ips, and 2-in. rotary-head 1,500 ips. Additionally, since very close head-to-tape contact is required, the heads are arranged so that they penetrate, or "dig," into the tape from 1.5 to 3 mils (Fig. 5-9). This head penetration (tip penetration), together with the high head-to-tape contact speed, causes increased temperatures and head and tape wear. The high temperatures, together with the physical distortion of the tape, have created special problems for the tape manufacturers. Under such conditions, ordinary tape binders break down and "clog" the head gap, causing the gap to short out and the tape to be lifted from the head surface. Thus special binders had to be developed that would hold up under such demands. The friction tended to wear the heads and tape badly; therefore special lubricants had to be used, particularly with tape that was used for helical-scan "stop action." Rough tape not only wears the heads but causes tape flap. Due to the moving heads, tape flap causes a dropout that lasts for some time. This long dropout produces a "shadow" in the video picture. For these reasons most video and rotating-head instrumentation tape is 100 percent tested.

Typical magnetic properties of video tape are:

Intrinsic coercivity H_{ci}:
- Longitudinal, 230–240 oersteds
- Transverse, 260–270 oersteds

Retentivity B_r:
- Longitudinal, 800 gauss
- Transverse, 1,000–1,300 gauss

Signal to dc noise:
- Longitudinal, —46 dB
- Transverse, —53 dB

7.5-mil-wavelength sensitivity:
- Longitudinal, —2.5 dB
- Transverse, +4.0 dB

The problems, critical areas, and magnetic properties of stationary-head video are the same as those already discussed for longitudinal instrumentation. The reader is referred to the preceding paragraphs on instrumentation tape for details.

REFERENCES

1. Spratt, H. G. M.: "Magnetic Tape Recording," The Macmillan Company, New York, 1958.
2. Haynes, N. M.: "Elements of Magnetic Recording," Prentice-Hall, Inc., Englewood Cliffs, N.J., 1957.

3. Mee, C. D.: "The Physics of Magnetic Recording," Interscience Publishers, a division of John Wiley & Sons, Inc., New York, 1964.
4. Lowman, C. E., and G. J. Angerbauer: "General Magnetic Recording Theory," Ampex Corporation, Redwood City, Calif., 1963.
5. Athey, S. W.: "Magnetic Tape Recording," National Aeronautics and Space Administration, Technology Utilization Division, Washington, D.C., 1966.
6. Kirk-Othmer: Encyclopedia of Chemical Technology, 2d ed., vol. 12, pp. 801–818, John Wiley & Sons, Inc., New York, 1964.
7. Weber, P. J.: "The Tape Recorder as an Instrumentation Device," Ampex Corporation, Redwood City, Calif., 1967.

CHAPTER SEVEN

The Recording and Reproducing Process

HOW RECORDING TAKES PLACE

Fundamentally, the process of magnetic recording is the magnetization of the magnetic particles in the tape. The exact pattern into which these magnetic particles are formed depends on the type of recording.

With direct recording, for example, the current which the record amplifier presents to the head has a frequency and an amplitude which are analogous to the frequency and amplitude of the incoming signal. In other words, it might be stated that the incoming electrical signal contains three parameters, amplitude, polarity, and time. When this signal is recorded on magnetic tape, the magnetic medium must be magnetized so that a given amplitude, a given polarity, and a given point on the tape represent the electrical signal at any instant in time. Figure 7-1 illustrates the resulting magnetic flux pattern in the tape for a changing sinusoidal input. Here it may be seen that the magnetic field intensity produced is proportional to the head current. It should also be noted that there is a small kidney-shaped magnetic pattern for each half-wavelength. Additionally, where the signal passes through its zero value, the polarities of the kidney-shaped magnetic fields are the same, i.e., north-north- or south-south-pole combinations. These

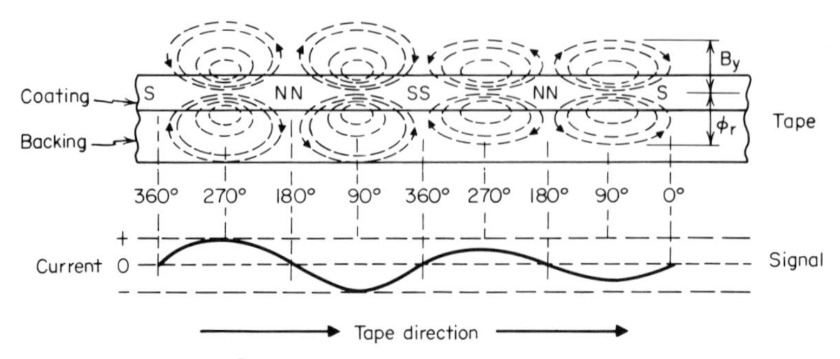

Fig. 7-1 Magnetic flux pattern on tape.

similar poles will cause demagnetization, a loss factor that will be discussed in detail later in this chapter.

The reader is also referred back to Fig. 4-2, showing the effects of a magnetizing force on a number of domains, for an illustration of the domain-vector alignment with varying amounts of head current. This figure also indicates the hysteresis effects of magnetic materials (nonlinear *B-H* curve, with its knee and instep).

Because of the hysteresis and demagnetization effects, the amount of remanent flux ϕ_r is not directly proportional to the magnetizing field strength H. As stated in Chap. 3, a conversion factor K must be added to the formulas for recording. To sum up the recording mode mathematically, it can be shown that the record current and the resulting remanent flux at any instant equals

$$i = I \sin \omega t \tag{7-1}$$

where t = time, sec
I = maximum current
i = instantaneous value
$\omega = 2\pi f$
f = recording frequency, Hz

The instantaneous magnetizing force H will be proportional to the instantaneous current i, less the loss due to the eddy currents,[1] etc. This means that the remanent flux ϕ_r and the remanent induction B_r, which are a function of the magnetizing force H, will also be proportional to the instantaneous current i, less hysteresis[1] and demagnetization losses. Thus, substituting in Eq. (7-1),

$$\phi_r = KI \sin \omega t \tag{7-2}$$

To represent the remanent flux at any point along the tape, it will be

[1] To be explained later in this chapter.

necessary to convert time t to distance x, and frequency f to wavelength λ.

Since time

$$t = \frac{\text{distance } x}{\text{tape speed } s}$$

and frequency

$$f = \frac{\text{tape speed } s}{\text{wavelength } \lambda}$$

then

$$ft = \frac{x}{\lambda}$$

and substituting in Eq. (7-2),

$$\phi_r = KI \sin \frac{2\pi x}{\lambda} \tag{7-3}$$

HOW REPRODUCTION TAKES PLACE

The fundamental process of reproducing from magnetic tape is based on the voltage induced in the reproduce head by the magnetic flux lines that emerge from the tape surface and pass through the reproduce-head core. The value of the voltage is a function of the speed at which a given number of flux lines cut a fixed number of turns of the windings of the reproduce head. Thus the amount of reproduce voltage is based on the rate of change of flux.

As may be seen in Figs. 7-1 and 7-2, the emerging flux is designated B_y. It will be proportional to and vary with ϕ_r, so long as the tape-coating thickness does not vary within the reel of tape. Tape-coating-thickness loss will be explained later in this chapter. The remanence flux ϕ_r at the center of the kidney-shaped magnetic fields is maximum, but its rate of change is minimum (Fig. 7-2), whereas at the end of the magnetic fields, the remanence flux is minimum but its rate of change is maximum. Thus the mathematics of the reproduce mode may be expressed as

$$B_y = K' \frac{d\phi_r}{dt} \tag{7-4}$$

where B_y = emerging flux
K' = proportionality factor (takes into account losses due to tape-coating-thickness changes, etc.)
d = symbol for change
ϕ_r = remanence flux
t = time, sec

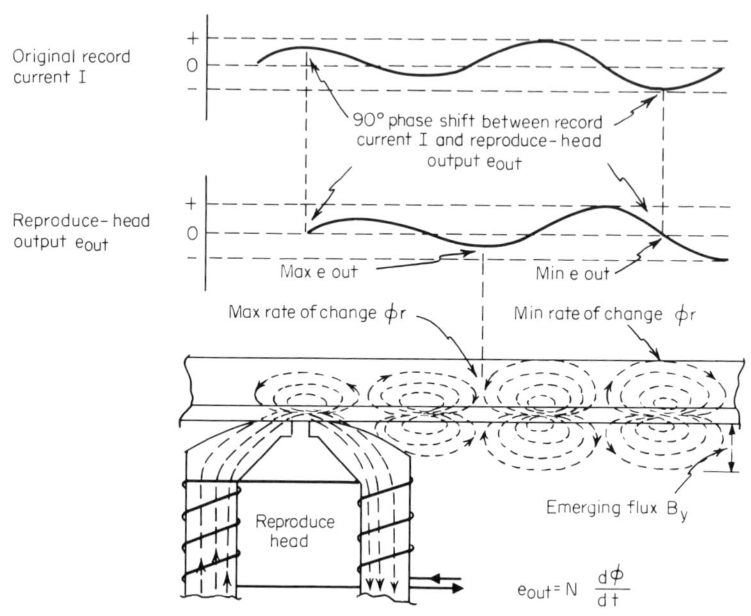

Fig. 7-2 Reproduce-head output and the phase relationship between record and reproduce processes.

Since $t = x/s$, $f = s/\lambda$, and $\phi_r = KI \sin \omega t$, by substitution in Eq. (7-4)

$$B_y = K''I \frac{2\pi}{\lambda} \cos \frac{2\pi x}{\lambda} \tag{7-5}$$

The instantaneous voltage developed in the reproduce head is directly proportional to the number of flux lines cut in unit time, thus:

$$e_{\text{out}} = K'B_y s = K''Is \frac{2\pi}{\lambda} \cos \frac{2\pi x}{\lambda} \tag{7-6}$$

More conventionally,

$$e = KIf \cos \omega t \tag{7-7}$$

These formulas have indicated that:

1. The output voltage is proportional to the record current.
2. If the tape speed is the same, the reproduce frequency is the same as the record frequency.
3. The output voltage is proportional to the frequency, and follows a 6 dB/octave curve (an octave is two times the frequency).
4. The change from the sine to cosine in the formulas indicates that the record current and the output voltage have a 90° phase difference for corresponding points along the tape.

5. The surface induction and the output voltage will increase linearly with frequency for a constant level of recording current (where head losses, etc., are taken into consideration).

DETAILS OF THE RECORDING PROCESS

The Hysteresis Loop

The hysteresis loop (Fig. 7-3) may be considered as nothing more than a visual indication that:

1. The magnetic intensity B lags behind the magnetizing force H.
2. The remanent flux ϕ_r or the remanent magnetism B_r induced in the magnetic medium does not vary linearly with the magnetizing force H.
3. The domain-vector alignment is not linear.

If the magnetic medium (magnetic tape, for example) is assumed to be in an unmagnetized state, and a gradually increasing and unlimited magnetic force H is applied, the value of the magnetic intensity B for any value of magnetizing force will follow the curve OP until saturation

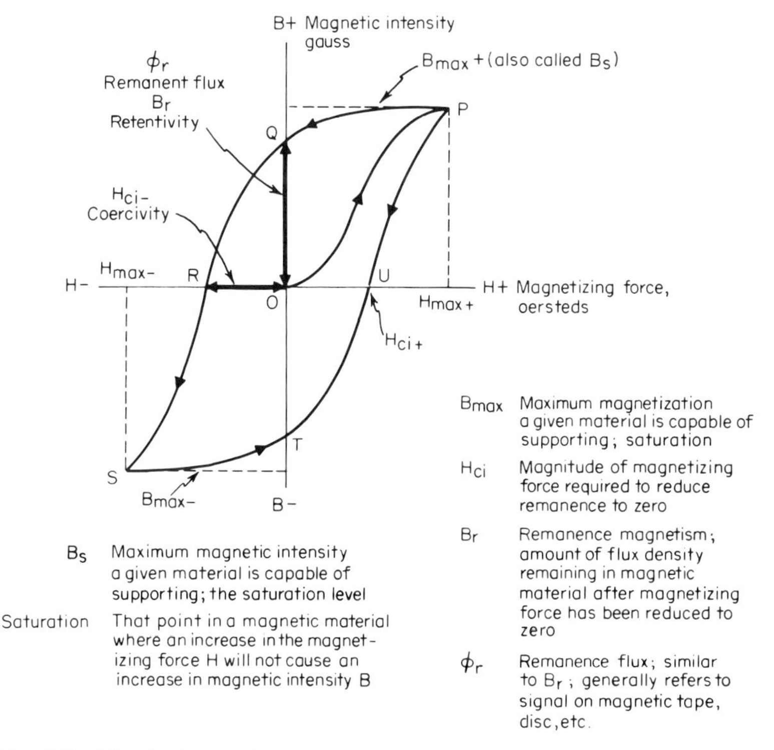

Fig. 7-3 The hysteresis loop.

is reached. The saturation point is designated in Fig. 7.3 as B_{max} or B_s, and the saturating force as H_{max}. If the magnetizing force is reduced to zero (point 0), the magnetic intensity will not follow the original path, but rather take the path *PQ*. The tape still contains a magnetic induction equal to B_r at this point *Q*. If a magnetizing force with a negative polarity is now applied to the medium, the curve *QRS* will be generated. Point *S* is the negative saturation point. The reduction of *H* to zero and its further application in the positive direction will complete the loop. *PQRSTUP* is called the hysteresis loop and illustrates the relationship of *B* to *H*.

With reference to Fig. 7-4, if the magnetizing force *H* is increased from zero to a value less than that required to saturate the magnetic medium, the magnetic intensity *B* will follow the curve *OV* for positive values of *H*, and *OX* for negative values. The reduction of the magnetic force *H* to zero and back to $+H'$ or $-H'$ would carry the magnetic intensity over the minor hysteresis loop *VWWV* or *XYYX*, depending upon the polarity of the magnetizing force. If the magnetizing force has a value of H' minus to H' plus, then a minor hysteresis loop of *XYVWX* will be formed.

To erase a recorded signal from the magnetic medium (tape, etc.), it is necessary to subject the medium to a slowly diminishing cyclic field. Referring to Fig. 7-5, it can be seen that if the magnetizing force *H* is cycled between the positive and negative saturation levels, while the medium is first permitted to be saturated in one direction and then slowly drawn from the influence of the magnetic force, the medium

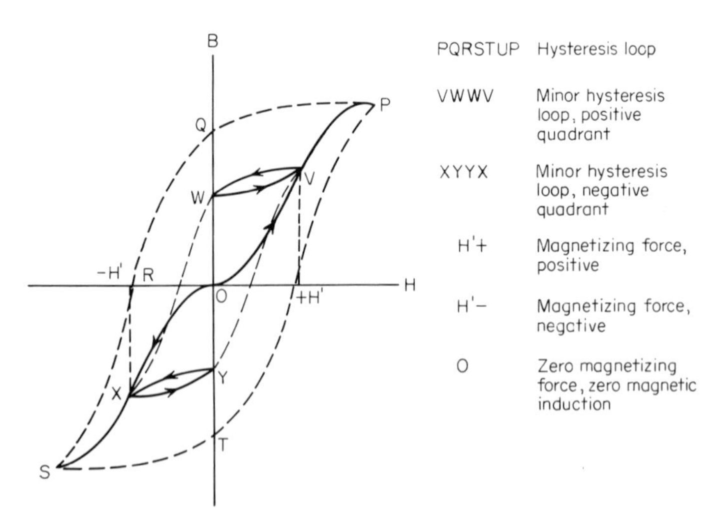

Fig. 7-4 Variations of magnetic induction *B* with magnetizing force *H* when *H* is less than that required to saturate the medium.

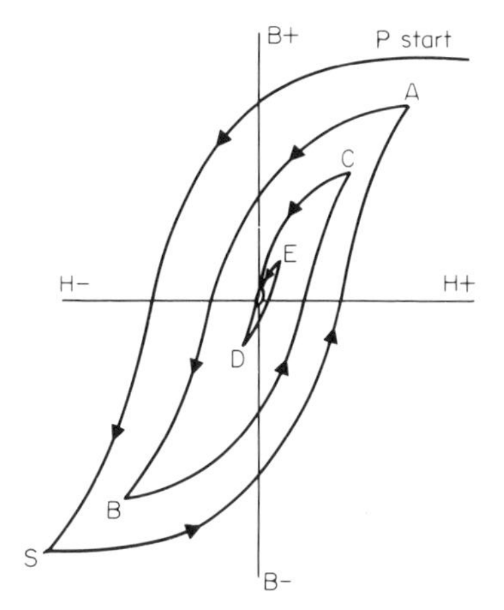

Fig. 7-5 The erase process.

will have been subjected to a slowly diminishing cyclic field, and the magnetic intensity B will have followed the path *PSABCDE,* etc. The magnetic induction has undergone a series of successively smaller hysteresis loops until it reached zero.

In modern commercial bulk-erasing devices called tape degaussers, it is normal that the reel of magnetic tape be placed on a moving turntable and a large 60-cycle electromagnet passed across both sides of the tape reel. This design creates the slowly diminishing cyclic field across and through the tape reel necessary for a minimum of —60 dB of erasure. Most audio recorders, instrumentation-loop recorders, and certain sections of a video-head assembly (Fig. 5-10) are fitted with a special erase head. These heads generally use a high-frequency current to provide the cyclic field. Erasing action could also take place if a rheostat or some similar device were used to bring the cyclic current through the erasing coil of a degausser to full saturation and back to zero. Using this technique, both the magnetic medium and the degausser are stationary; only the current changes.

The Need for Bias

If a graph of the remanence magnetism for each value of magnetizing force is plotted, it will form the curve shown in Fig. 7-6. The nonlinearity indicated is of great importance in magnetic recording. It is normal that direct and FM recording techniques use the relatively linear area of the B_r-H curve between the instep and the knee. Pulse-type recording, on the other hand (i.e., digital and PDM, etc.), use the portion of the curve that is above the knee (saturation). To place the direct and FM record signals in the linear area, it is necessary to add bias to the record current. Particularly with direct recording, a considerable

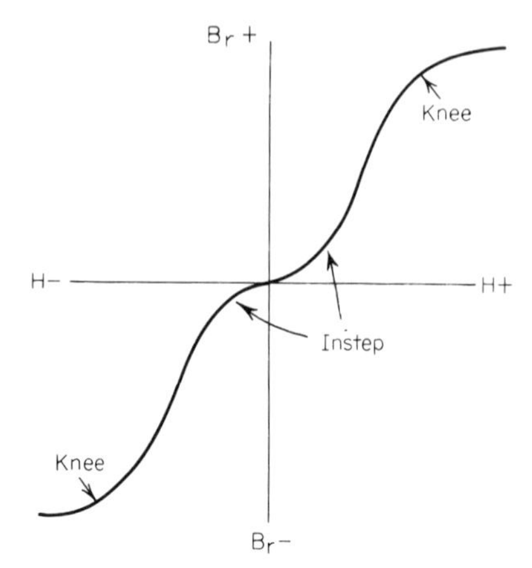

Fig. 7-6 Remanence flux vs. magnetizing force. B_r-H curve.

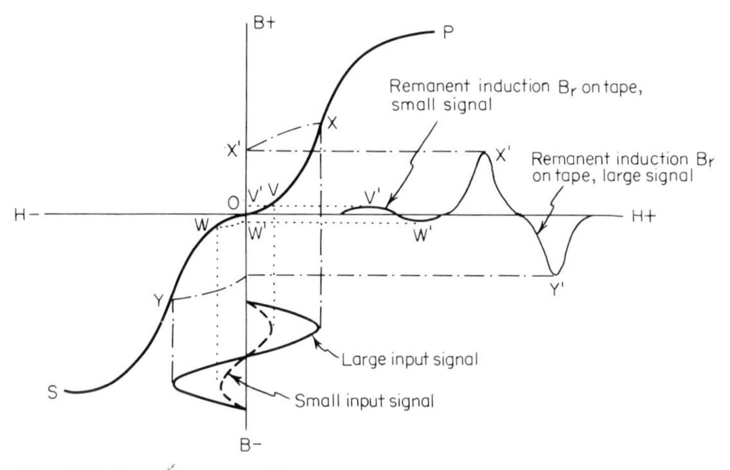

Fig. 7-7 Zero-bias recording.

amount of distortion would occur if the data signal were allowed to place the B_r at the instep or knee of the curve.

Recording without Bias Figure 7-7 shows what happens when a very small or a very large input sine wave is recorded without bias. With the large input, a large amount of distortion results. On the other hand, with the small signal, even though it is recorded on the linear portion of the curve (around the origin 0), the output is so small as to be unusable. Thus, with this type of recording technique, there are two choices, neither of them acceptable: the restriction on the maximum level of operating current results in a poor signal-to-noise ratio; a high operating level results in severe distortion.

Recording with DC Bias The first method of obtaining a linear recording was by using dc bias. This was done in one of two ways. One method is illustrated in Fig. 7-8. Here, starting with completely degaussed tape, enough dc bias is added to the data signal to shift it to the operating area between the instep and the knee. A good linear remanent induction B_r results. It should be noted, however, that only one-half of the *B-H* curve is used. Thus the remanent induction B_r has a relatively low level. Additionally, this method tends to magnetize the record head and tape with the dc component. Thus a low signal-to-noise ratio is obtained. The other method of using dc bias is to magnetize the tape to saturation in one direction before recording. This will leave a remanent induction B_r, as shown in Fig. 7-9. When an input signal with dc bias is applied, it will be placed on the relatively large and straight portion of the hysteresis loop, i.e., between *x* and *y*. The amount of dc bias should be nearly equal to the coercive force H_c and opposite in polarity to the saturation direction. With this dc-biasing method a larger portion of the B_r-*H* curve is used than with the first system. The output signal is greater, the signal-to-noise ratio is better, but as in the method of Fig. 7-8, dc noise is introduced.

Direct-current biasing can provide outputs with a signal-to-noise ratio of better than −30 dB and distortion levels of less than 5 percent. Since the circuitry used is relatively simple, most inexpensive tape recorders use one or the other of the dc-biasing methods described.

Recording with AC Bias As has already been discussed, recording without bias and recording with dc bias result in poor signal-to-noise

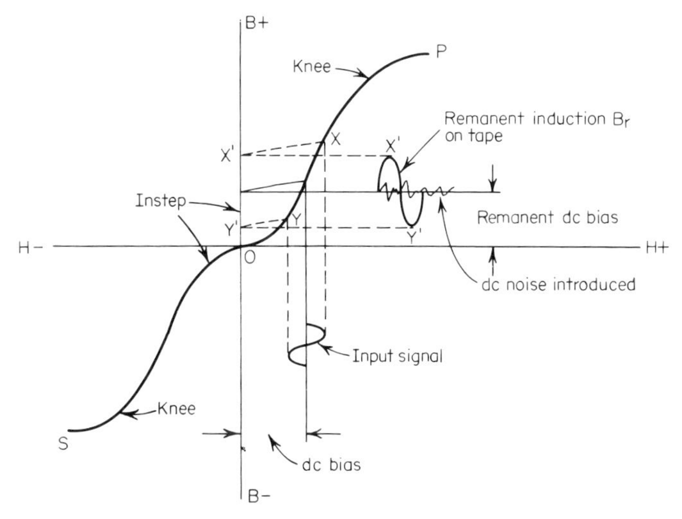

Fig. 7-8 DC-bias recording—method 1.

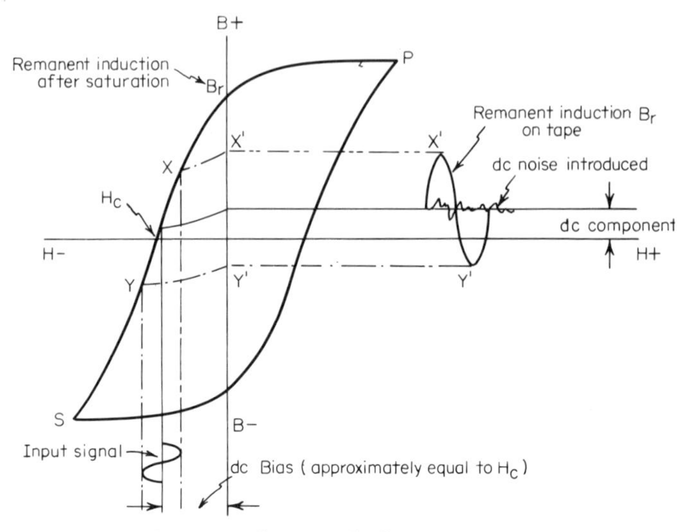

Fig. 7-9 DC-bias recording—method 2.

ratios and severe distortion. Many theories have been proposed as to why ac bias increases the output and reduces distortion; nevertheless, no single theory has yet accounted for all the phenomena that have been observed. Neither has any explanation been accepted by all the authorities.

From the studies of magnetism it is known that the domains are tightly locked together. To loosen these magnetic bonds and allow the signal to modulate the vectors (record a signal by changing the direction of the domain vectors—see Fig. 4-2), a large amount of magnetic energy is required. It would appear that until a particular threshold of response is reached, little or no signal recording can take place. This could be explained by a mechanical analogy; i.e., even though the atmospheric pressure has changed, and the needle of the aneroid barometer should change, it takes a slight tap on the glass to break the bearing friction before the needle is at its right reading. Like the barometer needle, the domain vectors, once unlocked, move with relative ease until the knee of the *B-H* curve (saturation) is reached. Although the nonlinearity of the knee area is fixed for the particular type of magnetic medium, the instep area can be influenced by the ac bias, the ac bias acting as the "tap on the glass." Both the shock of the tap on the glass and effect of the ac bias must die away (return to zero) before the small data signal (pressure, or magnetic flux) is recorded. Generally speaking, the ac-bias strength compared with signal size is anywhere from 5 to 25 to 1.

Referring to Fig. 7-10, it may be seen that the analogy of a push-pull circuit is being used to explain the action of ac bias. Many other valid analogies have been used by other authors (see References at the end of the chapter). However, we will stay with the push-pull explanation for the sake of simplicity. The data have been linearly mixed with the bias. Together they form the waveshape with the envelopes *ABC* and *A′B′C′*. The envelope is, effectively, the curve of the modulating data signal. The envelope *ABC* operates in the positive quadrant, and envelope *A′B′C′* in the negative quadrant, of the *B-H* curve. Examination of Fig. 7-10 shows that when envelope segment *AB* is in the positive straight-line portion of the *B-H* curve, segment *A′B′* is in the nonlinear negative portion. Conversely, when the envelope segment *BC* is in the nonlinear, positive portion of the *B-H* curve, segment *B′C′* is in the linear, negative portion. When using the push-pull analogy, it is reasonable to assume that the resultant remanence induction B_r is the difference between the two resulting curves *V′W′* and *Y′X′* at any instant. If this be the case, the resultant remanence induction *EFGHI* will be a faithful reproduction of the input signal *ABC*.

The ac-bias method of recording produces none of the noise that is inherent with dc biasing. It uses both quadrants, and therefore produces a greater amount of remanence induction. With a larger signal there is a better signal-to-noise ratio. The application of ac bias is

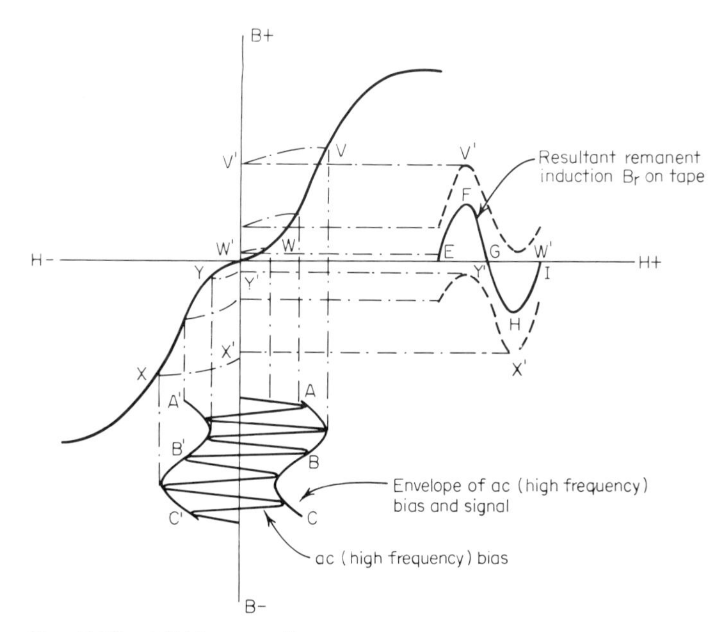

Fig. 7-10 AC-bias recording.

not without its pitfalls, however. To take full advantage of it, the following points must be observed:

1. The ac-bias frequency is chosen to be three to five times the highest signal frequency to be recorded. If this is not done, modulation effects taking place in the record system will give rise to beat notes that will increase the system noise. Audio systems commonly employ 50 kHz as the bias frequency, whereas frequencies as high as 7.7 MHz are used in wideband instrumentation systems.

2. The bias should contain no even harmonics. The even harmonics will make the positive and negative peaks of the bias current unequal. Any asymmetry will produce a dc component which will cause noise.

3. An optimum ac-bias current must be selected, one which is compatible with the heads and magnetic medium in use. This current should be the best balance between extended high-frequency response (a large amount of bias substantially reduces the high-frequency response), low distortion (too little or too much bias causes distortion), and high output.

The adjustment of the amount of bias, particularly with wideband systems, becomes extremely important. Audio systems and low- and intermediate-band instrumentation recorders normally use a fixed amount of bias. This amount has been set by the tape-recorder manufacturer as the value that gives the best balance of requirements by the heads and average reel of tape. For the direct mode of recording, this value will provide a good compatibility of recording from recorder to recorder. With wideband systems, however, the amount of bias must be set for the individual head and reel of tape. As can be seen in Fig. 7-11, the reel of tape has a saturation characteristic curve through the band of frequencies for which it was designed. If a high frequency is selected, i.e., the highest frequency of a particular recording system (in our example, 2 MHz), the amount of bias is set so that there is maximum output of the 2 MHz when it is recorded and reproduced.

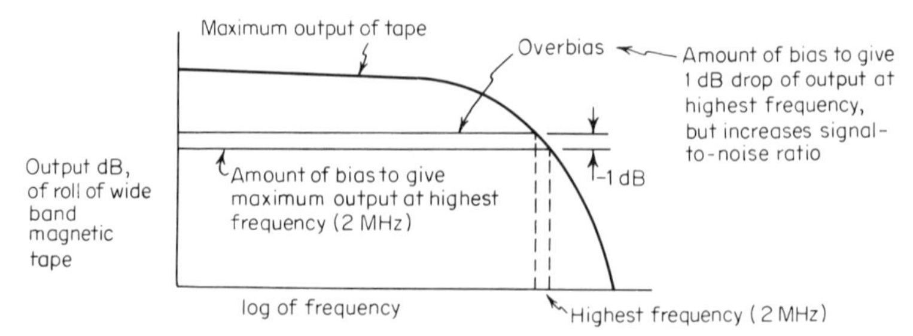

Fig. 7-11 Overbias setting for wideband recorders.

The ac bias is then increased so that the 2-MHz signal drops 1 to 1.5 dB. This is called overbiasing. This technique is used to improve the signal-to-noise of wideband systems while maintaining a wide frequency response.

RECORDING LOSSES

Refer to Fig. 3-2, where it was indicated that if a constant level of input was fed to the record amplifier and recorded on tape at increasing frequencies, the output of the reproduce head would not be constant; it would instead increase at a 6 dB/octave rate. The deviation from the straight line of the 6 dB/octave curve is caused by losses that are associated with head construction, magnetic-medium characteristics, and the speed of the tape, among other things. Some of the high-frequency dropoff of the 6 dB/octave curve is caused by the losses that take place in recording. These are principally due to demagnetization and bias erasure, but eddy currents also play a small role.

Demagnetization

Referring to Fig. 7-1, a series of recorded sine waves will develop a series of kidney-shaped magnetic patterns that are a half-wavelength long. If each pattern can be considered a bar magnet and it is agreed that each is made up of many particles, it will be seen that the length of the individual magnets depends upon tape speed and the frequency of the recorded signal. For very low frequencies the length of the magnet could be on the order of 0.25 to 0.5 in., and for very high frequencies, in microinches. These small magnets are located along the tape so that the like poles of the adjacent magnets are facing each other. The opposition between the like poles will cause a loss of magnetic induction (demagnetization). Of course, as the frequency increases, more and more like poles per inch will be formed and the demagnetism will increase. Also, the opposite poles of any magnet tend to cancel each other. Thus, as the frequency becomes higher and the length of the magnet smaller, the canceling effect of the opposite poles reduces the magnetic induction of the tape.

Bias Erasure

As the amount of bias is increased in a magnetic recording system, the magnetic induction will increase proportionately (to a point). When this bias value is increased further, the amount of magnetic induction decreases (Fig. 7-12). Large amounts of bias current exert an erasing effect similar to an erase head, the greatest effect being felt on the higher frequencies. The reason for this is as follows. The

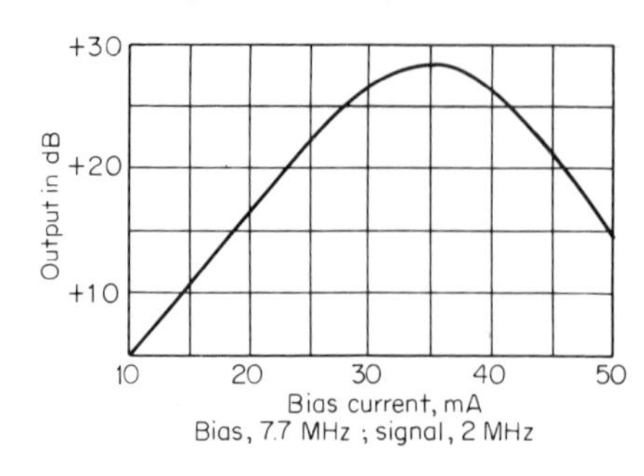

Fig. 7-12 Bias erasure.

prime, or optimum-level, point for the ac-bias field is, essentially, the surface of the tape. It is at this level that the magnetic fields for short wavelengths (high frequencies) are recorded. The longer wavelength uses more and more of the sublayers of the tape. Thus, if the bias field strength is greater than that corresponding to the optimum value, it will have a tendency to drive the signals with the shorter wavelengths below the surface of the tape. Maximum effect of bias erasure will be felt with short wavelengths, and minimum effect with long wavelengths.

Eddy Currents

Eddy currents may be defined as the circulating currents within the core of a magnetic head. They flow in the same direction as the core windings and, if permitted to circulate unhampered, would behave as if there were a number of shorted turns within the core. Such a short would dissipate a considerable amount of energy and decrease the efficiency of the head. Eddy currents increase with frequency. Thus, referring to Chap. 5 on Magnetic Heads, it can be seen in Fig. 5-1 that the number of laminations for the various head cores are different. As the frequency response of the system was increased, the number of head laminations was increased, in an attempt to cut down the amount of eddy-current loss.

DETAILS OF THE REPRODUCE PROCESS

In the earlier part of this chapter and in Chap. 3, it was stated that the basic function of the reproduce head is to change the magnetic field pattern found in the magnetic tape into a voltage e, and the reproduce head will act as a miniature generator following Faraday's law, i.e.,

$$e = N \frac{d\phi}{dt} \tag{7-8}$$

where e = instantaneous voltage
N = number of windings around the core
d = symbol for change
ϕ = magnetic flux
t = time, sec

It has also been stated that the voltage created in the reproduce head follows a 6 dB/octave curve (within certain limits). By this it was meant that as the frequency doubles, so will the output voltage from the head, within the limits set by the losses (Fig. 7-13).

REPRODUCE LOSSES

Gap Losses

It should be noted that the practical output curve of the reproduce head does not follow the theoretical straight line at the shorter wavelengths. This is due, neglecting all other losses for the moment, to the size of the reproduce-head gap l compared with the wavelength λ of the reproduced signal. The deviation from the straight line has been called the finite-gap-length loss, air-gap effect, or gap loss.

The physical and the effective magnetic gap lengths are somewhat different, due to manufacturing tolerances of the gap (size and contour of the edges) and the effects of the final polishing of the head. The polishing is done using very fine grinding tape; the grinding tends to harden the surface of the head at the gap, and this reduces the permeability of the head material in the gap area. Under these conditions, the gap tends to spread out on either side. The effective magnetic

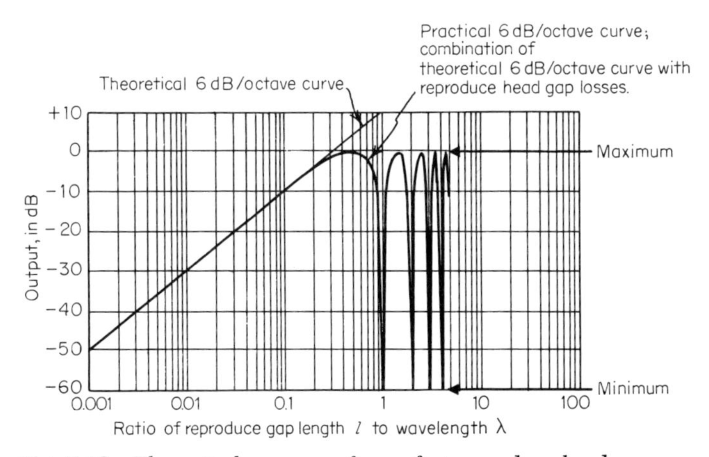

Fig. 7-13 Theoretical response of a perfect reproduce head.

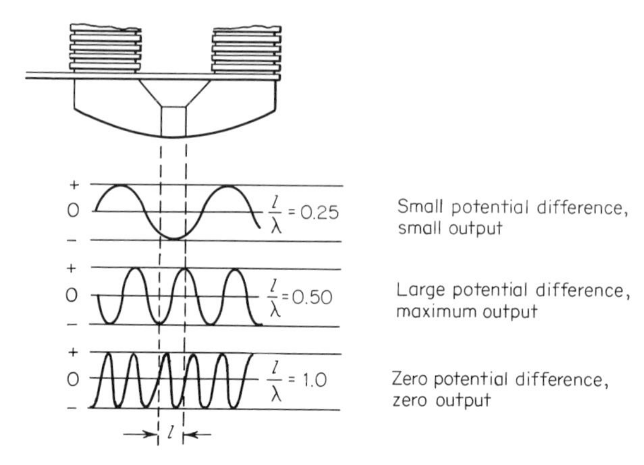

Fig. 7-14 Relationship of gap to wavelength.

gap length is 10 to 15 percent larger than the physical size. Thus l used in the formulas to follow, and as shown on Fig. 7-13, refers to the effective magnetic gap length.

In order for a voltage to be generated by the reproduce head, it is necessary that there be not only a change in the density of the magnetic flux, but also a magnetic potential difference across the head gap. The output level of the head will be determined by the formula 20 log sin $(180° \, l/\lambda)$. Thus, as can be seen by comparing the l/λ ratios of Fig. 7-14 with the corresponding points on the practical response curve of Fig. 7-13, the maximum output for a reproduce head, taking gap loss only into account, is the point where $\lambda/2 = l$. Maximum output will also occur where $3\lambda/2 = l$, $5\lambda/2 = l$, etc. Minimum outputs occur where $\lambda = l$, $2\lambda = l$, etc.

In addition to the reproduce-gap losses, there are also the losses that occur in the recording mode to take into consideration. These are demagnetization, bias erasure, and eddy currents. Thus Fig. 7-13 should be modified to look like Fig. 7-15.

Head-azimuth Loss

Incorrect head-azimuth alignment is a source of high-frequency loss. Correct azimuth alignment, on the other hand, means having the gap sides exactly perpendicular to the tape path. This is an extremely difficult thing to do, with all the manufacturing tolerances involved, particularly with multitrack heads (Fig. 5-4).

Figure 7-16 represents the relationship of a tilted head gap to long, medium, and short wavelengths recorded on magnetic tape. The north and south poles of the various recorded signal are shown by whole and dashed lines, respectively. The head gap is illustrated by two parallel lines. Referring to the long-wavelength portion of Fig. 17-16*a*, it can be seen that if the head is perpendicular to the tape path, both the upper and the lower gap edges will experience the same phase and there will be no azimuth error. On the other hand, if the gap is tilted, the upper and lower gap edges see a phase difference. This will result in magnetic flux lines flowing between the upper edge and the center line of the gap and the lower edge and the center line. These flux patterns will be slightly out of phase, and therefore a resultant flux will occur which produces an output voltage that will be lower than from a head with no azimuth error. With the medium λ (Fig. 17-16*a*), the upper edge of the gap is crossing the south pole, while the lower edge is crossing the north pole. This will mean that the flux lines from the upper edge to the center line will be passing around the core in one direction, while the flux lines from the center line to the lower edge will be passing around the core in the opposite direction. The total resultant flux is zero, and the voltage out will also be zero.

Referring to the short wavelength of Fig. 17-16*a*, it may be seen that the tilt corresponds to one complete wavelength. The total flux across the gap will be zero, as will the output voltage. Figure 17-16*b* shows the changes that will occur when the tape is moved a quarter-wavelength from the position shown in Fig. 17-16*a*. Very little change in output will be noted for long wavelengths. With the medium wavelength, however, the gap flux lines tend to pass around the core in the same direction. The maximum flux density will be at the center line, while the density at the edges will trail off to zero. The resultant voltage output will be much lower than it would be if the flux density had been placed across the length of the head gap, i.e., head gap perpendicular to the tape, but much larger than it was in Fig. 17-16*a*. The

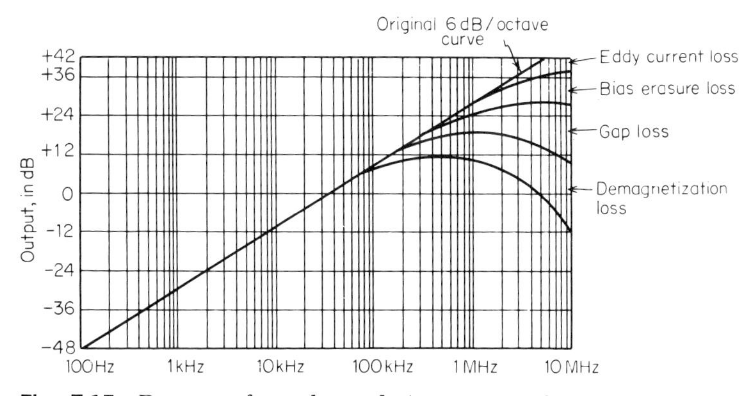

Fig. 7-15 Deviation from the 6 dB/octave reproduce curve due to losses.

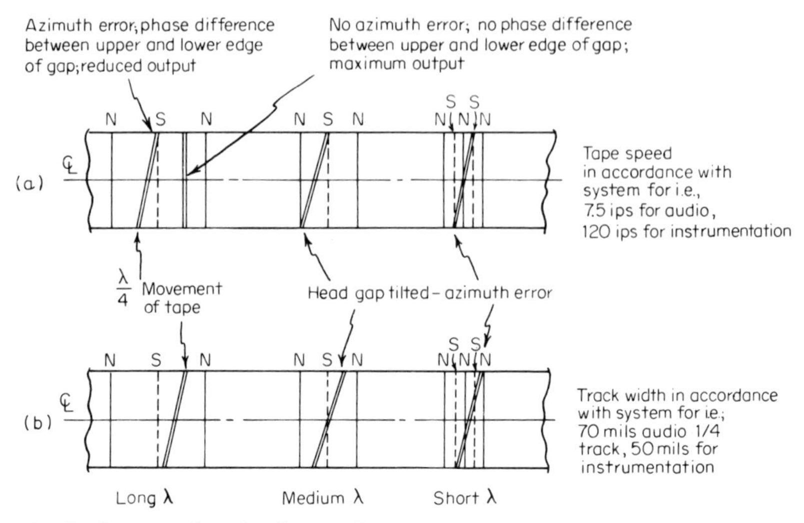

Fig. 7-16 Reproduce-head-azimuth error.

short wavelength of Fig. 17-16*b*, like that of Fig. 17-16*a*, will have an output of zero, since the phase of the flux patterns is such that they cancel each other out.

The mathematics of these losses are ably presented by S. J. Begun, H. G. M. Spratt, and W. E. Stewart (see Refs. 1, 3, and 4 in the list of References at the end of the chapter), but it should be noted that the final equations indicate that there is more than one output peak. Therefore the reader is cautioned, when making azimuth adjustments, to make sure that he has selected the correct setting. This is indicated in Fig. 7-17 and by the equation

$$\text{Alignment loss} = 20 \log_{10} \frac{\sin \dfrac{\pi W \tan \alpha}{\lambda}}{\dfrac{\pi W \tan \alpha}{\lambda}} \quad \text{dB} \tag{7-9}$$

where W = width of recorded track
α = angle of misalignment
λ = wavelength of recorded signal

Separation Loss

A serious loss of output, particularly at high frequencies, will take place in any magnetic recording system when there is separation between the magnetic medium and the head. R. L. Wallace (see Ref. 2 in the list of References at the end of the chapter) has shown that the losses will vary according to the formula

$$\text{Spacing loss} = 55\,\frac{d}{\lambda} \qquad \text{dB} \tag{7-10}$$

where d = distance between magnetic medium and face of head at gap, in.
λ = wavelength of recorded signal, in.

Any slight hump in the surface of the magnetic medium, such as redeposited oxide, lint, dust, etc., or any hollows due to scratches, creases, or oxide loss, etc., will cause a spacing loss and consequent reduction in output (Figs. 7-18 and 7-19). Since this loss is wavelength-dependent, the long wavelengths will be little affected and the short wavelengths greatly attenuated.

Eddy-current Loss

The reproduce head, like the record head, must be laminated to reduce eddy currents. Even the finest head construction does not prevent some loss due to eddy currents; therefore eddy-current loss must be taken into account and added to the balance of the reproduce losses.

Surface Loss—Magnetic Medium

Whenever the surface of the recording medium is rough (magnetic tape, discs, or drums, etc.), it will not only wear the heads, but will also cause an irregular separation. This bounce, or head bounce, as it is called, will cause separation losses similar to those already discussed. They are dynamic in nature, however, and almost impossible to compensate for. Therefore the reader is urged to make a careful choice of the type and quality of the recording medium. Wideband magnetic recorders, in particular, require extremely smooth tape to prevent high-frequency losses due to the medium surface.

When separation, eddy current, and surface losses are added to the

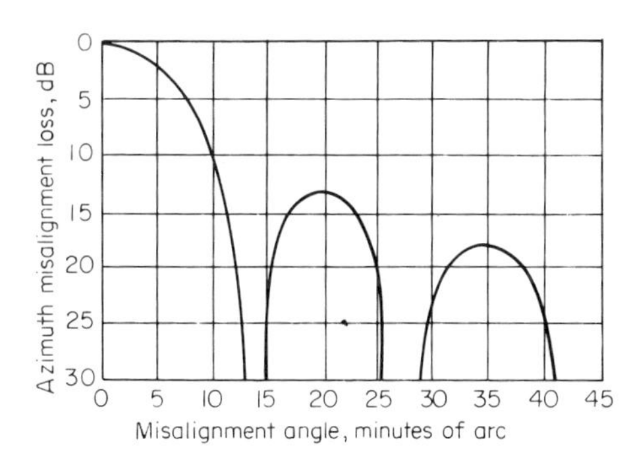

Fig. 7-17 Azimuth misalignment loss using 600 kHz signal and tape speed of 120 ips.

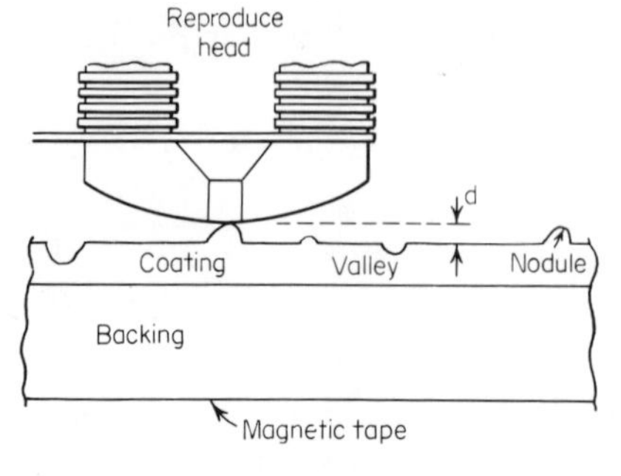

Fig. 7-18 Separation losses due to tape-surface imperfections.

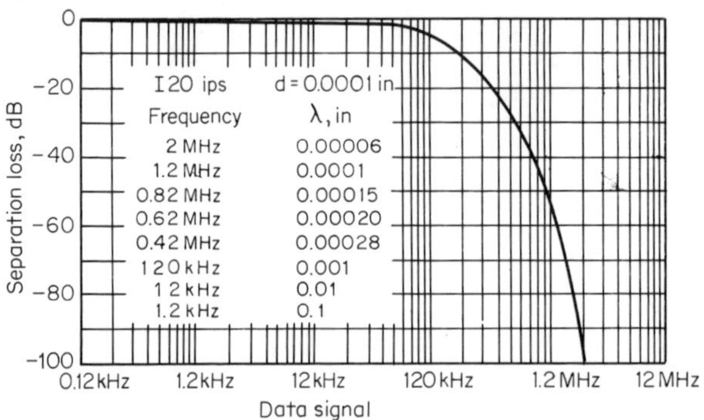

Fig. 7-19 Separation loss of an instrumentation recorder.

previously discussed reproduce losses, the 6 dB/octave curve must be further modified. Figure 7-20 shows the result of these additions. Each loss in turn has reduced the total output of the magnetic medium both for amplitude and frequency response.

Long-wavelength, Low-frequency Losses

Most of the losses so far discussed have been short-wavelength, high-frequency losses. There are some long-wavelength, low-frequency losses that should be taken into account, however, when considering the overall response curve of the system. These are:

1. Half-wavelength to head-contact-area ratio
2. Head bump
3. Tape-coating thickness
4. Faraday's law, $e = N\, d\phi/dt$

Half-wavelength to Head-contact-area Ratio Loss When the half-wavelength of the recorded signal is longer than the length of the head-to-tape contact area (ratio more than 1), the reluctance of the flux return path increases. As can be seen in Fig. 7-21*b*, there is a large air gap in the flux path between the head pole piece and the tape. Thus the magnetic flux does not follow a path that is entirely within the head core, as it does in Fig. 7-21*a*, with its shorter wavelengths and half-wavelength to head-to-tape contact area ratio of less than 1. It is possible, under very long wavelength conditions, that the flux lines may pass through only part of the head windings. If this should be the case, the output voltage of the head would be considerably less than the predicted 6 dB/octave characteristic curve.

Head-bump Loss When the recorded wavelength on tape approaches the overall dimension of the two head pole pieces, the pole pieces begin to act as a second gap. The additional flux lines will add to the normal output. Referring to Fig. 7-22 it can be seen that when a number of half-wavelengths is equal to the combined distance across the head pole pieces, there is an increase in the head output (bump). This increase in output gets larger as the number of half-wavelengths required gets smaller, until one half-wavelength equals the combined distance across the pole pieces. At this point the increase in output will be approximately +3 dB. Any further increase in the size of the wavelength will cause a rapid falloff in the head output. Referring to Fig. 7-22, it can be seen that at the 1, 2, 3, etc., wavelength points, the head output dips, the greatest dip being when one wavelength equals the combined distance across the pole pieces. The only way to alleviate these problems is by careful head design. There will always be a small amount of bump loss in any head.

Tape-coating-thickness Loss Both theoretical and experimental evidence (see R. L. Wallace, Ref. 2 in the list of References at the end of the chapter) would indicate that only the surface level of the magnetic medium contributes measurably to the recording of short wavelengths.

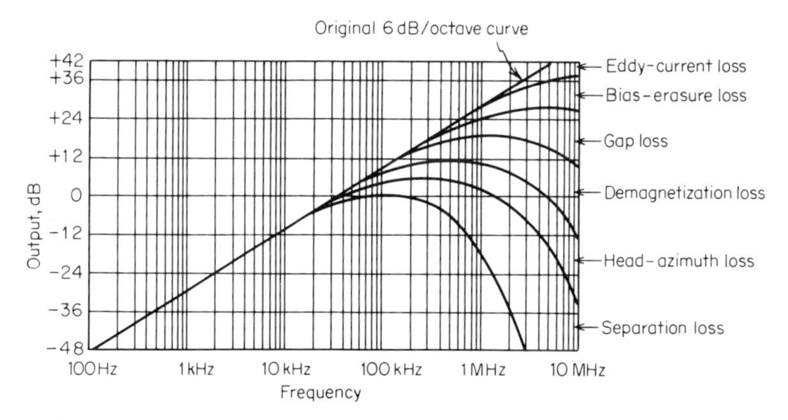

Fig. 7-20 6 dB/octave reproduce-head curve—high-frequency, short-wavelength losses taken into account.

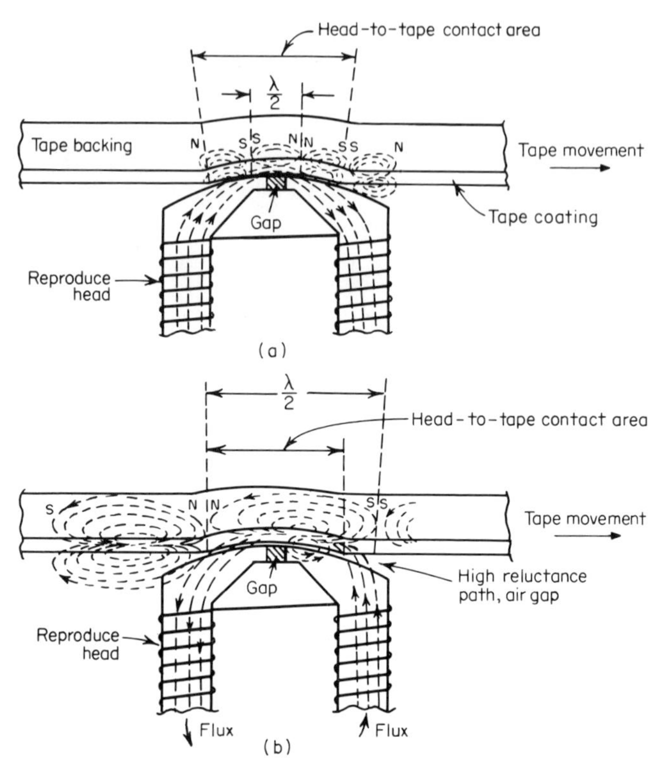

Fig. 7-21 Half-wavelength to head-contact-area ratio loss.

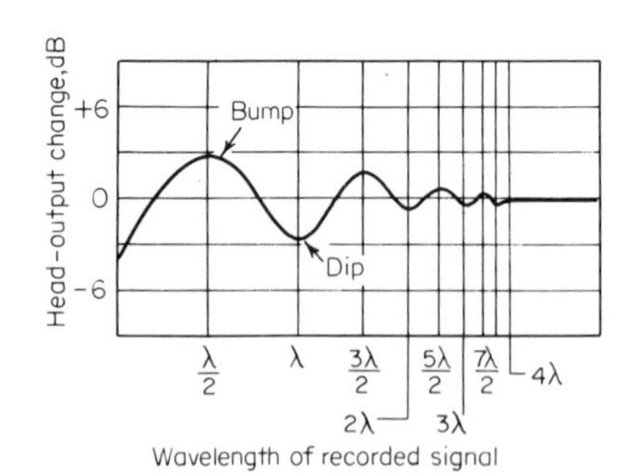

Fig. 7-22 Theoretical head-bump curve.

The long wavelengths use more and more of the sublayers. As the tape-coating thickness is increased, the various layers of magnetic particles are less and less subjected to optimum bias currents, and the recording sensitivity decreases. Since the tape must be chosen to suit a range of frequencies, the choice of coating thickness must be a compromise. It must suit both long and short wavelengths, while providing good signal-to-noise ratios, surface smoothness, wear qualities, etc. Generally, the coating thickness is chosen so that the long wavelengths will be limited by the thickness, since in effect the amplitude response of long wavelengths is proportional to the coating thickness and the effective penetration limit of the recording field.

Losses Due to Action of Faraday's Law As has already been stated, the reproduce head will act as a miniature generator following Faraday's law, $e = N\, d\phi/dt$ [Eq. (7-8)]. It is important to note that the voltage generated is not proportional to the magnitude of the flux, but rather to its rate of change. Thus, since the rate of change at long wavelengths is very low, the output voltage of the reproduce head will be very low. Under these conditions the output voltage will approach the inherent noise level of the system and be lost.

THE NEED FOR EQUALIZATION

From the discussions of recording and reproducing losses, it is obvious that if a band of frequencies were recorded at the same amplitude, the output of the reproduce head would not follow the same pattern, but instead, the output would follow a 6 dB/octave curve within limits. Beyond these limits there would be a falloff due to the long- and short-wavelength losses (Fig. 7-23).

Depending upon the recording technique used, an output characteristic curve with short- and long-wavelength (high- and low-frequency) losses is not always acceptable. Nor in fact can a 6 dB/octave curve always be accepted, even though it might be linear. A direct record and reproduce signal, shown in Fig. 7-23, was recorded at constant amplitude from one end of the bandwidth to the other. It must be fed from the reproduce amplifier in the same fashion. Figure 7-23*b* definitely shows that this is not the case in its uncompensated state (raw signal from the reproduce heads). For this recording technique (direct) some form of compensation is needed, i.e., amplitude equalization.

On the other hand, if the type of recording technique required only two-state operation, i.e., one and zero, or high and low potential, etc., such as digital and PDM, any slight variation in the amplitude would cause no serious problems. Generally speaking, neither would the lack of equalization pose problems with low-, intermediate-, and wideband group I FM, since in these modes of recording, the center-carrier frequencies and the related sidebands are located on the relatively linear portion of the 6 dB/octave curve. With wideband group II FM, however, the center carrier and its sidebands operate in the region of the reproduce curve between 500 kHz and 1.2 MHz. As can be seen in Fig. 7-23*b*, this is a relatively nonlinear section of the curve. For this reason it has been found necessary to include equalization of some sort.

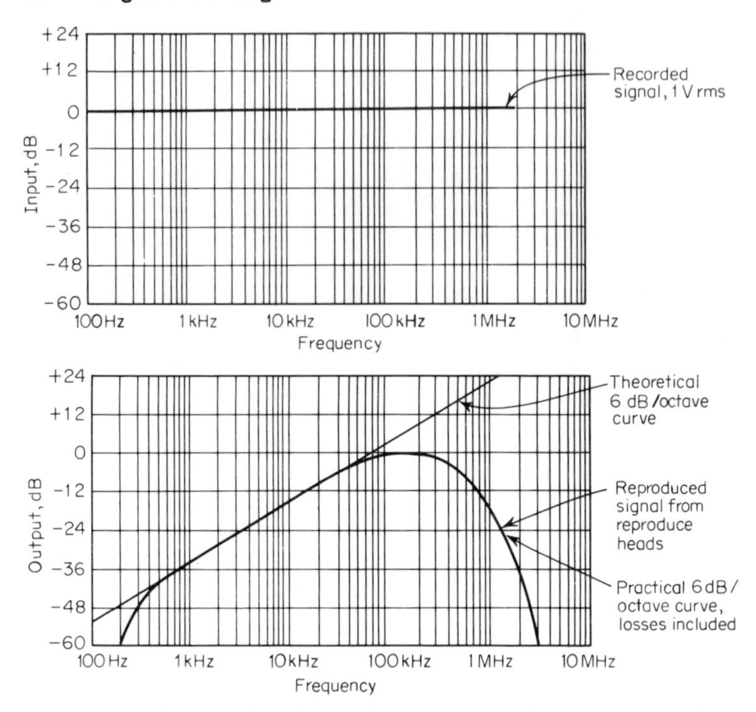

Fig. 7-23 The record and reproduce curves, indicating the need for equalization.

Amplitude Equalization

Instrumentation The curves involved in equalization for instrumentation recording are shown in Fig. 7-24. It should be noted that the resultant output, although low in value, is flat from one end of the passband to the other. This is quite satisfactory for most normal instrumentation readout devices which are also quite linear. In some rare instances, the user of instrumentation equipment has some special requirements calling for special response curves that are not flat, but emphasized at the high or low end. In most of the instrumentation recorders the equalization circuits are placed in the early stages of the reproduce electronics, the one exception being in the older intermediate-band direct systems. In these 500-kHz systems there was considerable loss in the record heads due to eddy currents, particularly above 125 kHz. A small amount of record equalization was included (a boosting of the record currents above 125 kHz) to maintain a constant amount of flux on the tape. As a result this type of technique was called constant-flux recording. With the newer heads record-current boost is no longer needed.

Audio (CCIR, NAB, AME) Audio equalization poses different problems from instrumentation. The readout device (the human ear) is not linear. Thus, if the same level of low-frequency sound were fed to the speakers as that for the mid-frequencies, the music would sound flat, unbalanced, or lacking in bass, because the human ear follows the curve shown in Fig. 7-25. The curves indicate that, in order to hear the bass notes at the same level as the mid-frequencies, the bass notes must be increased in level by some 30 dB. The higher frequencies must be treated in the same way. For these reasons, the response curves for audio recorders have very different shapes from those used for instrumentation. It is not recommended, therefore, that an audio recorder be used for instrumentation purposes where a flat response is required and where time-base errors must be kept down to nanoseconds. Nor should an instrumentation recorder be used for audio purposes where high-fidelity music or speech is required.

Consideration must be given to the placement of the equalization circuits in the audio systems, since the location can make a substantial difference with respect to the signal-to-noise and the distortion figures.

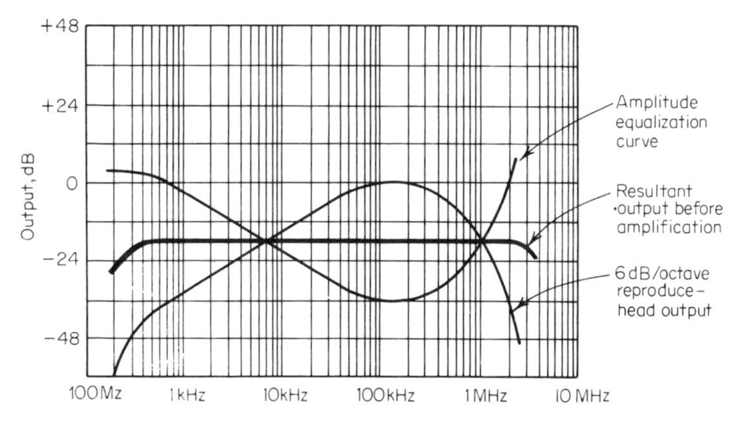

Fig. 7-24 Instrumentation amplitude equalization.

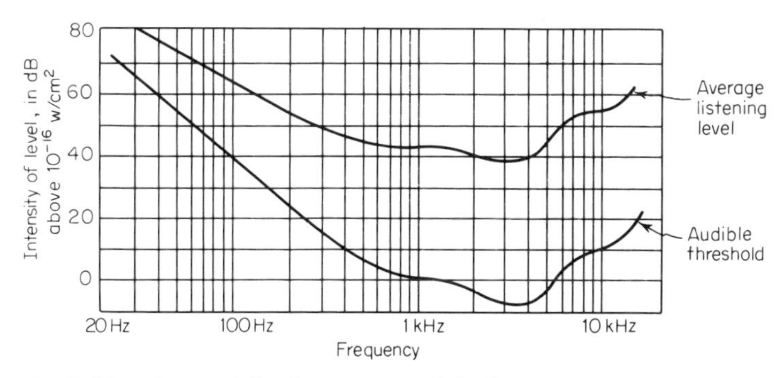

Fig. 7-25 The equal-loudness curves of the human ear.

For example, the bass notes must be boosted by some 30 dB to be heard at the same level as a 3-kHz note. To put this amount of bass energy into the record circuits, considering the amount of energy contained in low frequencies, would overload the record driver circuits and cause excessive distortion. On the other hand, if all the bass boost were placed in the reproduce circuits, it could contribute to noise by amplifying the hum picked up by the reproduce head. Consequently, it is general practice, particularly when using NAB (National Association of Broadcasters) Standards, to boost the low end (low frequencies) by a small amount in the record side (approximately +6 dB) and compensate for the rest of the boost needed in the reproduce electronics.

Audio treble boost is done primarily in the record section, to avoid amplifying tape hiss and reproduce preamplifier noise. Since the short-wavelength losses are so large, it is not possible to compensate for all the loss in the record stage, or the record current would be high enough to saturate the tape and cause marked distortion. Figure 7-26 shows some of the common equalization curves in use with professional-type recorders. Notice that the record CCIR curve, which is a European equalization standard, reduces the bass boost. This is done to overcome the large amounts of high energy that are produced in organ music, so popular in Europe. The NAB and AME record curves show a boost of about +4 dB at the low end. Notice that the center of the band of the NAB and CCIR curves is relatively flat, although the CCIR curve introduces considerable boost at the high end. The AME record curve (Ampex mastering curve) is a professional equalization curve used for making master tapes. A relatively large amount of boost is used in those areas that are most responsive to the normal hearing range of the human ear. This type of equalization can be extremely dangerous in the hands of the uninformed user. The center frequencies are boosted close to saturation level with this system. Care must be observed in its use, or the recordings will be severely distorted. The normal equalization curves used with the majority of home recorders is the NAB system. With these standards (NAB), the record equalization, together with the reproduce equalization and the characteristic response curve of the record head, will result in an output that will be within the following tolerances:

Flat within ±1 dB from 100 Hz to 7.5 kHz

Not more than 1 dB up or more than 4 dB down at 50 Hz and 15 kHz

It has often been said, in the audio recording trade, that the NAB standard has a "50-μsec time constant (transient response)." This is simply

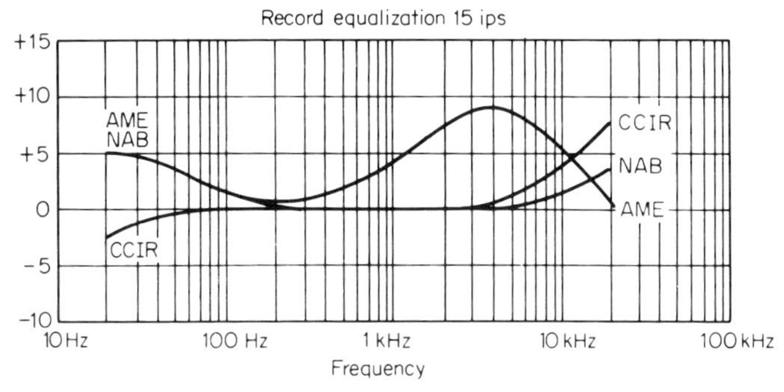

NAB - National Association of Broadcasters
CCIR - Consultant Committee for International Radio (Europe)
AME - Ampex Mastering Equalization (professional use only)

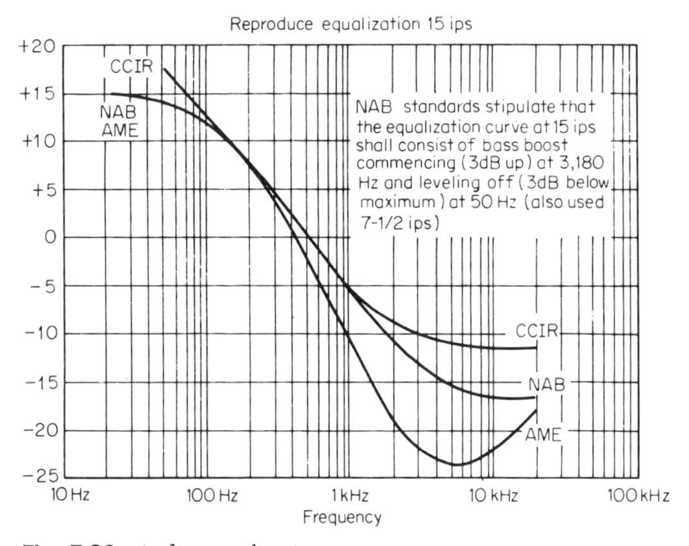

Fig. 7-26 Audio equalization curves.

another way of stating that the +3-dB point is at 3,180 Hz. For more information on the other equalization standards, the reader is referred to J. G. McKnight, Ref. 12 in the list of References at the end of the chapter.

Phase Equalization

There are many magnetic recording applications, in the instrumentation field in particular, where the phase response of the recorded/reproduced signal is extremely important, especially since the introduction of longitudinal wideband recording has permitted recording of direct signals to 2 MHz and FM to 500 kHz. This broadening of the bandpass has made possible expanded FM/FM telemetry multiplex systems and the recording of high-frequency square waves and repetitive pulse trains. However, in multichannel FM/FM telemetry systems, if the phase response across the telemetry channels is not reasonably linear, the data will be distorted and unusable. The same is true with pulse-type recording. A square wave, for example, consists of all the odd harmonics of the fundamental frequency. If any of the harmonics are delayed in phase, the reproduced pulse will not be square, but would be distorted, as shown in Fig. 7-27.

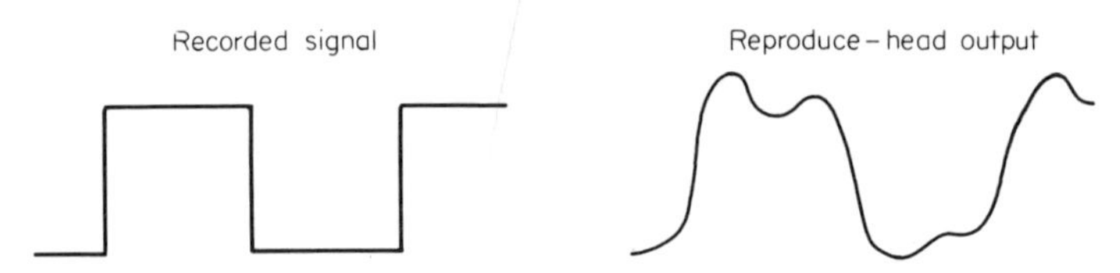

Fig. 7-27 Distortion of square wave due to nonlinear phase response.

As has been mentioned several times earlier in this book, there is a 90° phase shift between the recorded and reproduced data. This is due to the current-voltage relationship between the two processes. To this 90° shift must be added the phase shift produced by the other components of the system, i.e., the phase shift introduced by the L and C components of the amplitude equalizers and the inductive reactance of the heads.

Phase equalization may be accomplished by using specially built delay lines and/or constant-time-delay low-pass filters. In addition, selection of the proper portion of the bandpass for operation will help to overcome phase errors.

REFERENCES

1. Begun, S. J.: "Magnetic Recording," Rinehart & Company, Inc., New York, 1949.
2. Wallace, R. L., Jr.: The Reproduction of Magnetically Recorded Signals, *Bell Syst. Tech. J.*, pt. II, vol. 30, no. 4, p. 1145, October, 1951.
3. Spratt, H. G. M.: "Magnetic Tape Recording," The Macmillan Company, New York, 1958.
4. Stewart, W. E.: "Magnetic Recording Techniques," McGraw-Hill Book Company, New York, 1958.
5. Lowman, C. E., and G. J. Angerbauer: "General Magnetic Recording Theory," Ampex Corporation, Redwood City, Calif., 1963.
6. Haynes, N. M.: "Elements of Magnetic Tape Recording," Prentice-Hall, Inc., Englewood Cliffs, N.J., 1957.

7. Weber, P. J.: "The Tape Recorder as an Instrumentation Device," Ampex Corporation, Redwood City, Calif., 1967.
8. Burstein, H.: Tape Recording, *Radio Telev. News,* September–December, 1954, January–February, 1955.
9. Davies, G. L.: "Magnetic Tape Instrumentation," McGraw-Hill Book Company, New York, 1961.
10. Mee, C. D.: "The Physics of Magnetic Recording," Interscience Publishers, Inc., a division of John Wiley & Sons, Inc., New York, 1964.
11. Daniel, E. D., P. E. Axon, and W. T. Frost: A Survey of Factors Limiting the Performance of Magnetic Recording Systems, *J. Audio Eng. Soc.,* vol. **5**, no. 1, p. 42, January, 1957.
12. McKnight, J. G.: Flux and Flux-frequency Measurements and Standardization in Magnetic Recording, *J. SMPTE,* vol. 78, pp. 457–472, June, 1969.
13. Lowman, C. E.: "The Magnetic Tape Recorder/Reproducer and Concept of Systems Used for Recording and Reproducing FM Analog Test Data," vol. 1 of "Fundamentals of Aerospace Instrumentation," Instrument Society of America, Pittsburgh, Pa., 1968.

CHAPTER EIGHT

The Transport

SELECTING A MAGNETIC RECORDING SYSTEM—CONSIDERATIONS

The selection of the proper magnetic recording device is often difficult, because many factors must be taken into consideration. These include:

1. Cost: Does it fit the budget?
2. Performance: Will it record and reproduce without error?
3. Bandwidth: Is the bandwidth wide enough?
4. Reliability: Will it perform for reasonable periods without failure?
5. Compatibility: Is it compatible with existing systems? Will it reproduce tapes, cassettes, and discs from older equipment?
6. Standardization: Is it standard with the industry? Are spares, tapes, cassettes, and discs readily available? Does it have standard speeds, tape widths, and disc, reel, and cassette sizes?
7. Flexibility: Can it be expanded, modified, or updated?
8. Maintainability: What type of personnel is needed for repair?

No single magnetic recorder will satisfy all users. Nor, in fact, will a magnetic recorder from one category fit well into another; i.e., although computer operation requires fast start-stop time, tape-speed stability is relatively unimportant. The instrumentation user is not particularly concerned with fast start-stop, but is greatly interested in tape-speed stability (low flutter and wow). Where audio requires relatively narrow bandwidths at slow speeds and nonlinear equalization (to make the music sound better to the human ear), the video user is most concerned with very wide bandwidths and time-base stabilities in the nanoseconds. Where one user wants 1 or 2 tracks, another may require 14.

THE IDEAL TRANSPORT

Surprisingly enough, in spite of all the seemingly opposing standards just mentioned, the basic design principles of most magnetic transports are the same. The tape transport should all ideally meet the following standards:

1. It should move tape across the recording and reproducing heads at a perfectly constant rate and be completely free of short- and long-term speed errors.
2. It must move the tape accurately across the heads even with a bad tape pack or bent reels.
3. It must be capable of recording information so that it may be reproduced in exact time interval.
4. It must be capable of passing tape across the heads hundreds and thousands of times with minimum wearing of heads, tape, or disc.
5. It should be able to accelerate the tape or disc from stop to full speed instantaneously and stop it instantly.
6. It should be capable of being controlled locally or remotely.
7. It should be capable of operating in a wide variety of temperatures, altitudes, shock conditions, etc.

Of course, there is no single transport that meets all these requirements. In fact, there is no transport that comes even close. For the most part a trade-off must be made for every increased demand. Where portability is required, maintainability becomes difficult due to the compactness. Where multitrack recording is required, the cost increases. Where time stability is required in the nanoseconds, complicated tape-speed and head-drum (video) servo systems are required. When very wide bandwidths are desired, high head-to-tape contact speeds are required. These quite naturally decrease the amount of recording time available from a reel of tape.

It must be concluded from all this that there is no ideal transport that can be used for all purposes. It would seem, therefore, that the best way of making a choice of a magnetic recording system is to know the principles behind transport design, and using this knowledge, to select a magnetic recording system that fulfills as many of the better design principles as possible, while satisfying the initial considerations of cost, performance, bandwidth, etc.

Cassettes and Cartridges—Discussion Delayed

Although the basic operation of transports using cassettes or cartridges is similar to those that will be discussed in this chapter, because of the great popularity of these devices, the author will discuss them separately in Chap. 12.

TRANSPORT COMPONENTS

The Baseplate Assembly

In order to move the tape across the heads at exactly the right angle (90° to the head gap), the components that are used to guide the tape and mount the head assembly must be referenced to some form of precision surface. The exact form that this reference surface takes varies from machine to machine. In general, however, it is provided by a precision ground plate or a series of reference points established by milling or grinding. This provides for easy interchangeability of tapes between machines without introducing guiding errors.

Some of the components that are mounted on the precision surface are capstan assembly, pinch-roller assembly, turnaround idlers, tape guides, and magnetic heads. Where zero-loop and optimum-loop drive systems are used, pinch rollers and turnaround idlers are not included in the tape path. Nevertheless, the reference surface retains its importance.

Longitudinal Drive Systems

Many types of drive systems have been used to move the magnetic recording medium across the heads. The most common of these are the open-, closed-, optimum-, and zero-loop drive systems, as well as the rotating-head (transverse-scan), helical-scan, and disc-recorder drive systems.

Each of these drive systems has been used in one or more of the audio, video, instrumentation, or digital recording fields. Each has its advantage and disadvantage. Each will be discussed in turn in the following pages.

Open-loop Drive System This is the simplest and most inexpensive of all the drive systems. It pulls the tape across the heads via a capstan and a single pinch-roller assembly. The capstan is driven by a hysteresis-synchronous motor. When the tape is clamped to it by the pinch roller, the tape is pulled at the speed of the capstan. It can be seen, in Fig. 8-1, that there are long unsupported lengths of tape. These create flutter (tape-speed instability). To overcome that problem, a series of fixed and rotating guides are located in the head area, and an inertia idler is installed between the supply reel and the heads. Extremely high grade bearings are used for the low-mass rotating guide to prevent the introduction of flutter due to bearing irregularity. The inertia idler is used to isolate the supply reel and any tape-pack irregularity from the heads. In the older versions, a large-mass flywheel is attached directly to the idler shaft. Unfortunately, this large mass had to be brought up to speed, or slowed down by the tape, during starting or stopping. This caused long start-stop times and some distortion of the tape, particularly with ¼- and ½-in. widths. Additionally, when a tape-speed servo system was used, the large mass made it difficult to achieve a compensated state quickly. For this reason, in the later versions, a viscous-damped flywheel was used. The attachment of the flywheel mass to the idler shaft is through a viscous fluid (silicon grease). Using this type of flywheel, the tape has only to turn the idler itself during the starting periods and the flywheel will catch up later through the energy supplied via the viscous fluid. When the flywheel is up to speed, it provides the large mass to reduce the flutter components in the tape movement. Using the viscous-damped flywheel results in servo systems reacting faster, since only the idler need be speeded up or slowed down immediately, and the flywheel can catch up later. The viscous flywheel acts as a damping device for large changes of tape speed. Generally speaking, the action of the viscous system reduces the flutter by 50 percent.

The open-loop drive system is the most popular drive system in the audio field. It plays a prominent role with cassette and cartridge recorders.

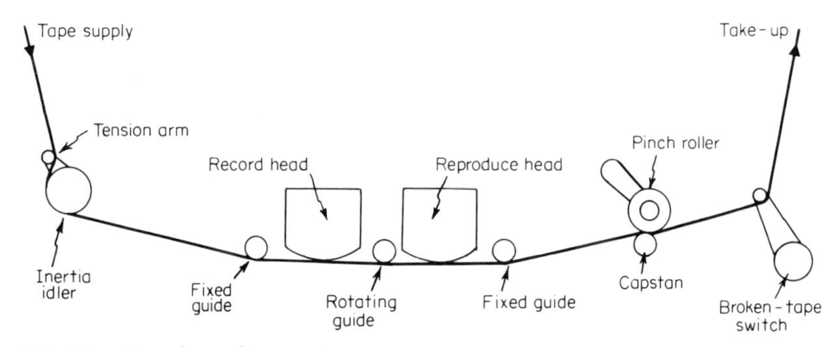

Fig. 8-1 Open-loop drive system.

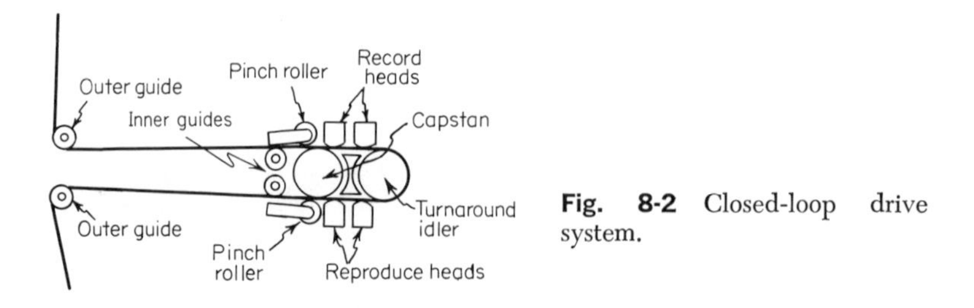

Fig. 8-2 Closed-loop drive system.

Closed-loop Drive System In the closed-loop system, as may be seen in Fig. 8-2, the unsupported lengths of tape are kept to a minimum to eliminate any vertical displacement of the tape across the heads. The capstan is driven at a constant speed via belt or direct drive. Two pinch rollers are used which, together with the turnaround idler, heads, and capstan, complete the closed-loop configuration. The heads contact the tape in the relatively short span of tape between the capstan and the turnaround idler.

The capstan has been driven by a variety of motors. Initially, hysteresis-synchronous motors were used; now, however, the trend is toward the use of dc printed motors, or ac split-phase induction motors. The capstan speed is kept very constant during the record mode (errors in speed are quoted in the microsecond range) by either synchronizing it with the power source or by using some form of servo system. These may take the form of slotted skirts or photoetched disc tachometer. These and other servo systems will be discussed under separate headings. The outer and inner guides provide isolation from tape-reel irregularities for the head area.

Optimum-loop Drive System The optimum-loop drive system (Fig. 8-3) uses a directly driven, all-metal, hollow capstan, a dc printed-circuit motor, and a highly sophisticated capstan servo system to achieve a time-base stability of ± 0.5 μsec. The hollow aluminum capstan hub has two sets of three rows of perforations that mate with a vacuum manifold over a 67° area. As a vacuum is applied to the capstan hub, the tape entering and leaving the capstan area will be drawn against it by the vacuum, and the tape will be moved at the same speed as the capstan. One set of perforations is used with ½-in. tape, and both sets with 1-in. tape. Since the capstan hub has a relatively large diameter (approximately 4.5 in.), extremely close matching tolerances can be maintained during the manufacture.

The capstan assembly is fitted with a photoetched tachometer disc. It is attached directly to the capstan shaft, and thus its output truly represents the capstan speed (see capstan servo systems).

A bidirectional, dc printed-circuit motor provides the high-torque and

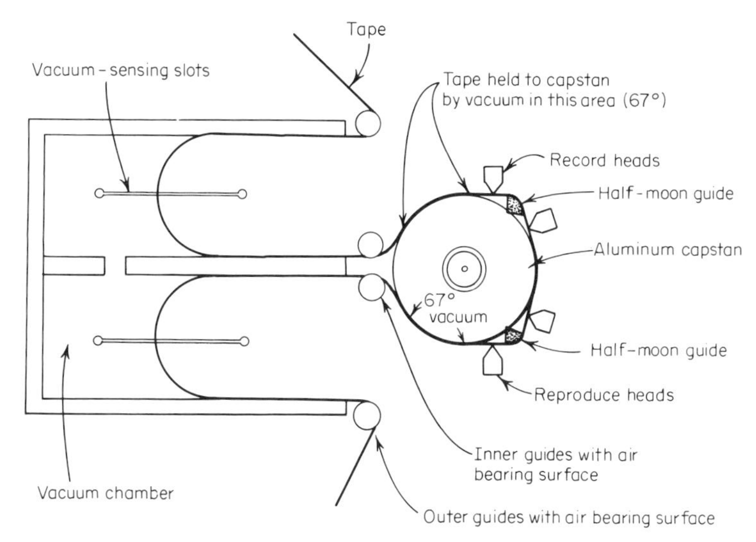

Fig. 8-3 Optimum-loop drive system.

low-mass drive system which, together with direct drive (no belts, gears, pinch rollers, or pulleys), provides maximum reliability.

The tape movement into and out of the capstan area is closely controlled by the use of a vacuum chamber (Fig. 8-3) and the air-lubricated guides. The vacuum chamber provides:

1. Low-mass storage. Weight of tape in the vacuum chamber is negligible.
2. Tension of tape. This is a combination of the radius of tape arc in the vacuum chamber and the vacuum level. Since both are held constant, the tape tension will remain constant.
3. Guiding. Since the chamber is the same width as the tape, it provides a reasonably long tape-edge guide. The air-lubricated guides hold the tape in alignment with the tape path while providing a friction-free air bearing to reduce the wear on the tape and guides.

Zero-loop Drive System Referring to Fig. 8-4, it may be seen that the tape is clamped to the capstan by the spring-loaded heads. The surface of the capstan has been coated with a polymer to provide the necessary resilience so that the heads are not damaged as they squeeze the tape against the capstan surface. Additionally, the polymer assists in preventing tape slippage.

An earlier zero-loop system used vacuum to hold the tape against the capstan surface. This vacuum was applied to precision-machined slots in the capstan surface via the hollow capstan. As the tape was

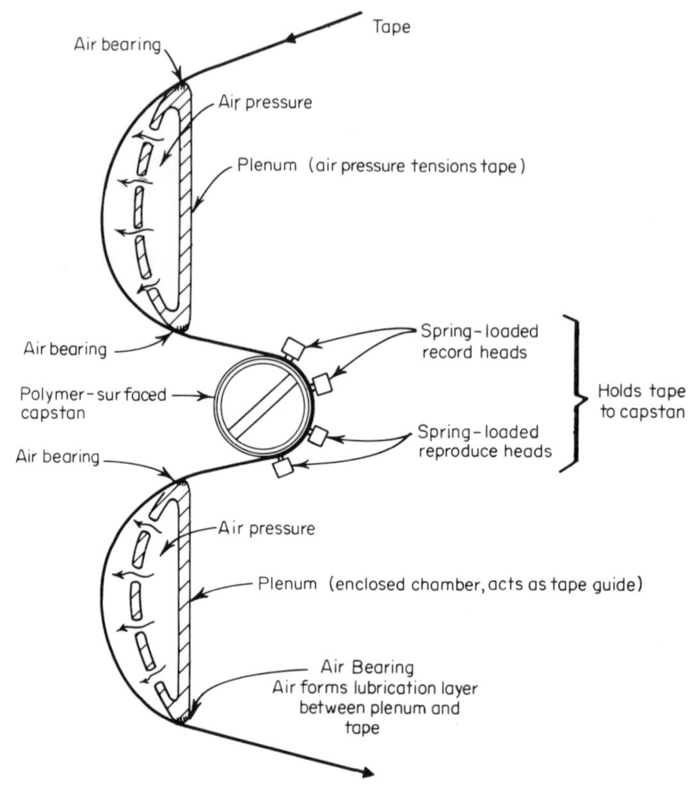

Fig. 8-4 Zero-loop drive system.

pulled into the slots by the vacuum, it formed small bridges of tape across the sides of the slots (similar to that shown in Fig. 5-9). The heads were advanced against the tape at these points.

Modern machining methods and availability of plastic materials for capstan surfacing have revived the zero-loop systems. With the zero-loop drive system the unsupported length of tape is held to an absolute minimum, and time-base error and flutter are considerably reduced.

One type of motor used as the drive for the zero loop is an air-cooled, ac, split-phase induction motor. Normally, it is capable of bidirectional rotation and provides full power continuously during tape motion. The rotational speed of the capstan motor is governed by the amount of braking force applied to it via a servo-controlled disc brake. Attached to the capstan shaft is a flywheel (to smooth out the effects of motor cog), and electrically controlled brake assembly (to provide capstan-speed control), and a photoetched tachometer assembly (to sense capstan speed and control the brake assembly).

Drive Motors—The Printed-circuit Motor

Since the printed-circuit motor is a relatively new device, and is now being used with many of the drive systems, we present here some of the major construction and operational details.[1]

The printed-circuit motor is a dc device in which the conventional wire windings of a high-mass cylindrical armature have been replaced by flat conductors formed on a disc by modern printed-circuit techniques (Fig. 8-5).

The armature can be considered as being a coil of wire with relatively few turns and having poor inductive coupling with the surrounding field structure. The absence of any magnetic attraction, together with the light weight of the armature, means that the motor bearings are very lightly loaded, and frame-conducted motor noise is extremely low.

The generated torque of this type of motor is smooth at any speed. The accuracy of the armature position, when driven with an appropriate servo system, is excellent. Because the construction is basically in the form of a pancake, it is well suited to applications where it becomes an integral part of the driven device (capstan system).

Since the magnetic field in the motor is supplied by a structure that does not rotate with the shaft (permanent magnet of Fig. 8-5), the rotation of the armature does not tend to modulate the permeability of the magnetic path. The only factor to cause any change in the torque, therefore, is the slight shift in area and distribution of the current sector on the armature disc when one of the brushes engages a new conductor. Thus the torque developed in response to a constant current is constant within approximately 0.5 percent for any armature position, and there is negligible cogging.

The lack of armature inductance will result in an extremely fast response

[1] Extracted from texts supplied by and reproduced with the permission of Printed Motors, Inc., Glen Cove, N.Y.

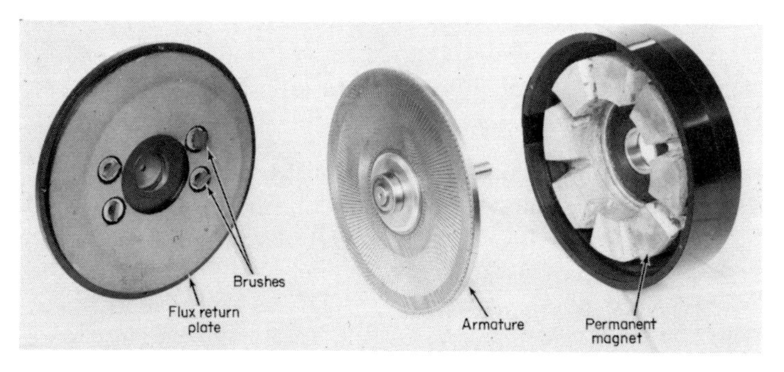

Fig. 8-5 Exploded view of a printed-circuit motor. (*Printed Motors, Inc.*)

to applied armature voltage. The developed torque will be controlled only by the back emf and the series resistance of the motor at the instant of change in the terminal voltage.

In general, it may be stated that the printed-circuit motor is useful where direct drive (no gear or pulley trains) is desirable. It has rapid response where load inertia is small, and generates smooth torque at any speed. It is smaller in size and lighter in weight than the conventional wire-wound motor. It has freedom from preferred armature position. But it should not be considered as the direct replacement for any and all conventional motors, under all conditions.

Capstan Servo Systems Used with Longitudinal-tape Transports

Real and Apparent Errors A servo system is used to correct inaccuracies in the reproduced signals resulting from physical changes in the magnetic tape, changes in power-line frequency, and nonconformity of mechanical components from one tape transport to another. These types of errors can be listed under two categories: real and apparent errors. They occur in the following ways:

1. Magnetic tape has a dimensional instability due to its temperature and humidity coefficients. If either the temperature or humidity changed between the time of recording and the reproduction, the signal on tape would have increased or decreased in size. This is called an *apparent error.*
2. The rotational speed of a synchronous motor used as the capstan drive source is directly proportional to the input-power-source frequency. Thus, if the line frequency changes (most commercial power sources have a ±0.5-Hz frequency tolerance), the speed of the capstan will change. These errors are called *real errors.*
3. The mechanical components within the drive systems of any two transports may have slight differences in size due to tolerance allowances during manufacturing. Thus, if a signal is recorded on one transport and reproduced on another, slight differences in tape speeds may occur. These errors are also called *real errors.*

Compensation for such errors may be accomplished by one of several types of servo systems. They may be mechanical, photoelectric, crystal oscillator, Wein bridge oscillator, or tuning-fork based, but they can also be a combination of two of these.

Speed Sensing

The Slotted Tachometer: An optical tachometer is often used to sense the speed of the capstan, and thus the tape. Figure 8-6 shows one of these. A skirting with a precise number of slots is attached to the capstan flywheel. An exciter lamp is placed on one side of the skirt, and a photocell on the other. As the capstan revolves, the photocell receives alternately pulses of light and no light. Since the skirting is attached to the flywheel, which is itself attached to the capstan, the pulses of light bear direct relationship to the speed of the capstan.

The Photoetched Tachometer: A second type of optical tachometer uses a photoetched disc attached directly to the capstan. This disc contains a large number of opaque lines (that is, 2,000 to 10,000) and an equal number of transparent spaces. When an exciter lamp is placed on one side of the disc and a photocell on the other, alternate pulses of light and dark will be produced as the capstan moves (Fig. 8-7). The output of the tachometer disc is representative of the capstan speed. It is compared with a frequency from a crystal source. If there is any frequency or phase difference between the two signals, an error voltage is generated which is used to control the speed of the capstan via the capstan servo system.

18.24- and 17-kHz Direct Servo Systems One of the problems of servos used with synchronous motors is the frequency limitation. Since a 60-Hz synchronous motor makes a complete revolution by 60 separate steps of 16.7 msec each, it is not possible, by ordinary methods, to completely correct for all speed errors, particularly when they have a frequency higher than the motor step rate. Essentially, correction cannot be made for errors that happen faster than the step rate; it is like trying to climb a set of stairs half a step at a time.

To overcome as much of the line-error problem as possible, the older capstan drive systems used a very accurate internal frequency standard as the source of power. Referring to Fig. 8-8, it should be noted that the 60-Hz oscillator is used to drive the capstan in record and also to provide a 60-Hz precision signal to be recorded on tape. Since the 60 Hz cannot be recorded via the direct record technique directly, a

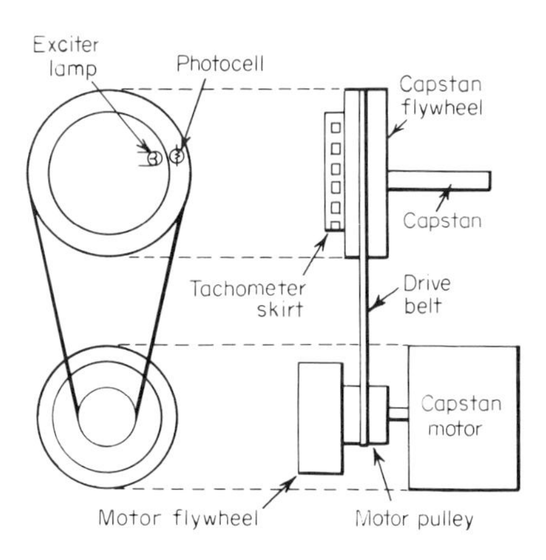

Fig. 8-6 Capstan tachometer system using slotted skirt.

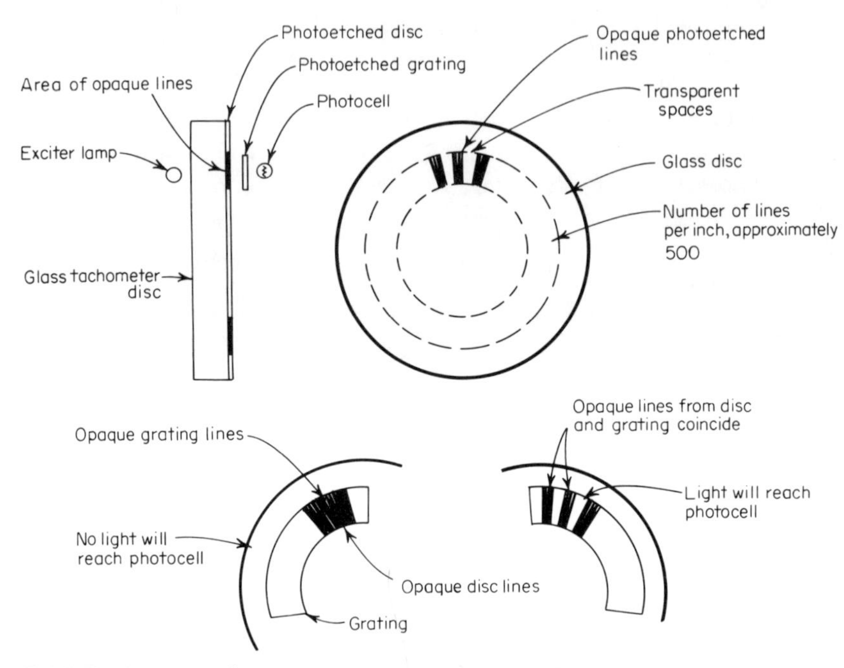

Fig. 8-7 Capstan tachometer system using photoetched disc.

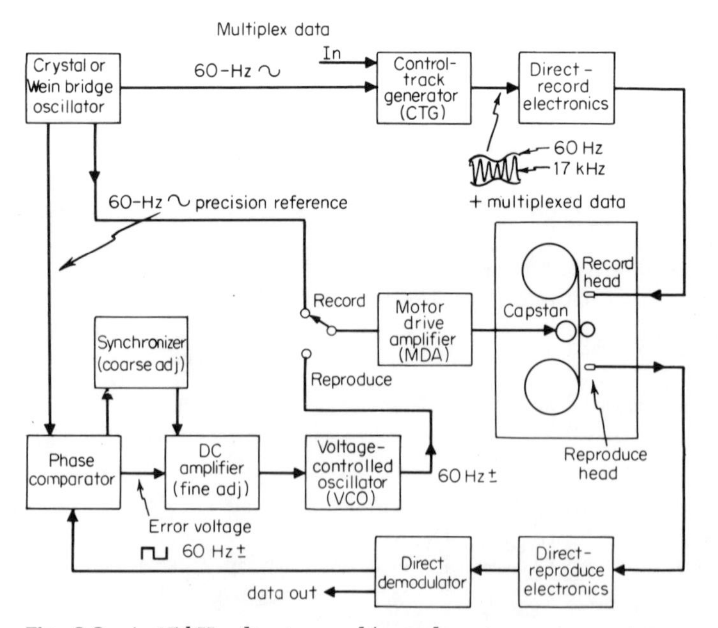

Fig. 8-8 A 17-kHz direct record/reproduce tape servo system.

modulation process was developed whereby a high-frequency control-track-generator (CTG) frequency was modulated 50 to 70 percent by the 60 Hz; i.e., essentially, the CTG frequency acts as a carrier for the 60-Hz data, placing it within the frequency band of the reproduce system.

Two separate frequencies are used as the CTG:

1. 18.24 kHz—referred to as the Old Standard, rarely used now.
2. 17 kHz—referred to as IRIG Standard, in common use.

The 17 kHz fits nicely between the ±7.5 percent deviation channels 13 and 14 of the IRIG telemetry standards, and does not interfere with the ±15 percent deviation channel A, whereas the 18.24-kHz frequency does.

The modulated CTG signal, placed on tape during the record mode, is retrieved from the tape during the reproduce mode. After amplification it is passed to a demodulator, where the CTG signal is extracted from any other data that have been multiplexed with it. The 60-Hz signal is then extracted from the CTG by detection and filtering. Finally, the 60 ± Hz (± if any speed error) is squared up and passed to a phase comparator, where it is compared with a square wave from the frequency standard. Taking into account the natural 90° phase shift between recorded and reproduced signals, the phase comparator will produce an error voltage if phase or frequency differences are found between the signals from the tape and the frequency standard.

The error voltages from the phase comparator will be used to control the frequency of a voltage-controlled oscillator (VCO). The error signal from the phase comparator is amplified in a dc amplifier and used to change the output frequency of a saturable reactor oscillator to 60 ± 3 Hz (Fig. 8-8). For initial correction, if the errors are large, a synchronizer stage is used. It feeds a large correction signal to the VCO via the dc amplifier until a ball-park correction state is reached. At that time the synchronizer is phased out of circuit, leaving the dc amplifier to take over vernier control. As the frequency is changed, the capstan motor will speed up or slow down and the tape-speed errors will be corrected.

FM Servo System One of the problems with the system just described is that, if the tape should be recorded at one speed and reproduced at another (time-base expansion and contraction), the control-track signal could no longer be used, and reproduce servo would be unavailable. To overcome this, an FM servo system was developed whereby a number of precision frequencies were mixed together in a mixer stage (this replaces the CTG of a direct system) and recorded through the FM record electronics. If the recording speed were 120 ips, for example, the frequencies mixed together would be 60, 120, 240, and 480 Hz. If, during reproduce, the tape speed were reduced to 30 ips, or $^{120}/_{4}$, then the frequencies that were recorded at 120 ips would also be divided by 4 and become 15, 30, 60, and 120 Hz, and there would still be a 60-Hz signal for the servo system to use. An FM demodulator is programmed to pick out (via high- and low-pass filters) only the 60 ± Hz signal, square it up, and feed it to the phase comparator. From there on it is treated in the same fashion as the direct signal of Fig. 8-8. Frequency-modulation servo provides time-base expansion and contraction, but the channel on which it is recorded can be used for servo signals only (no multiplexing).

200-kHz Servo System To overcome the obvious disadvantages of both systems just described, a new set of frequencies was evolved that could be used for both direct and FM servo systems and also provide time-base expansion and contraction. These frequencies have become the accepted standards of the industry. They are constant-amplitude 400-, 200-, 100-, down to 3.125-kHz speed-control frequencies. 200 kHz is normally used at a tape speed of 120 ips, and 3.125 kHz for a speed of 1⅞ ips.

In Fig. 8-9, the frequency standard can be an internal crystal oscillator or an external reference source. Its accuracy should be 2 parts in 10^5 or better in a 24-hr period. The 400 kHz is fed to a series of binary dividers, where the selection of the required control-track frequency for the tape speed in use will be made. Diode gates are used to select the proper frequency. These are enabled by a voltage taken from the transport speed-selector switch.

To simplify the explanation of the capstan servo system of Fig. 8-9, it will be assumed that the tape speed selected is 120 ips, the required control-track frequency is 200 kHz, and the frequency required by the phase comparator is 50 kHz.

The output of the first set of diode gates will feed a 200-kHz square wave to a stage of fundamental filters (control-track filters), where a 200-kHz sine wave will be produced. The sine wave requires less bandwidth than a square wave when fed to the record electronics and recorded on tape.

During the reproduce mode, the 200-kHz signal will be retrieved from the tape, passed through the control-track filters to extract the control signal from any other multiplexed signal, and fed to the phase comparator. There it will be compared with a signal from the crystal reference frequency. If any frequency or phase differences occur, they will produce an output from the phase comparator indicative of a speed error. This output will be used to correct the speed of the capstan.

To control the capstan speed during recording, a photoetched tachom-

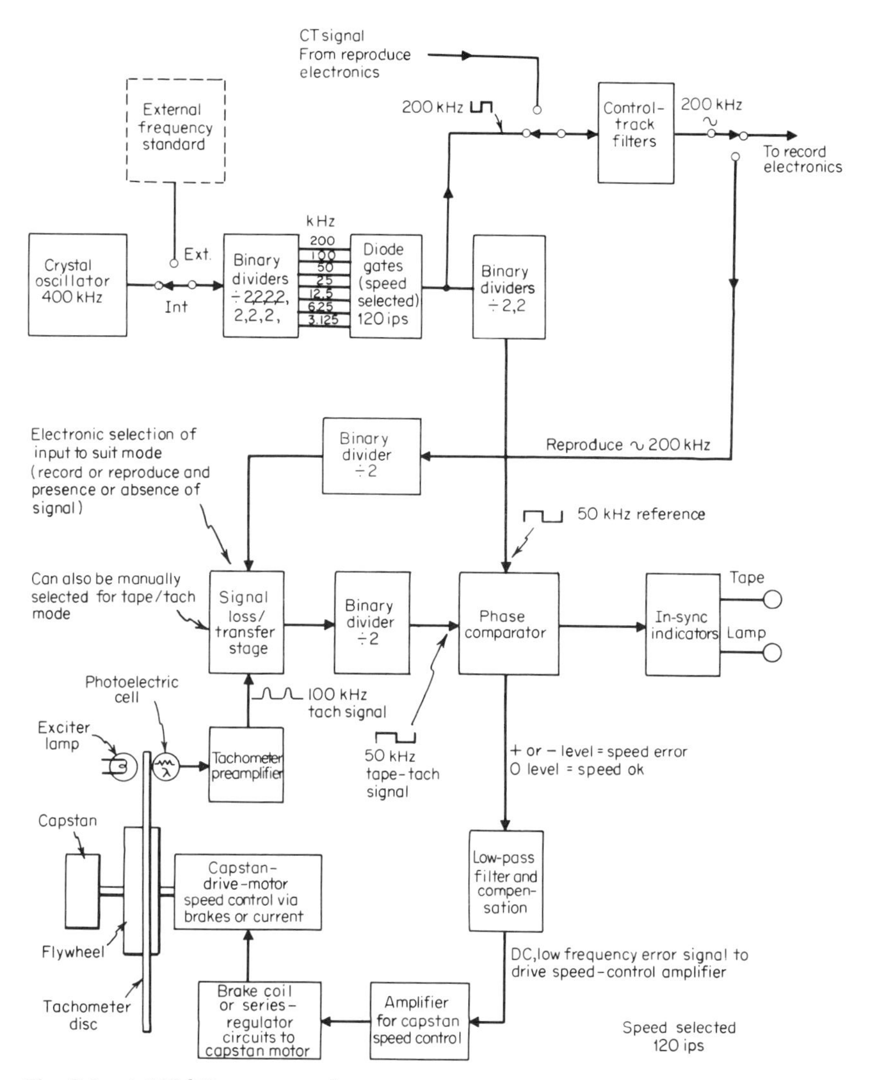

Fig. 8-9 A 200-kHz tape-speed servo system.

eter disc is attached to the capstan shaft (Fig. 8-7). A fixed number of pulses will be produced from the disc for each revolution of the capstan. Assuming that the circumference of the capstan is 12 in. and the number of lines on the tachometer disc 10,000, then with a tape speed of 120 ips, the capstan must turn 10 times per second and the output of the tachometer disc will be $10 \times 10{,}000$, or 100 kHz. After these pulses are shaped and passed through a set of binary dividers, they form a train of 50-kHz square waves. These, like the control-track signal, will be compared against the crystal reference in the phase comparator. Any difference in frequency or phase represents a capstan speed error.

The output of the phase comparator will be either positive or negative if an error exists, and essentially zero potential if no error is present, the particular output (+ or − representing fast or slow state) being based on the choice of positive or negative logic circuits used in the phase comparator (also called the forward-backward counter and digital comparator). These levels are biased; so they are always one polarity (all + or −), and may be used to control the large power transistors of the final stage. Additionally, the high-frequency components that result from the comparison processes are removed. This is done by the use of 240-Hz low-pass active filters in record and 800-Hz filters in reproduce. The larger bandpass is required in reproduce because part of the errors in speed may be due to wow (low-frequency flutter), tape stretch, and shrinkage, as well as the capstan speed. Thus a wider range of control is needed in the reproduce mode.

The dc level from the phase comparator and the low-pass filter and compensation stages is used to control the current through a large brake coil. The more current through the coil, the greater the braking and the slower the capstan speed.

In another system, the dc level is used to control a series regulator which applies more or less power to a dc printed circuit, capstan motor, or standard dc motor. With nominal power, these capstan motors turn at a nominal rate.

In the record mode, the capstan speed is kept accurate by referencing its speed to a crystal frequency. In the reproduce mode, the capstan is made to follow the tape and compensate for any changes that real and apparent errors introduce. There is one problem in systems of this nature, however, that occurs when the signal from the tape disappears for one reason or another. This means that the phase comparator will have only one input, an impossible operating state. Therefore a signal-loss and transfer stage is included in most of the modern servo systems. Both the tachometer and the tape signal are fed to the stage, and if the reproduce mode (tape) were selected, the stage would permit the tape signal (control track) to go through to the phase comparator and inhibit the passage of the tachometer signal. If the record mode (tach) were selected, the stage would permit the tachometer signal to go to the phase comparator and inhibit passage of the control-track signal. If the system is in the tape mode and the control-track signal disappears for more than a few milliseconds, this stage will automatically revert to the tach mode. Then, when the tape signal reappears, it will automatically revert to the tape mode. The control of this loss-and-transfer stage can also be manual. Thus, if multiplexed data were interfering with the control-track signal, the system could be manually switched to the tach mode, cutting off the interference and maintaining a constant capstan speed. The introduction of the printed-circuit motor has improved the response time, so that present capstan servo systems are quoted with time-base errors of ±300 nsec at tape speeds of 120 ips.

TAPE DRIVE AND SERVO SYSTEMS FOR ROTARY-HEAD TRANSPORTS

Capstan Drive Systems[1]

The drive system for the rotary-head recorder is designed to move magnetic recording tape from the supply reel, past the record and reproduce heads, to the take-up reel, as in a conventional longitudinal recorder. It must in addition, however, include a drive system for the rotary head. Therefore it could be stated that the drive system must control the tape speed, the rotary-head speed, and the tape contour and tension within extremely close tolerances.

The capstan drive is a servo-controlled system, used to move 2-in.-wide magnetic recording tape longitudinally across the head drum at a speed of 12.5 ips. For two-channel machines, the speed would be 25 ips. A spring-loaded pinch-roller assembly is used to clamp the tape to the capstan. The capstan shaft is driven by a two-speed hysteresis-synchronous motor, coupled to the shaft by a series of pulleys such that, as the motor turns at 7,320 rpm, the capstan shaft, with a diameter of 0.5 in., is caused to rotate at 472 rpm (12.5 ips). When the system is used for 25 ips, the motor speed is doubled to 14,640 rpm (244 rps).

Rotary guides which have precision flanges top and bottom guide the tape from reel to reel, as shown in Fig. 8-10. The bearings used for this type of guide are extremely high grade. Two guides in the tape path do not follow this rule, the first being the guide used in the stationary-head area. Normally, this guide is fixed, to reduce the possibility of passing on any tape perturbations from the canoe area to the auxiliary channel heads. The second is part of the female-guide assembly used with the rotary-head drive. See Fig. 8-11 for details.

Rotary-head Drive Systems

The rotary-head drive, like the capstan drive, is also a servo-controlled system. The head drum is rotated at 244 rps by a hysteresis-synchronous motor which causes the four heads (mounted on the head

[1] These speeds and drive systems are somewhat different in the rotary-head recorders used for television and closed-circuit productions. The reader is referred to Chap. 11 for details.

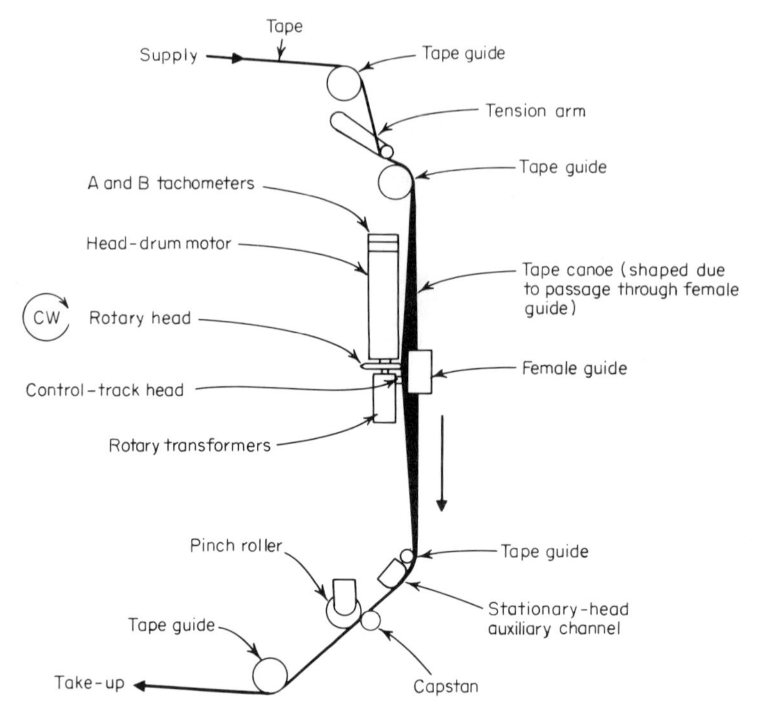

Fig. 8-10 Rotating-head (transverse-scan) drive system.

drum) to traverse the width of the tape sequentially. The heads are coupled to the record and reproduce electronics via rotary transformers and changeover relays. The rotary transformer includes an individual rotary and stationary winding for each head tip. For a dual-channel recorder, since there are eight heads, the output of two head tips goes through each winding.

Referring to Fig. 5-7, the head-drum motor is secured to the assembly base, while the head drum itself is mounted on the motor shaft. The rotary-transformer rotor is mounted to the face of the head drum and thus turns with the motor shaft. The stationary windings (stator) are supported by the base of the head assembly. Two tachometer rotors are attached to the end of the head-drum motor shaft opposite to that on which the head drum is mounted. The tachometer pickup windings are part of the motor housing.

The female guide (vacuum guide) performs a variety of functions. Referring to Fig. 8-11:

1. It contours the tape so that it fits the head-tip path (circular).
2. It holds the tape (via two vacuum slots) so that a tent of tape is formed across the guide groove. This tent provides an area where

the head tips make intimate contact with the tape as the head drum spins, without damage to either head tips or tape.

3. It provides a means whereby the tape may be tensioned or the head tip may penetrate the tape by a precise amount. Essentially, this is a form of tape tensioning combined with control of head-to-tape contact.

4. It edge-guides the tape by means of a fixed guide mounted on the lower portion of the female-guide assembly.

During the record/reproduce modes, the female guide is pulled toward the head drum by a solenoid. For other modes the solenoid holds the guide away from the head drum. Manual control of the guide position is also provided via a vernier adjustment so that fixed amounts of movement toward and away from the head drum may be made. This takes care of stretch or shrinkage of the tape due to temperature and humidity or the nonconformity between various recorders and reproducers. The nonconformity comes about when the female guide is moved closer to the head drum, causing the head tips to push against the tape harder,

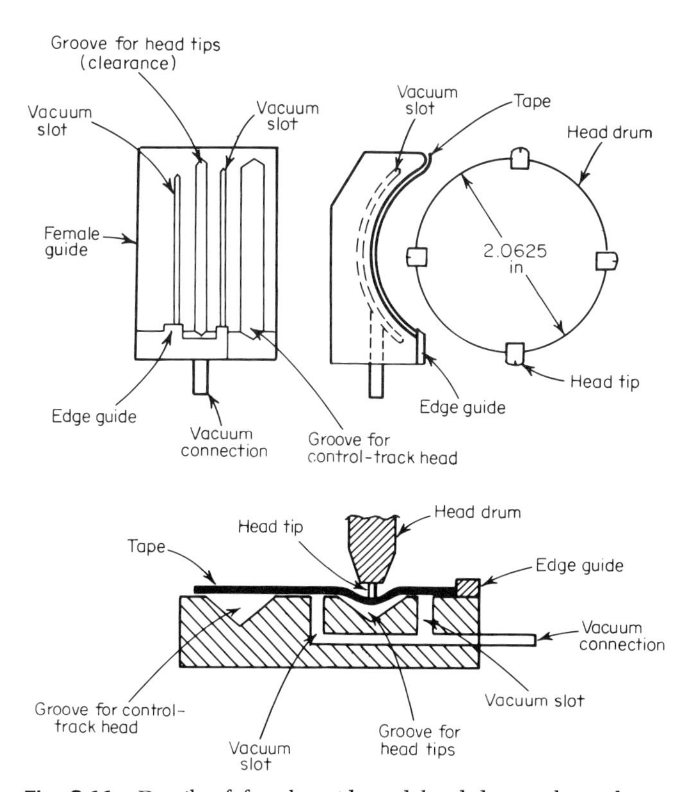

Fig. 8-11 Details of female-guide and head-drum relationship.

increasing contact and also lengthening the tape by stretching it a little. This tape stretch results in a timing change between tracks. The tape is formed into a canoe-shaped section by the female guide and the guides used for entering and leaving the rotary-head area. Thus it is referred to by the trade as the canoe. The vacuum level used to hold the tape to the female guide is approximately 40 in. of water.

Capstan-drive Servo Systems

During the record mode, the capstan motor speed is controlled by a frequency produced in a variable frequency oscillator (VFO) that is locked to a very accurate crystal-controlled frequency standard. The VFO frequency is used to produce a series of three-phase 244-Hz square waves in a motor-drive amplifier. These crystal-synchronized three-phase square waves will drive the delta-wound capstan motor at a very accurate speed (Fig. 8-12, dotted lines).

At the same time that the capstan speed is being accurately controlled, a control track is being recorded via a stationary head on the bottom edge of the .tape. The control-track head is mounted 0.5 in. in front of the rotary-head drum, on the stationary portion of the rotary-head assembly. One pulse (approximately 20 μsec long) is recorded for each revolution of the head drum. This pulse is derived from the tachometers located on the end of the head-drum motor shaft. Their position on the tape reflects the rpm of the head drum in direct relationship to the longitudinal speed of tape (capstan speed).

In the reproduce mode (Fig. 8-12, solid lines), the control-track pulses are retrieved from the tape and compared with the head-drum-tachometer outputs. Since the control-track pulses were originally derived from the head-drum tachometers, any differences in frequency or phase indicate a tracking error; that is, the relationship of the capstan speed to the head-drum speed is not correct. The frequency or phase differences produce a nonsymmetrical output from the phase comparator, which is sampled once each head-drum revolution. The sampling produces a varying dc level, which in turn is used to change the output frequency of the VFO. The VFO frequency changes will speed up or slow down the capstan as required to correct the longitudinal tape-speed error. Short-term errors are detected by using a discriminator in place of a phase comparator and comparing the timing between successive pulses. Any deviation from nominal produces an ac error output which, like the dc, affects the output of the VFO.

Manual adjustment of the rotary-head tracking permits the operator to center the head tips on the recorded transverse tracks. This allows the selection of the maximum reproduced signal level and the best signal-to-noise ratio.

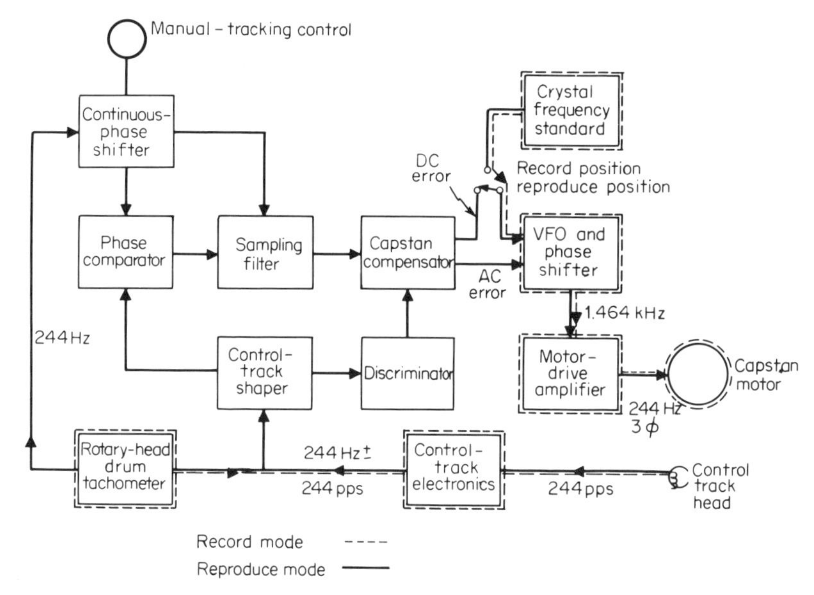

Fig. 8-12 Capstan servo systems—record and reproduce modes.

These systems correct for any variations in tape speed due to the belt drives of the capstan, differences between the drive components of the record and reproduce transports, and the dimensional differences of the tape due to temperature and humidity conditions.

Rotary-head-drum Servo Systems[1]

During the record mode and reproduce modes where there is no control signal (pilot) available, the head-drum motor is driven at a constant speed by a frequency derived from a VFO that is locked to a crystal frequency standard. This standard mode of operation is controlled like the capstan drive system; i.e., the VFO is locked to a crystal frequency standard and the VFO output changed into three-phase 244-Hz square waves in a motor-drive amplifier and used to drive the head-drum motor at 14,640 rpm. There is one difference, however: there is a servo feedback loop provided to stabilize the motor drive's natural resonances. In addition, the loop compensates for any phase shift introduced in the circuits. As can be seen in Fig. 8-13, tachometer A and B pulses are fed via a shaper circuit to a discriminator. Since these pulses are a direct function of the head-drum rotation, any changes of drum speed will result in changes of time (phase) between succeeding pulses, and the discriminator will produce an error signal. This signal will be used

[1] The reader is referred to Chap. 11 for details of the servo system used with the television and closed-circuit video recorders.

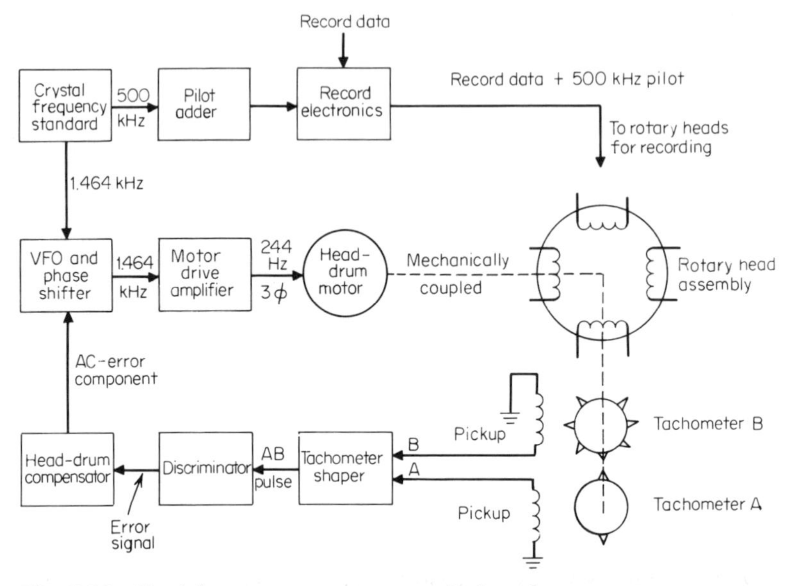

Fig. 8-13 Head-drum servo system—standard mode.

to change the VFO output, so that the head-drum motor speeds up or slows down as required to correct for the rotational errors. Such changes are small and brought about by changing the phase of the 244-Hz pulses that drive the head-drum motor.

Where a control signal is being retrieved from tape (Fig. 8-14), the mode is called the servo mode. In this mode of operation, the VFO and phase-shifter frequency will be controlled by the error voltages produced by comparing the 500-kHz pilot signal with the frequency standard. Since the pilot signal recorded on tape was originally derived from the crystal frequency standard, this control arrangement ensures that the head-to-tape speed is kept at precisely that speed required to produce the 500-kHz pilot signal, or precisely the same speed as that used during the record mode.

HELICAL-SCAN DRIVE AND SERVO SYSTEMS

Why Single and Dual Heads?

The basic design features for helical-scan magnetic recorders must satisfy the requirements of a very wide market, as follows:

1. They must have a frequency response wide enough to handle the standard television signal, specialized closed-circuit video, high-bit-rate digital, radar-type signals, and high-frequency PCM.

2. They must be economical enough to comply with the budget restrictions in the educational and medical fields, as well as those of the small industrial user.

3. They must be portable, to provide easy handling in operating rooms, classrooms, mobile situations in TV studios and field situations, and a wide array of closed-circuit applications.

4. They must be simple to operate, with a minimum amount of maintenance, and have little downtime.

These recorders may be subdivided into two classes, single-head and dual-head machines.

The single-head recorder (Fig. 8-15*a*) with the omega wrap format has a point in the head path where it must leave the tape at the end

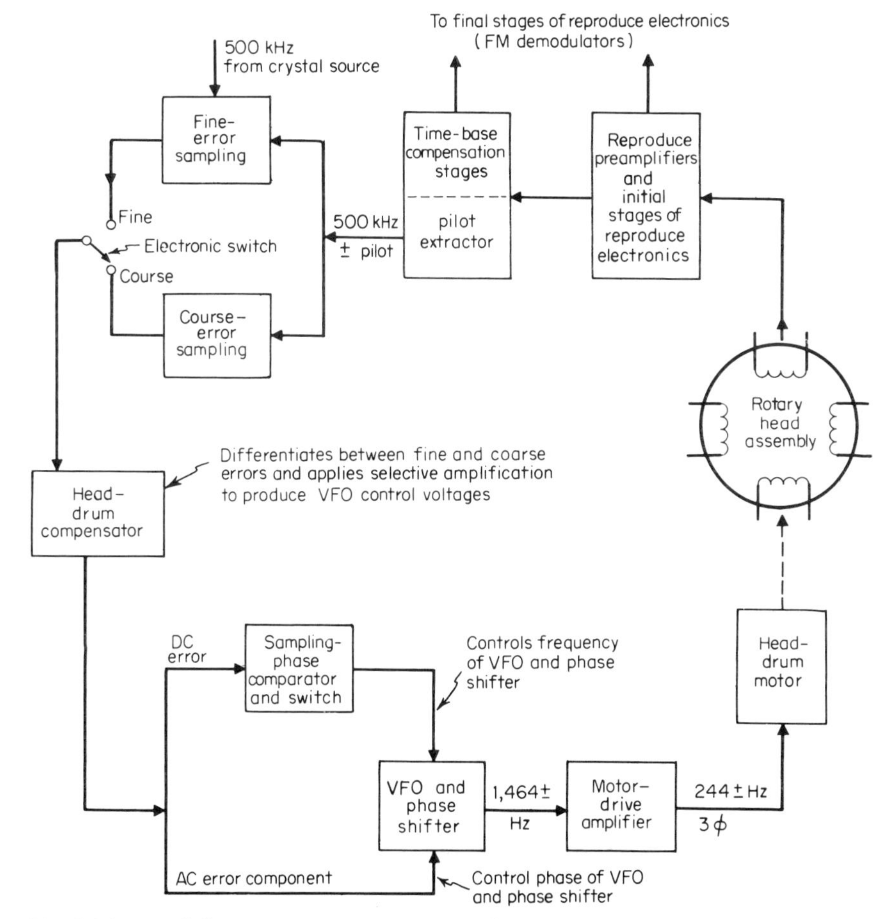

Fig. 8-14 Head-drum servo system—servo mode.

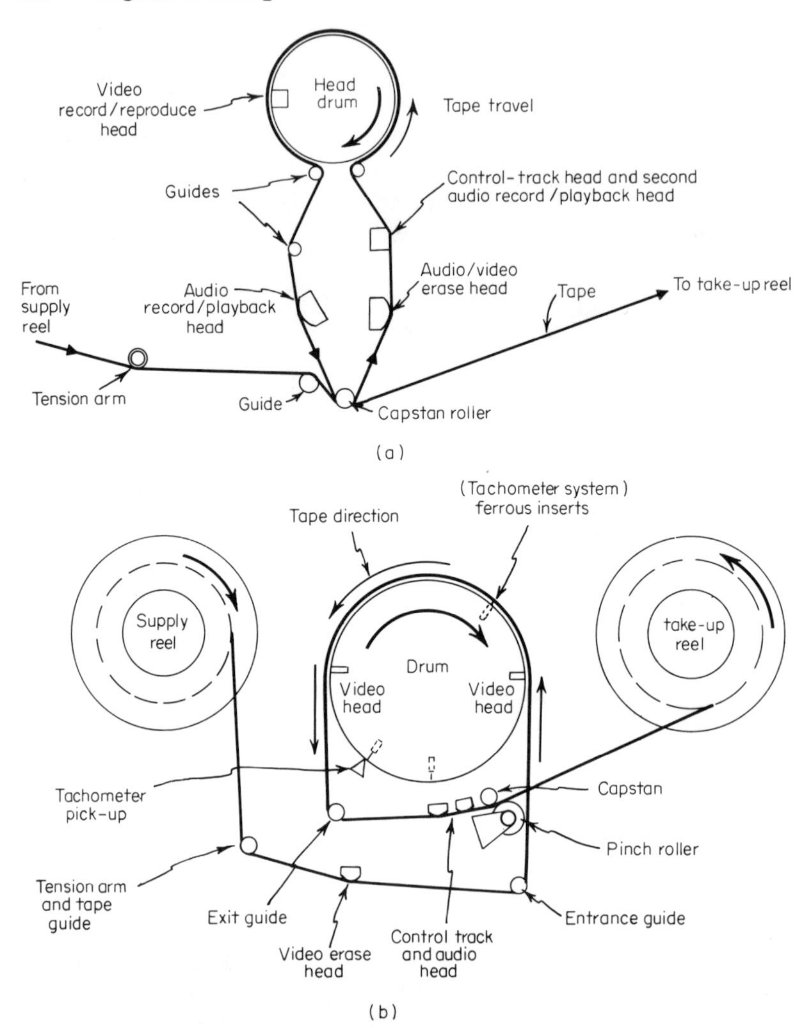

Fig. 8-15 Helical-scan drive systems. (*a*) Single-head system; (*b*) Dual-head system.

of one track on one tape edge, to begin the next track on the other tape edge. Since there cannot be a perfect overlap, there will be a loss of information for a few microseconds. Therefore the single-head recorder is used with a processing amplifier when a break in the information path can be tolerated.

The dual-head recorder (Fig. 8-15*b*) provides continuous information. This method solves the problem of missing information, since when one head leaves the tape, the other remains in contact. Whenever two heads are used, extreme care must be taken to place the heads accurately with respect to each other. Mechanical tolerances are critical, and placement errors are easily introduced. When the heads are not exactly 180° apart, the error is called dihedral error. Tape interchange between machines is impossible with large dihedral errors. The dual-head recorder requires some form of electronic switcher to switch from one head to the other.

Figure 8-16 shows an example of another type of wrap format, the D wrap. This format, in which dual heads are also used, is prevalent among Japanese manufacturers of helical-scan recorders.

Compatibility between Helical-scan Recorders

Unfortunately, no standard was established in the early days of the development of the helical-scan recorder. Thus the individual manufacturers chose drum speeds, longitudinal tape speeds, and tape widths that suited their own purposes. This meant that the helix angle varied between machines, and a tape made on one recorder could not be played back on another that was not made by the same manufacturer and

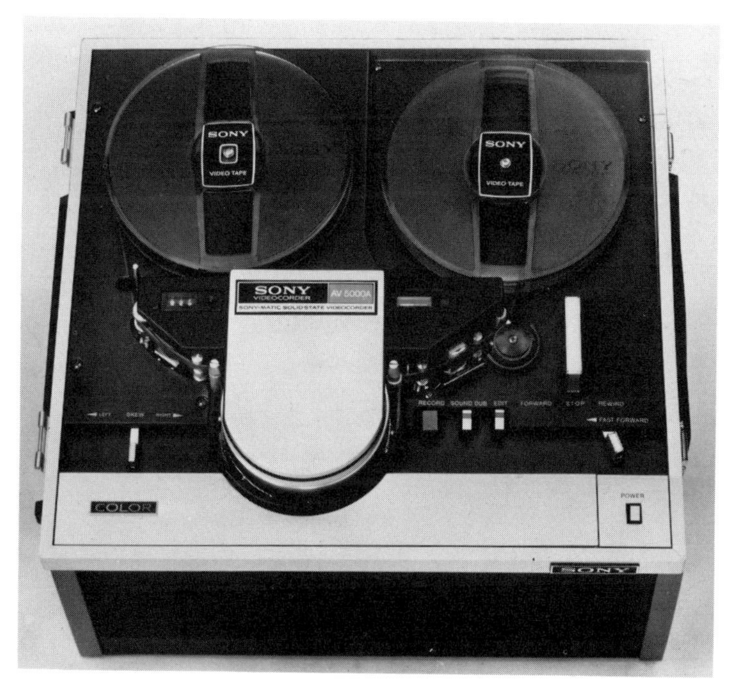

Fig. 8-16 AV 5000A Videocorder. This solid-state helical-scan color recorder uses 1-in. tape and a longitudinal tape speed of 7.5 ips and produces a tape pattern that conforms to the EIAJ Type I VTR Standard. (*Sony Corp. of America.*)

sometimes also not of the same series. Recently, Japanese manufacturers have established a common standard, based on the EIAJ Type I VTR tape pattern. The longitudinal tape speed is 7.5 ips, and the tape width is normally ½ in. ¾ in. tape has also been used. A recent model brought out by the Ampex Corporation also follows this standard. There will be compatibility between machines that conform to the same standard. Figure 8-17 indicates typical helical-scan track patterns. Note the change in helix angle. This sort of change could also occur using the same tape widths if head drum or longitudinal tape speed were different.

Due to their specialized recording format, helical-scan recorders require a relatively high writing (head-to-tape) speed. The exact speed will depend on the machine configuration and the width of the tape used. With single-head helical-drive systems, the writing speed is relatively high, approximately, 1,000 ips. Dual-head recorders, on the other

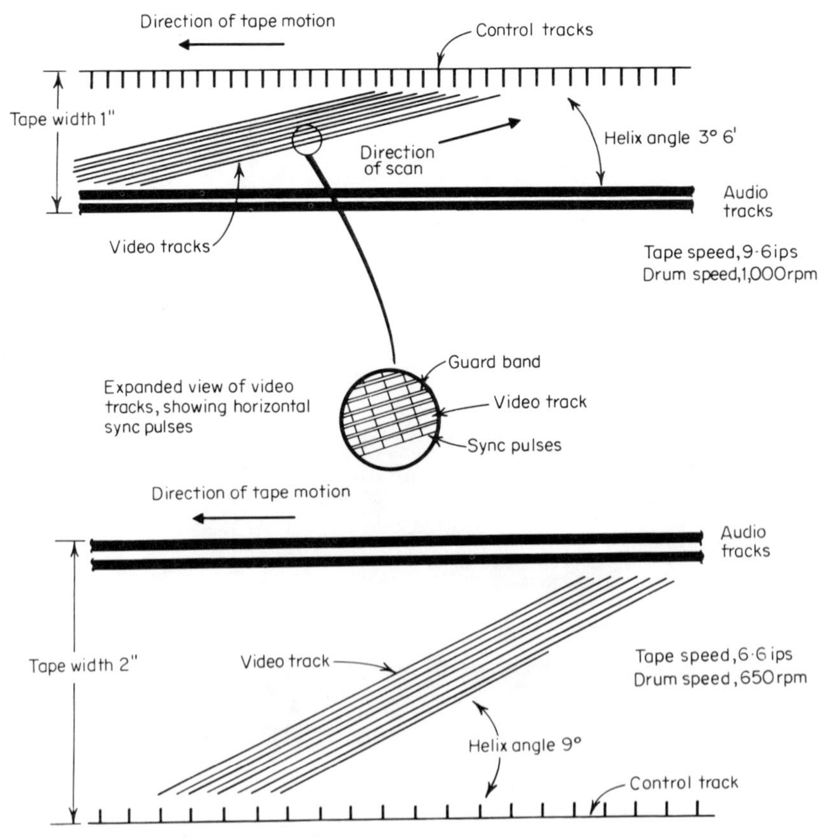

Fig. 8-17 Typical 1- and 2-in. helical-scan track patterns.

hand, have writing speeds that are much lower, approximately two-thirds of the single-head speed. To accomplish this speed, a scanning assembly consisting of a rotary drum fitted with one or two video heads is used. A capstan drive assembly provides the necessary longitudinal tape velocity.

Capstan Drive Systems for Single-head Helical Scan

Figure 8-15*a* shows the path the tape takes from the supply reel on the left to the take-up reel on the right. After passing over the first tape guide, on the left of the capstan roller, it goes over the bottom half of a rubber-sleeved capstan. From this point, the tape passes across the audio and video erase heads (active only in the record mode), and then over the control track and audio 2 heads. The tape moves counterclockwise around the clockwise rotating drum. During this maneuver, the tape is wound in a helix. The video record and reproduce head (same) is mounted on the upper rotating half of the drum. The tape then passes over the audio 1 record and playback head and finally across the top half of the rubber-sleeved capstan. This routing makes up the closed-loop system. From there the tape is fed to the take-up reel.

Capstan drive systems vary somewhat from machine to machine. Some use hysteresis-synchronous motors, others dc printed-circuit motors which are speed-controlled through sophisticated servo systems. With the hysteresis-synchronous motor a solenoid is used to pull the motor-and-drive pulley, which is mounted on its shaft, against the capstan-drive-wheel rim. Drive ratios between the pulley and the drive wheel are so arranged that the rubber-sleeved capstan will pull the tape around the head drum at a speed necessary to provide the proper writing speed for the frequency range, and a guard band between tracks. See the 1-in. track pattern of Fig. 8-17. No pinch rollers are used with this closed-loop drive system. The wrap of the tape around the lower half of the capstan before entering the drum helix, and the upper half of the capstan after leaving the helix, is sufficient to respond to capstan drive.

Capstan Drive Systems for Dual-head Helical Scan

Referring to Fig. 8-15*b*, it can be seen that the tape path varies from that of the single-head machine. This system employs a capstan and pinch-roller drive, along with a different arrangement of the heads. Figure 8-17 illustrates the differences in the track pattern that occur with the configuration changes. The longitudinal tape speed is generally lower for dual-head machines than it is for the single-head versions. The helix angle is generally larger also.

When the capstan motor is a hysteresis-synchronous motor, it is effectively geared down by a pulley and flywheel to 10:1. A solenoid is used to pull the motor-and-drive pulley against the capstan-flywheel rim. The pinch roller is pulled in by a solenoid during the record and playback modes.

As has been previously stated, the video heads for both the single- and dual-head recorders are mounted on the upper and rotating portion of the head drum. Signals to and from the heads are coupled to the signal electronics via slip rings and brushes or by rotary transformers. In addition to the video heads, the head drum is often fitted with small magnetic inserts, or a magnetic blade is attached directly to the drum. A coil pickup (tachometer head) senses the movement of these magnetic components and produces a "tachometer" pulse for each one of them. The pulses may be related directly to the speed of the drum rotation. In dual-head machines the output of two of the magnetic inserts will be used to control the head-switching circuits that select the head which is in contact with the tape. Once each revolution of the head drum, one of these tachometer pulses will be recorded on the edge of the tape. Such recordings are called the control track and are an accurate indication of the position of the video head.

Many types of motors are used to drive the rotating-head drum and the capstan. These are the hysteresis-synchronous, the induction, and the dc printed-circuit motor. Their speed is a function of the accuracy of the frequency fed to them (for synchronous motors) or the servo system controlling the current through them. Since helical-scan recorders are used in conjunction with television and specialized closed-circuit operations, frequency standards that are used for control are generally locked to the power line or the vertical sync signal of the television station.

Servo Systems Used in Helical-scan Recorders

An essential requirement of helical-scan recorders is that the tape tension and the relative speed of the head-drum-to-tape movement be controlled during record and playback modes. The accuracy of the timing information of the video signal is dependent upon the accurate and stable tape and scanner dimensions and head-to-tape writing speed.

Tape-tension Servos Tape-tension systems use both manual and automatic operation. Fixed tension is used during the record mode, and electronic servo tension during playback. In the record mode, as the tape is passed over a tension arm, its pressure against the arm is transmitted through a linkage system to a brake band. As the tape tightens against the arm, the brake band will be loosened and the brakes released. With less braking, more tape will be allowed to feed into the system, releasing the pressure on the arm, and the brakes will be tightened. Quite naturally, a nominal position will be found by the system, and the resulting tape tension will be nominal. This is known as holdback tension.

In the better-quality recorders, during the playback mode, a solenoid is used to move the brake linkage to change the amount of braking. This solenoid is controlled by the amount of current flowing through it. Some of the less expensive models rely on mechanical servo action for holdback tension for both record and playback.

In the electronic servo systems, a VCO, simulating a TV monitor's horizontal oscillator, is used to produce a fundamental frequency which will be compared with the horizontal sync pulses that are recorded as part of the TV signal on the helical track. Any change in frequency between the horizontal sync pulses and the fundamental frequency is routed back to the frequency standard (VCO) and controls its frequency. Any change in the VCO frequency represents an error in tension. It will be used to control the automatic tensioning of the tapes by controlling the amount of current through the solenoid. (See Fig. 8-18.)

Rotating-head-drum Servos Quite naturally, the speed of rotation of the rotary head will also control the laying down (spacing) of the horizontal pulses on the tape. Where the head-drum motor is a hysteresis-synchronous type (older version), its speed is controlled by the frequency of a voltage-controlled oscillator. The frequency and phase of the VCO is compared with the frequency and phase of the tachometer

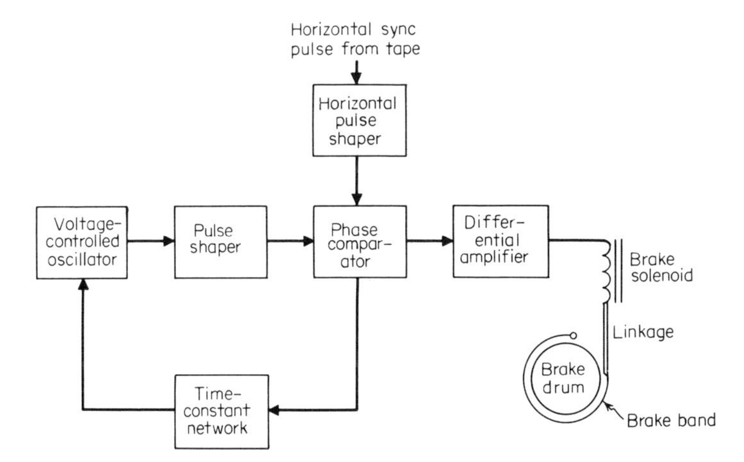

Fig. 8-18 Typical tape-tension servo system.

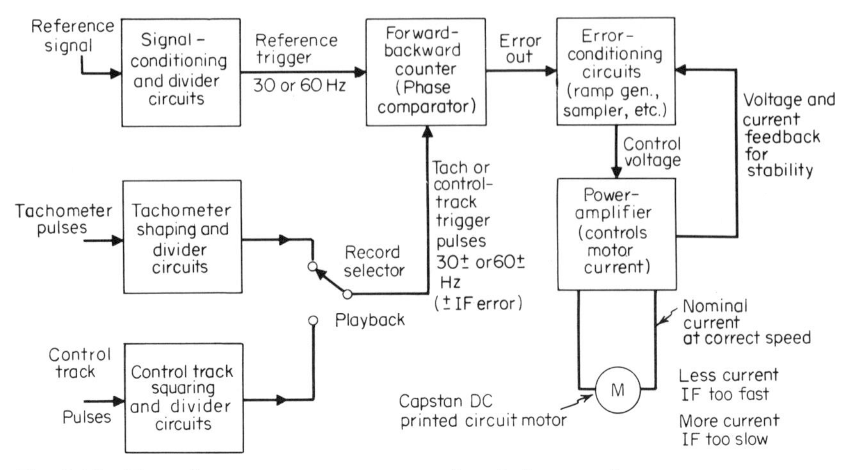

Fig. 8-19 Typical capstan servo system used with dc printed-circuit motors.

pulses. If any difference occurs, this represents a head-drum speed error, which will be used to change the frequency of the VCO, and thus the speed of rotation of the head drum. In the playback mode, control-track pulses will be used for comparison in place of the tachometer. Additionally, when the helical-scan recorder is being used with TV, the VCO will be locked to the TV station sync signal or to the power-line frequency. In the newer versions the head-drum motor is normally a dc printed-circuit motor. The servo system for such a motor will be used to control the amount of current through the motor in order to control its speed.

Capstan-drive Servos The speed of the capstan, like the speed of the rotating-head drum, is the function of the frequency of the VCO (in the older versions) or the amount of current fed to the dc printed-circuit motor in the newer models. This servo circuit must maintain the proper capstan speed with reference to the rotating-head assembly and the pulses it produces from its tachometer components. In the newer versions, during the record mode, the capstan servo phase-locks the capstan to a reference signal. The tachometer signal of Fig. 8-19, which is an accurate indication of the drum-capstan speed, is phase-compared with a precision reference frequency. Any phase difference represents a capstan speed error. The error is used to control the current through the dc printed-circuit motor. The tachometer pulses are also recorded on the tape to act as the control track during the playback mode. During playback, the control-track signal is compared with the precision reference in place of the tachometer signal. Again the phase differences are used to control the speed of the capstan motor by controlling the current through it.

DISC RECORDING AND REPRODUCING SYSTEMS

The disc recorder has been used for many years in the computing field, particularly in the disc pack recorder for digital data-storage systems. The disc was made of a polyester base, coated with gamma ferric oxide. To prevent damage to the oxide surface of the disc and excessive wear to the heads, the more advanced systems used a moving head (flying head) that was air-supported a fixed distance above the surface of the disc. Unfortunately, when the need arose for recording the high frequencies of television and instrumentation radar or similar pulse-type data, the very short wavelengths involved precluded the use of any system that had large amounts of separation between the heads and the disc. This problem was handled differently by the various manufacturers. One company, Data Discs, Inc., designed Micro-space heads which skim the magnetic disc with a separation of only 10 to 15 μin. Another designed an in-contact head-to-disc system. Both systems have merit so long as they are properly serviced and maintained.

In either the spaced-head or the in-contact head-to-disc system, the oxide-coated-polyester disc was replaced by an all-metal disc. These discs vary in size from a diameter of 12 to 24 in., the commonly used sizes being 12, 14, and 16 in. Their construction, although somewhat varied from manufacturer to manufacturer, follows that shown in Fig. 8-20. The recording can take place on both sides of the disc. The average weight of a 16-in. disc is 5.5 lb.

In-contact Head-to-disc System

In Fig. 8-21 the heads are mounted on four carriages, located at each corner of the assembly, one carriage per disc surface. The head carriages move in a set of guide rails that extend radially across the disc. (See Fig. 8-22.) The carriages are moved in controlled steps across the disc or discs while they are being rotated at 3,600 rpm by a servo-controlled dc printed-circuit motor. The head assemblies consist of a ferrite head and two small ferrite pads, which together form a stable

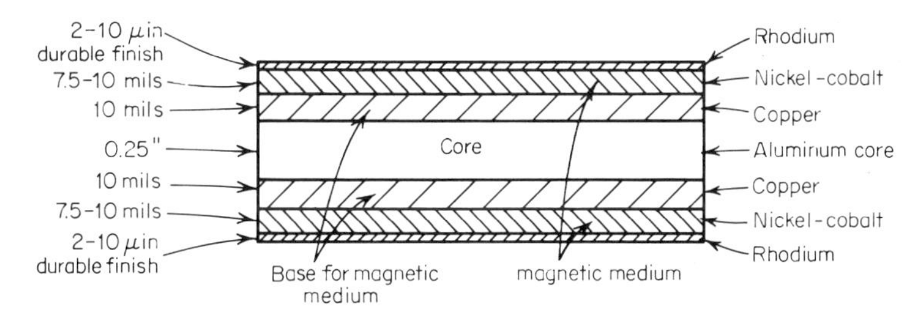

Fig. 8-20 Disc construction.

Fig. 8-21 HS 100 color slow-motion disc recorder. (*Ampex Corporation.*)

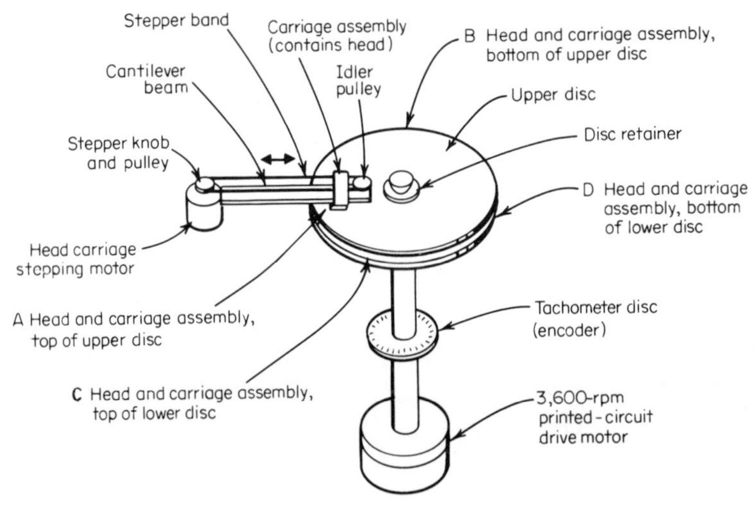

Fig. 8-22 Disc-recorder drive system.

three-point contact with the surface of the disc. The assembly is normally held against the disc surface by a cantilever spring that bears against a ruby jewel in the center of the head platform. The ferrite pads, head, and ruby form a tetrahedron. The amount of spring pressure is extremely light (approximately 5 to 6 g). The driving power for the carriage assembly is provided by a dc digitally controlled stepping motor, which has 200 steps of 1.8° per revolution. The stepping distance across the disc surface is 10 mils, 7.5 mils being used for the record track and 2.5 mils for a guard band. The motor is coupled to the sliding carriage by means of a metal belt.

Referring to Fig. 8-23, the first track is laid down ⅛ in. from the edge of the disc, after which the stepping motor moves twice before a second track is recorded. This means that one track has been skipped. In this fashion, a series of 225 tracks will be recorded as the head is stepped from the outer edge toward the center of the disc. There will also be 224 blank tracks. When the head (and carriage) has reached track 225, a logic circuit takes over and changes the stepping pattern so that no space is provided beyond track 225, and track 226 is recorded. After this action, the stepping motion is reversed, and the head will traverse toward the outer edge of the disc in the blank areas between the previously recorded tracks, laying down tracks 226 to 450.

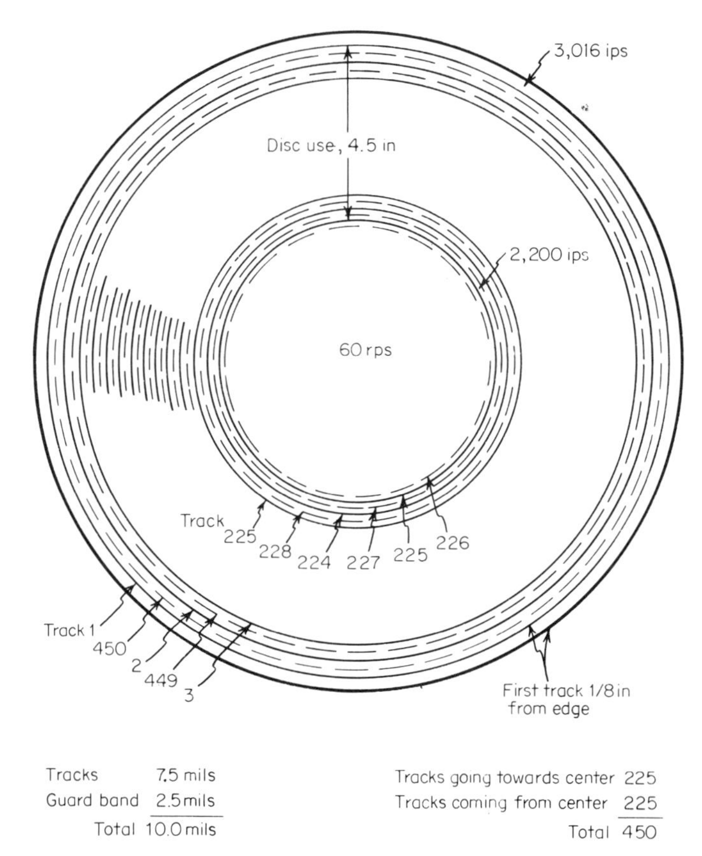

Fig. 8-23 Track numbering—HS 100/HS 200.

The rotational speed of the disc and its size (diameter) create some rather interesting head-to-disc speeds; i.e., the outer rim of a 16-in. disc, rotating at 60 rps, will be traveling at a speed of 175 mph, or 3,016 ips. The inner track, which is 4⅝ in. from the rim, will be moving at 1,320 ips. These speeds will limit the wavelength that can be recorded and reproduced.

Television Disc Systems

Servos A television frame consists of two vertical fields. Each field is a complete picture unto itself, but half the vertical resolution of the interfaced components. When a disc recorder is used with television, the primary purpose of the drive system is to lock the rotation of the disc in phase with a signal derived from the reference vertical sync of the television station. This ensures that each complete revolution of the disc corresponds exactly to one television field, beginning and ending with the vertical blanking period. An optical tachometer forms part of the servo system. The tachometer disc is similar to those used with the longitudinal instrumentation recorders. They are modified so that there are two encoder sections. One is a continuous ring of approximately 12.6-kHz segments, and the other contains a single transparent slit that is used to produce one pulse per disc revolution. The output of the single slit indicates one complete disc revolution and is called "the once-around tach." The 12.6-kHz tachometer section is used to bring the disc drive motor to the approximate speed (the velocity section). The "phase section" uses the once-around signal to correct for the balance of the speed errors (vernier control). Since it would be incorrect to attempt to record or reproduce when the disc is not up to speed, the velocity servo system is used to inhibit the movement of the head carriages whenever the discs are stationary or at very low speed. Head movement across the surface when the disc is stopped could result in damage to the disc.

Slow-motion and Still Pictures During playback the heads can be stopped anywhere and the result would be a still picture with one-half the resolution of the original. Slow motion can be produced by repeating the scan of each track a number of times before moving to the next track. Repeating each frame three times would produce the effect of one-third speed. However, due to the makeup of a television picture (see Chap. 11, The Television Recorder), it is necessary to produce slow-motion effects by repetition of fields rather than frames, and produce interlaced frames artificially, using the information contained in a single field only. For fast motion, every other field may be recorded. This would provide, in addition, two times the recording time.

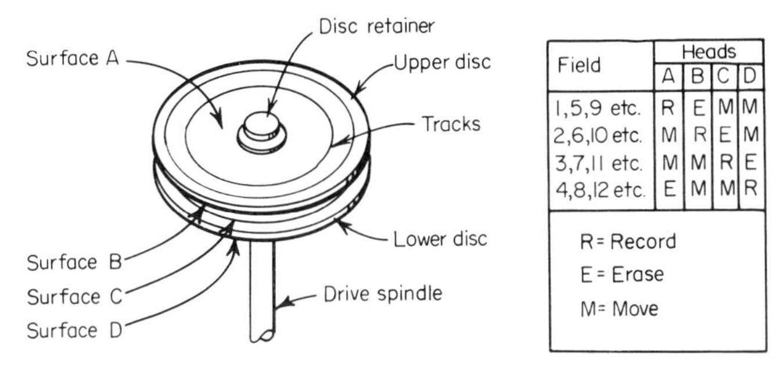

Field	Heads			
	A	B	C	D
1,5,9 etc.	R	E	M	M
2,6,10 etc.	M	R	E	M
3,7,11 etc.	M	M	R	E
4,8,12 etc.	E	M	M	R

Fig. 8-24 Disc-surface lettering and switching format.

Recording Pattern In the Ampex HS 100 and HS 200 disc recorders, four heads are used, one for each surface of the two discs. Each head combines three functions: record, reproduce, and erase. Referring to Fig. 8-24, when head A has completed recording field 1, head B records field 2. After head D records field 4, head A is now in place to record field 5. Each head records every fifth field. During the record mode one head is always recording, one is erasing, and the other two are moving into new positions. Using this technique, the machine can be operating in a continuous updating mode, making the disc recorder ideal for continuous monitoring or for recording short data sequences. For reverse motion the carriages are stepped in the opposite direction and the sequence of head switching is changed from A, B, C, D, to D, C, B, A.

Instrumentation Disc-recorder Servo Control

Since television data are separated into two fields, the space between the fields can be used for switching between the tracks without any loss of data. However, for instrumentation applications, where the signal is continuous, some means must be found to switch between the tracks without any loss of data. This is accomplished by recording overlapping tracks, as illustrated in Fig. 8-25. In the overlap period both tracks carry the same data. Transient-free recombining is achieved by soft (slow) switching from one track to the next during reproduction. The result is continuous and uninterrupted data output. (See Fig. 8-26.) For successful recombining there must be extremely small time-base errors between the two signals. For this reason a 500-kHz pilot signal is added to the RF carrier in the same fashion as was done for rotating-head instrumentation recorders. By use of the tachometer discs plus electronic time-base correction circuits (pilot signal + equalizers + time-base correction), time-base errors are kept below 25 nsec peak to peak.

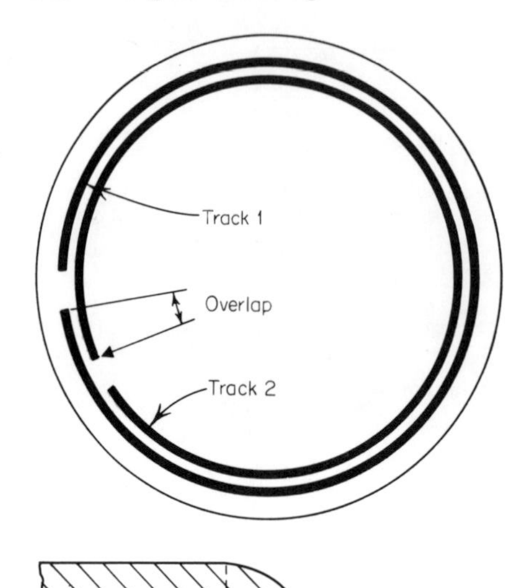

Fig. 8-25 Instrumentation disc-recording overlapped tracks.

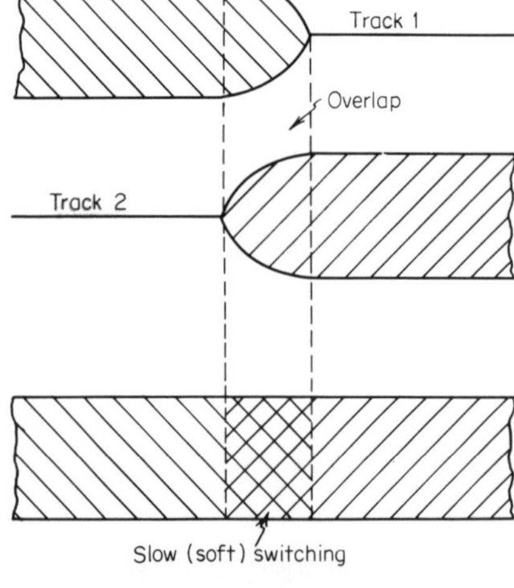

Fig. 8-26 Slow-switching, soft-switching, continuous-data reproduction.

Spaced-head Systems

The spaced-head systems use a head mount that keeps the head separated from the disc by a minute distance (10 to 15 μin.). Both fixed- and movable-head assemblies are available using the Micro-space design. (See Fig. 8-27.)

Using the fixed-head technique, each track on the disc is recorded and played back by its own head. There can be as many as 72 individual heads recording at one time across the disc. Most of these systems are available in increments of eight heads. (See Fig. 8-28.) The disc is normally driven by either a hysteresis-synchronous motor or a dc printed-circuit motor and a solid-state servo drive system. Disc speeds may vary from 1,800 to 3,600 rpm.

Fig. 8-27 Close-up view of Micro-space heads. (*Data Disc, Inc.*)

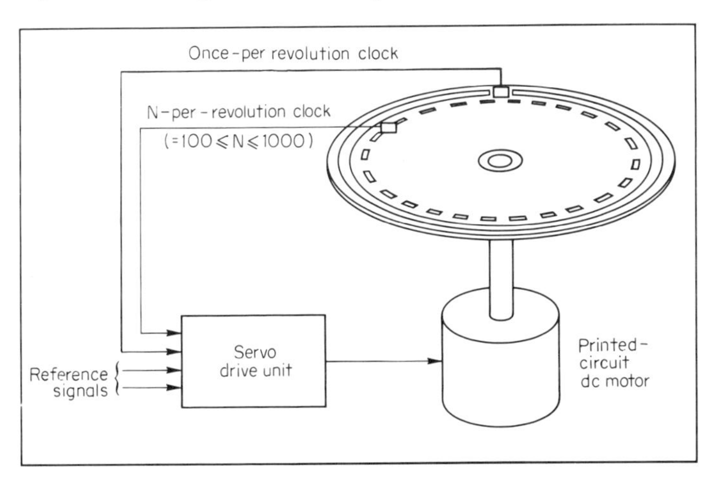

Fig. 8-28 Servo drive system. (*Data Disc, Inc.*)

Spaced-head Servos[1]

The servo system used to control the speed of the disc synchronizes the disc's speed in-phase with respect to an external composite sync signal for video applications or to a phased high-frequency signal (N-per-revolution clock with a once-per-revolution phase reference) for other applications. As illustrated in Fig. 8-28, there are two recorded tracks

[1] Extract from "Servo Drive System," Data Disc, Inc., Palo Alto, Calif.

on the disc used for servo. One track provides a once-around mark (or REF), while the other provides *N* marks per revolution (or *N*). Synchronization of the disc speed is achieved by comparing similar signals (*N* or REF) from a reference signal source with those from the disc. Comparison is made via two separate phase comparators. If the disc and the reference sync signals are within 1 part of *N* of being phase-locked, the servo will be in the high-frequency operating mode. In this mode, error information from the *N*-input-phase comparator is used for continuous phase locking in a very high gain servo loop. If the disc and reference signals are not within 1 part in *N* of each other, the servo system automatically switches over to the reference-phase detector for appropriate adjustment of the loop gain. As soon as the servo is within 1 part in *N* again, the *N*-phase control is reestablished by the automatic switching. Any phase error measured by the comparators will cause the servo to accelerate or decelerate the disc drive motor as required for phase sync.

REEL- OR TAPE-STORAGE SYSTEMS

Functions

The reel- or tape-storage assemblies provide the following functions:

1. Storage of tape on either side of the capstan and heads.
2. Development of necessary holdback tension so that a constant amount of stress is applied to the tape as it enters the head and capstan area, despite the changing mass.
3. Development of the tension required to take up the tape as it is fed from the capstan.
4. Fast movement of the tape in those systems that do not use the capstan for this purpose, i.e., fast forward, fast reverse, or rewind modes.
5. Guiding the tape into and out of the head area.

Reel Servo and Tensioning Systems

Bin Loop and Bin-loop Adapter When continuous playing time is required, a single storage chamber is often used and the tape is in the form of a continuous loop. These systems are normally called bin loops. They may be adapted to fit on a reel-to-reel transport so that the modified transport operates as a continuous-loop recorder/reproducer system. One such system, developed by Consolidated Electrodynamics Corporation (CEC), consists basically of a housing having a bin compartment to store the tape loop, a drive mechanism having a free-running differential-drive pulley, and a pinch-roller system to transmit power from the capstan drive to move tape out of the bin and return it under tension. There is also a system of guides to direct tape into and out of the bin. The front side of the bin compartment is a transparent plastic door which allows the operator to see the tape loop during operations. In this system, tape tension outside the capstan drive is maintained by a differential-drive pulley which attempts to feed tape into the bin faster than it is removed.

Another bin-loop system, the Ampex FB 450, uses a different system to maintain the holdback tension in the closed-loop area. Figure 8-29 shows that there is a tape-tension capstan and a stuffing capstan. The tape-tension capstan consists of a capstan, a solenoid-actuated pinch roller, a tension-sensing arm, a brake drum, and a brake cord. In operation the pinch-roller solenoid is actuated, clamping the tape against the tension capstan. The tension-sensing arm detects the change in tape tension and tightens or loosens the brake cord around the capstan brake drum. In this manner, the tension capstan regulates the amount of tape leaving the bin and develops the proper holdback tension. Take-up tension is developed by the tape-stuffing mechanism. It consists of a pinch roller and a capstan which is attached to the shaft of a dc motor. The torque of the stuffing motor produces the tape tension, and the tape is fed into the bin.

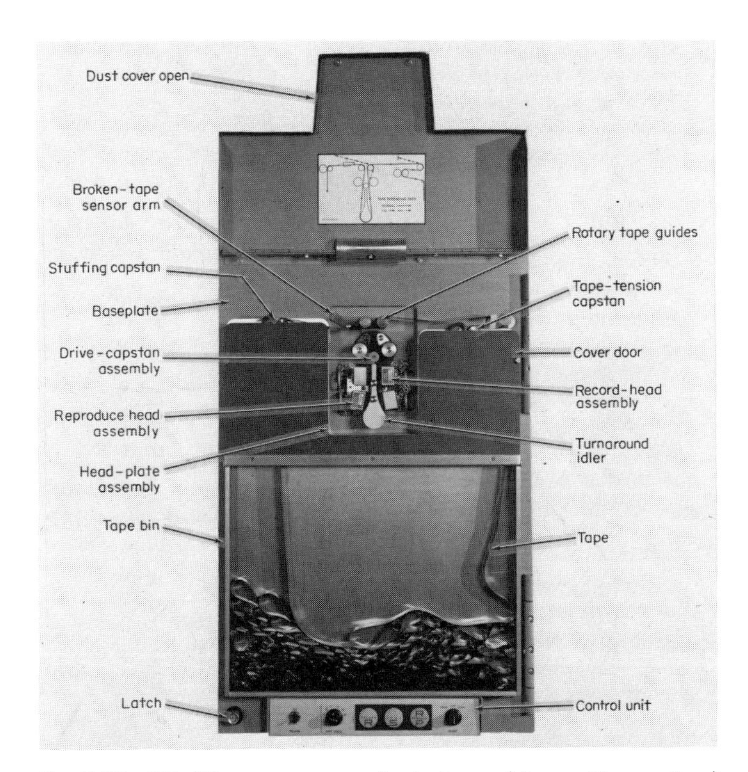

Fig. 8-29 FB 450 tape transport, front view. (*Ampex Corporation.*)

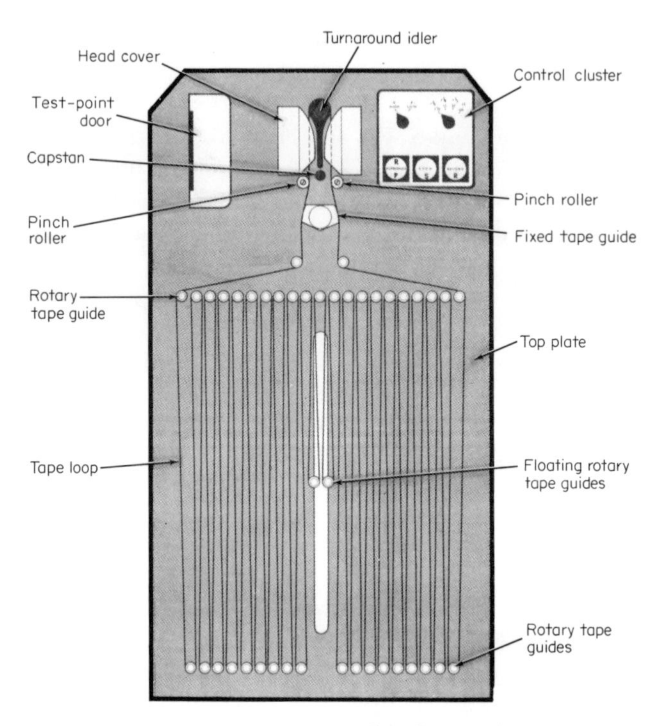

Fig. 8-30 Loop transport—tape-threading path.

The Continuous Loop A second type of loop system is shown in Fig. 8-30. The tape is stored on a series of precision rollers. The distance between the upper and lower sets of rollers and the way in which the tape is threaded over them determine the total loop length. The capstan provides all the driving power to move the tape. This is possible because of the extremely small mass of tape to be driven in the loop. Most of the load is provided by the friction of the tape against the head, guides, and bearings of the rollers. Tape tension is set by introducing a standard weight into the loop; i.e., the weight is attached to two of the rollers (floating rotary tape guides), over which the tape is threaded. The weight is vertically mobile, riding on two rods attached to the transport frame.

In systems such as these there are no supply or take-up motors or turntables. Movement of the tape takes place when pinch rollers press the tape against a precision-driven capstan. Stopping of the tape is controlled by the braking of the capstan. This is usually some form of solenoid-applied brake pad (asbestos or cork, etc.) that is pulled or pushed against the capstan flywheel. Guiding of the tape is maintained by using flanges on the precision rotating guides (rollers) in the loop and via precision stationary guides in the head and capstan area.

Tape Storage and Servoing Using Vacuum and Photoelectric Sensing

Combination Systems Figure 8-31 shows a system in which a combination of reel and vacuum storage is used. Supply and take-up tension will be a function of the position of the tape in the vacuum chamber. As the position of the tape changes, the amount of light that reaches a series of photocells changes. The output of the photocells is used to control the amount of tape fed into or taken out of the chambers.

Vacuum Chambers and Vacuum Sensing Another method of vacuum control is shown in Fig. 8-32. Here the take-up and supply motors are dc motors that contain two field windings; one for clockwise and the other for counterclockwise rotation. Direction and speed of the motor rotation are controlled by gating on a pair of silicon controlled rectifier (SCR) circuits. When tape is fed into the vacuum chamber, it divides the chamber into two portions. The right-hand side will be at atmospheric pressure, and the left-hand portion under vacuum. A sensing slot is common to these two portions of the chamber. The position of the tape along the slot will determine the ratio of atmospheric pressure to vacuum that is felt by the vacuum-sensing transducer. The resultant value will control the movement of a set of bellows which is part of the transducer assembly. A moving core attached to the bellows determines the amount of signal coupling from the primary to the secondary of the transducer windings. The output of the transducer is a sine wave of an amplitude that is a function of the position of the core (the position of the tape in the chamber). This is fed via a half-wave rectifier to the circuit that controls the turn-on time of the SCRs. Whenever an SCR is turned on, it completes the circuit for the series field winding to which it is attached; i.e., if the clockwise

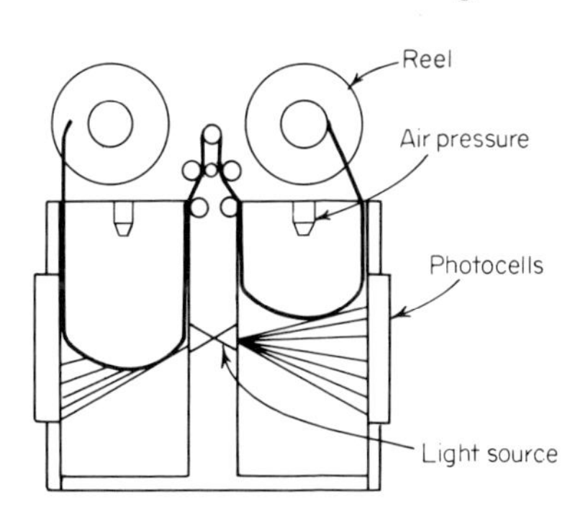

Fig. 8-31 Tape storage using vacuum and photoelectric sensing.

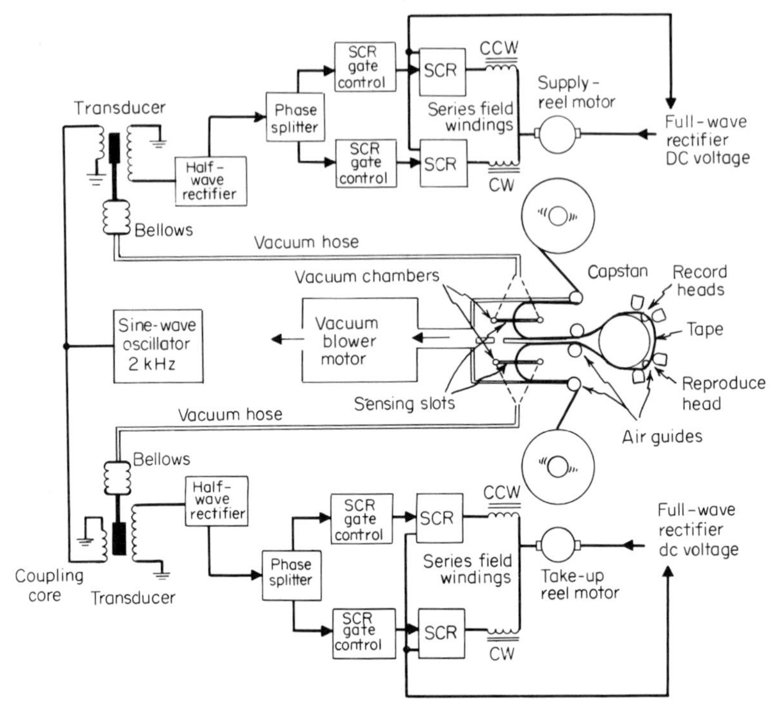

Fig. 8-32 Block diagram of reel servo using vacuum chamber.

SCR was turned on and the counterclockwise SCR was not, the motor will move in a clockwise direction. However, if the clockwise and counterclockwise SCRs were both turned on, the motor will be given equal command to move clockwise and counterclockwise and will thus not move at all. Therefore the principle of operation of this type of circuit is the amount of time that each of the SCRs is on. If there is a difference, the SCR that comes on first will command the motor to move in that direction until the turn-on command is countered by an equal and opposite turn-on command of the other SCR.

Using a Plenum In systems which use a plenum, as shown in Fig. 8-33, when the tape changes its position in the plenum, more or less light reaches the reel-servo solar cell. The amount of light controls the amount of positive or negative drive signal from the preamplifier stage and provides the controlling force through the reel-servo amplifier to a brake assembly mounted on the reel motor. The motors in systems of this sort are 115 V, ac, split-phase, stall-torque induction motors, capable of continuous duty and having an ability to accelerate relatively large and variable inertial loads.

The motor brake is disc-type and consists of a segmented lining, a stator coil and housing, a cast-iron armature, a support diaphragm, and a fixed cover plate. (Refer to Fig. 8-34.) The cover plate is keyed to the reel motor shaft via the hub and turns with the reel motor. The cast-iron armature is coupled to the cover plate through a flexible diaphragm. The diaphragm causes the armature to rotate with the cover plate, and also permits it to be drawn against the brake lining by the magnetic attraction of the stator coil. Friction provided by these contacting surfaces produces a braking force on the motor shaft. This force is controlled by the amount of current through the stator coil. The speed of the reel motor is controlled by the braking force.

In Fig. 8-33 assume that there is too little tape in each plenum and that the tape is being moved from the upper to lower plenum. In systems of this type both motors rotate in the same direction. The reel motors move opposite to the capstan rotation. To increase the amount of tape in the upper plenum, the upper brake must be released.

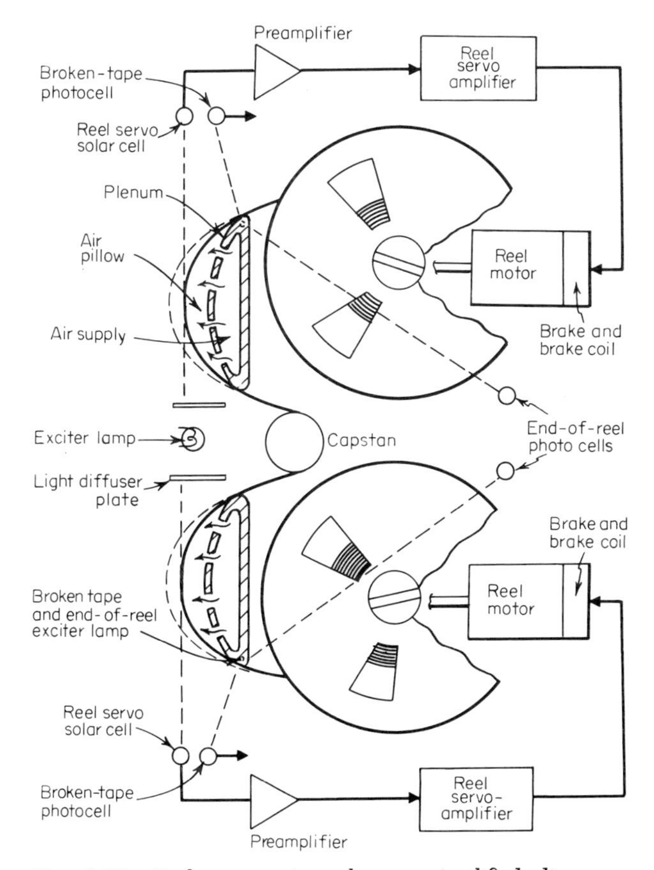

Fig. 8-33 Reel servo using plenum—simplified diagram.

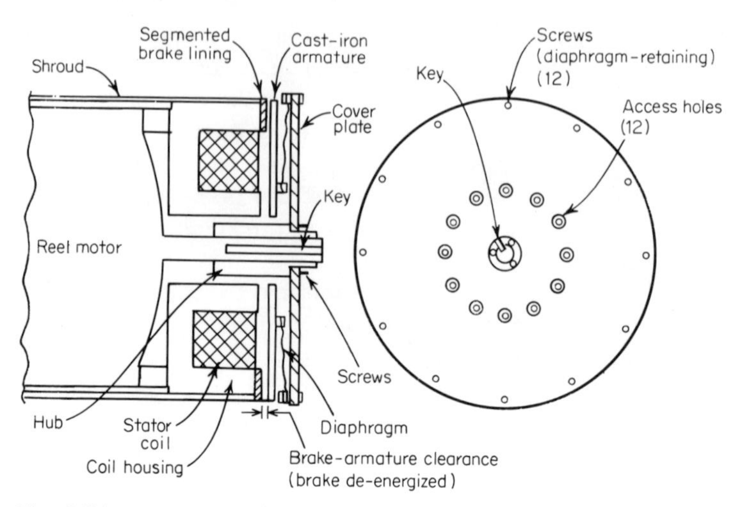

Fig. 8-34 Disc-type-reel motor brakes.

To increase the tape in the lower plenum, the lower brake must be tightened. Thus it may be seen that the reel-servo amplifiers must be mirror images of each other. With more light hitting the upper solar cell, less current will be passed through the upper brake coil and the brake will be released. This permits the upper reel motor to turn faster and feed more tape into the plenum. With more light falling on the solar cell in the lower plenum, the brakes will be tightened and the motor will slow down. This being the case, less tape will be pulled out of the plenum with respect to the amount of tape being fed in by the capstan, and as a result the amount of tape will increase in the plenum.

Reel Servo and Tensioning Systems Using Mechanical Control

One of the oldest forms of tape-tensioning systems used a mechanical brake on the supply reel to provide the tape tension. Referring to Figs. 8-35 and 8-36, it can be seen that if the tape is too tight, the servo-tension-arm tape guide will be pulled in such a direction that the supply tension arm will loosen the brake band from around the supply turntable and brake drum. This means that the drum will turn more easily and the tape tension will be relieved. Take-up tension is provided by a constant torque on the take-up motor. The band brake and the take-up tension arm are not used in the forward mode. A modified version of this system uses string brakes in place of band brakes. The string brakes are made of 130 strands of beryllium copper wire covered with a woven nylon jacket.

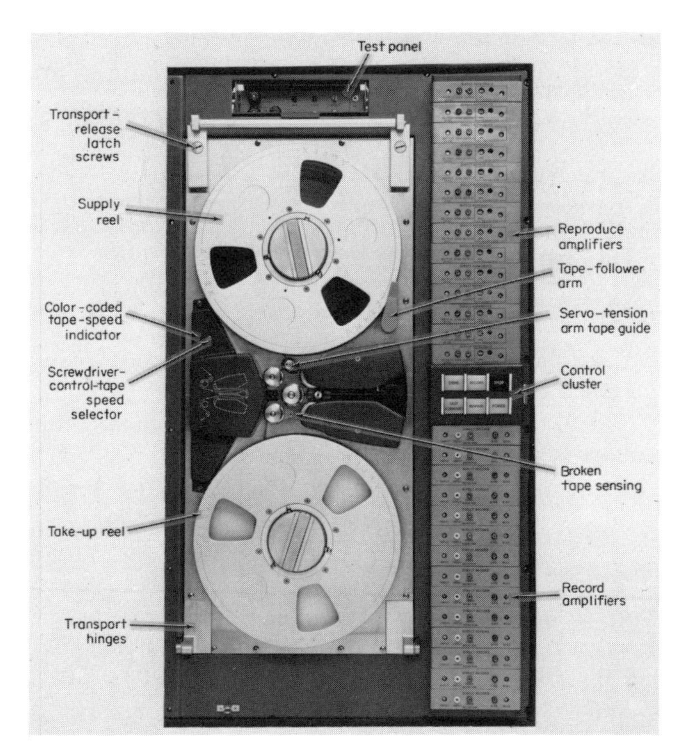

Fig. 8-35 CP 100, top view. (*Ampex Corporation.*)

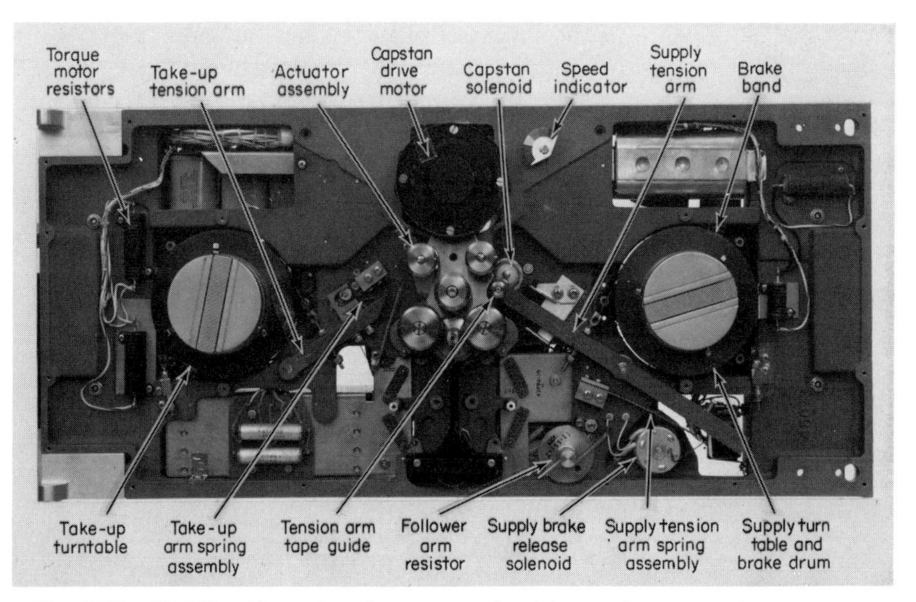

Fig. 8-36 CP 100 with overlay plates removed. (*Ampex Corporation.*)

Although this type of tape tensioning is relatively old, it is still being used in modern airborne recorders where light weight and space are of prime consideration.

SENSING THE END OF THE TAPE

Mechanical, pneumatic, photoelectric, and vacuum designs have been used to sense the end of the tape.

Mechanical

In Figs. 8-35 and 8-37 a follower arm senses the amount of tape remaining on the reel. Attached to this follower arm is a potentiometer which adjusts the current fed to a remote readout device that indicates the amount of tape remaining. When the end of the tape is reached, a microswitch, also attached to the follower arm, is closed and breaks the ground return to the mode relays, turning off the transport.

Pneumatic End-of-tape Sensing

Figure 8-38 shows a means of pneumatically sensing the end of the tape. Figure 8-38*a* shows the angle of tape when there is a full reel. Under this condition the pressure in the two air chambers is equalized and the rubber diaphragm with its metal tip will be in the rest position. The active contacts of the microswitch are open. As the end of the reel is approached, the angle of the tape changes, as shown in Fig. 8-38*b*, to a point where it closes off the mushroom-head aperture. This builds up a pressure in the lower air chamber and expands the rubber diaphragm, placing the microswitch in the position where the active contacts are closed. When this occurs, the end-of-tape circuit will be energized and the tape transport will be stopped.

Photoelectric End-of-reel Sensing

Figure 8-33 indicates that as the amount of tape on the reel is reduced, there will be a point at which light from the broken tape and end-of-reel

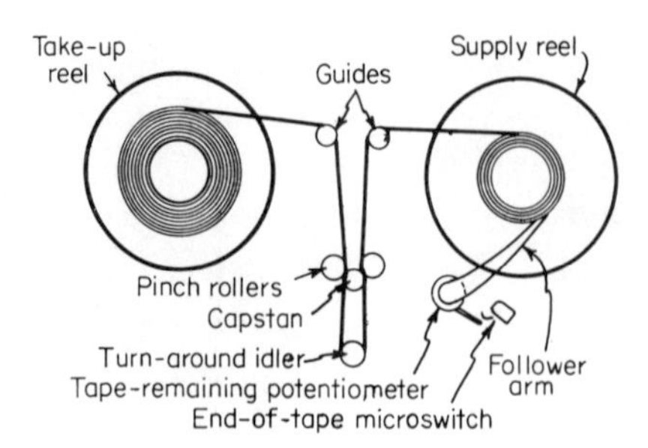

Fig. 8-37 End-of-tape sensing using follower arm.

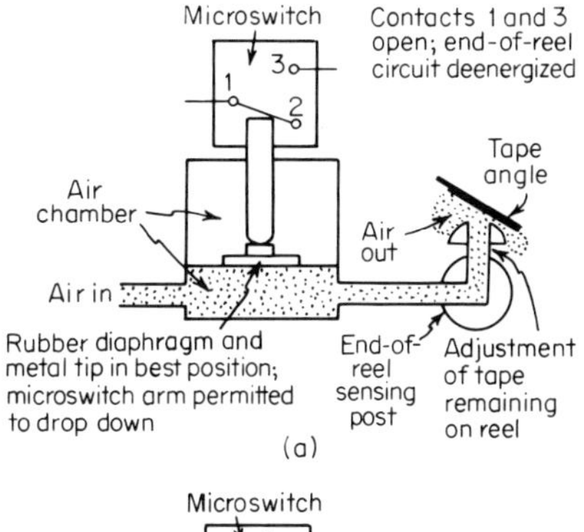

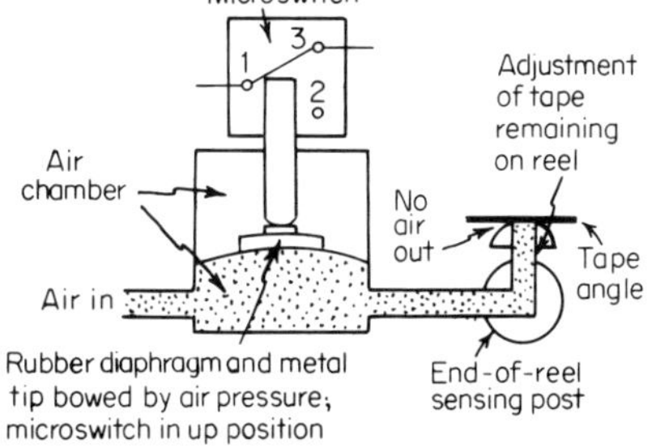

Fig. 8-38 Pneumatic end-of-tape sensing. (*a*) Full reel of tape; (*b*) end of tape reached.

exciter lamp will reach the end-of-reel photocell (see dotted lines). When light hits the photocell, a circuit is energized which breaks the ground return for all the mode relays, and tape-motion stops. This type of end-of-reel sensing is used in a great number of modern recorders.

Vacuum End-of-reel Sensing

Most systems that use vacuum chambers contain sensors for sensing the absence of vacuum. For the most part, they would indicate not only the end of the reel, but also broken tape, and therefore will be discussed under the heading of broken-tape devices.

BROKEN-TAPE SENSING

Broken-tape sensing, like end-of-tape, may be classed under several groups: mechanical, pneumatic, photoelectric, and vacuum.

Mechanical Broken-tape Sensing

Mechanical broken-tape sensing is accomplished by putting a small spring-loaded tension arm in the tape path. If the tape breaks, the arm automatically swings in such a fashion that it opens the contacts to the mode relays. This turns off the transport and applies the brakes. (See Figs. 8-35, 8-36, and 8-39.) This type of broken-tape sensing is used in a number of instrumentation recorders, most of the audio recorders, and a great number of the rotating-head instrumentation and video recorders. It is simple and relatively foolproof.

Pneumatic Broken-tape Sensing

Figure 8-40 shows broken-tape sensing using a pneumatic switch. The mechanical action of the broken-tape sensing switch is illustrated in the (*a*) and (*b*) sections of the figure. The switch consists of two air chambers separated by a rubber diaphragm which has a metal plate and tip mounted on it. As long as the tape contacts the guide, the lower air chamber will be under pressure and the rubber diaphragm will be bowed. The microswitch will be in the position shown in Fig. 8-40*a*, and the contacts 1 and 2 are open. When the tape breaks, the air pressure in the lower chamber will escape and the diaphragm will return to its rest position. Thus the microswitch will drop down and the contacts 1 and 2 will be closed (Fig. 8-40*b*). With contacts 1 and 2 closed, the broken-tape circuit is energized and the tape transport is shut off.

Photoelectric Broken-tape Sensing

Another system for energizing broken-tape circuits is shown in Fig. 8-33. In the figure it is seen that if the tape breaks, light will reach the broken-tape photocell either on the supply or the take-up side. The increase of current through the photocell will energize a circuit that

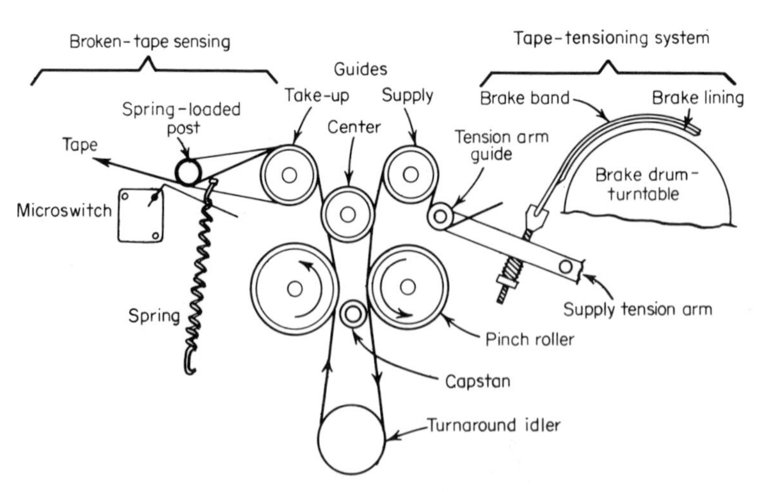

Fig. 8-39 Mechanical broken-tape and tape-tensioning systems.

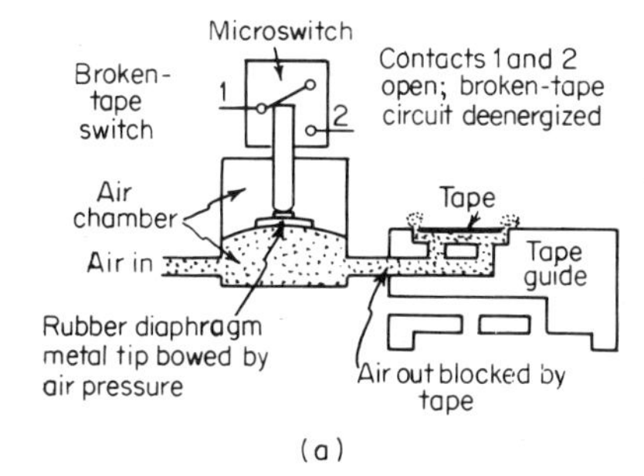

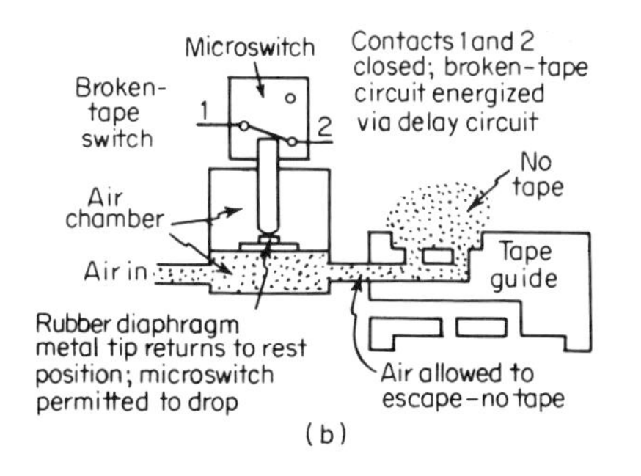

Fig. 8-40 Pneumatic broken-tape sensing.

will provide ground to the broken-tape relay. This in turn will shut down the transport by opening up the ground return line for the mode relays and put on the brakes.

Vacuum Broken-tape Sensing

Referring to Fig. 8-32, a secondary vacuum sensing switch is placed into the vacuum chamber. If the tape breaks, the amount of vacuum pressure will drop and the vacuum switch will open, in so doing cutting off power to the mode relays, so that the tape transport will come to a halt.

Quite naturally, not all the types of systems used with all tape transports have been discussed. However, this text has attempted to cover this broad field as fully as possible, using all the general categories of control and design. Illustrations and descriptions of each system have been included so that the reader will have a ready reference to the overall techniques used.

REFERENCES

1. Athey, S. W.: "Magnetic Tape Recording," National Aeronautics and Space Administration, Technology Utilization Division, Washington, D.C., 1966.
2. Begun, S. J.: "Magnetic Recording," Rinehart & Company, Inc., New York, 1949.

3. Data Disc, Inc., Display Division: "Servo Drive System," Palo Alto, Calif., 1970.
4. Davies, G. L.: "Magnetic Tape Instrumentation," McGraw-Hill Book Company, New York, 1961.
5. Haynes, N. M.: "Elements of Magnetic Tape Recording," Prentice-Hall, Inc., Englewood Cliffs, N.J., 1957.
6. Lowman, C. E.: "The Magnetic Tape Recorder/Reproducer and the Concepts of Systems Used for Recording and Reproducing FM Analog Test Data," vol. 1 of "Fundamentals of Aerospace Instrumentation," Instrument Society of America, Pittsburgh, Pa., 1968.
7. Lowman, C. E., and G. J. Angerbauer: "General Magnetic Recording Theory," Ampex Corporation, Redwood City, Calif., 1963.
8. Printed Motors Inc.: "Printed Circuit Motors," Glen Cove, N.Y., 1965.
9. Stewart, W. E.: "Magnetic Recording Techniques," McGraw-Hill Book Company, New York, 1958.
10. Weber, P. J.: "The Tape Recorder as an Instrumentation Device," Ampex Corporation, Redwood City, Calif., 1967.

CHAPTER NINE

Direct Record and Reproduce Signal Electronics

GENERAL

This chapter will be devoted to discussion of the block and simplified diagrams of typical circuits that are used for direct recording and reproducing systems. It will also lay down some of their limitations and advantages over the other types of modulation processes. Discussion will be restricted to the modern transistorized and integrated circuits. For the older vacuum-tube circuits, the reader is referred to Refs. 1 to 4 in the list of References at the end of the chapter.

An increase of the bandwidth of direct recording systems has been made possible primarily by improvements in head technology; but better tape, improved tape transports, and solid-state circuitry have also played their part. All these improvements have permitted a wider range of frequencies while using the same or lower tape speeds.

LOW-BAND DIRECT RECORD/REPRODUCE ELECTRONICS—TO 100 kHz

Audio recording and video auxiliary track recording fit into this class. Low-band systems for instrumentation are usually involved with voice log or monitoring and provide a combined record/reproduce function to decrease cost and limit space requirements.

Typical Voice-log System

Figure 9-1 is the block diagram of a typical instrumentation voice-log system. It is capable of 300- to 3,000-Hz response. The voice-log track is an auxiliary edge track 5 mils wide for ½-in. tape and 10 mils wide for 1-in. tape. It is normal practice to restrict the bandwidth in order to improve the signal-to-noise ratio of these systems. Since a large amount of noise is found in normal recording environments (hum of motors, footsteps of operating personnel, etc.), they contain attenuation circuits to remove the low frequencies picked up by the microphone. This causes the signal to sound a little high-pitched, but it does improve the understandability.

Referring to Fig. 9-1 the relay *K*1 is shown in the reproduce (deenergized) position. In this mode, the reproduce head is connected to the dc-stabilized amplifier stage. It also has an additional feedback path (not shown in the figure) that operates to reduce the gain at high frequencies. This type of feedback is used to remove any bias feed-through.

After amplification and low-end limiting (attenuation network), the signal is fed to the emitter-follower driver amplifier. The output stage is normally a dc-coupled class A amplifier with transformer coupling to the speaker or earphones.

In the record mode, relay *K*1 is energized and the microphone is connected to the first amplifier stage; the output of the driver is con-

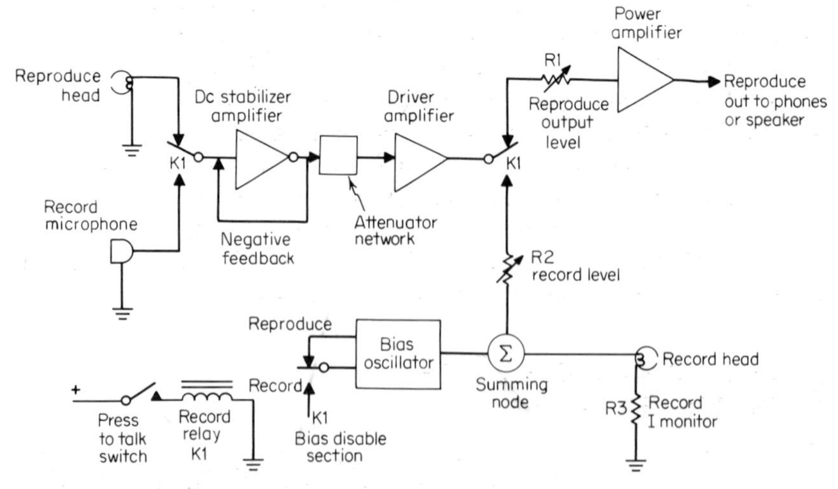

Fig. 9-1 Block diagram of a typical voice-log circuit.

nected to the record head via the record-level potentiometer $R2$ (bias is added to the voice signal at the summing node Σ); the bias oscillator is permitted to free-run. The frequency of the bias will depend on the system, but is normally 50 kHz or so. The oscillator may be one of a number of types. The most common is a modified form of the Colpitts oscillator. Signal currents of 2 to 4 mA and bias current of 20 to 25 mA are mixed together (linearly) at the summing node before being passed to the record head. Resistor $R3$ is used when checking or setting the signal or bias currents.

Typical Audio- and Cue-track Systems

Figure 9-2 is a typical block diagram of the circuits used in video and helical-scan recorders for audio and cue tracks, and in the instrumentation rotating-head recorder for auxiliary tracks. The bias current provided by these systems will also be used as erase current for the audio and cue tracks. The bias frequency varies somewhat with different systems. But with recorders used for closed-circuit and standard television, 100 kHz is the most common. Instrumentation rotary-head recorders tend to use 250 kHz as their bias frequency.

Since the output of the microphones used with these systems is fairly small, a microphone preamplifier is used to boost the signal level by 40 to 46 dB prior to feeding it to the record amplifier. The difference between the audio, instrumentation, and video systems is shown in Fig. 9-2*a* and *b*.

There are other differences in the two systems that are not evident in the block diagrams, the types of record amplifier circuits, for example. The instrumentation version uses a totem-pole power amplifier to develop the head current. It also has two adjustable bias traps to prevent the bias from entering into the signal amplifier stages and causing distortion. Degenerative feedback, which varies inversely as the frequency, is used to provide a constant-current level to the heads. The video version uses a single nonadjustable bias trap and a push-pull power amplifier for head current. The degenerative-feedback path has been expanded to include equalization circuits so that the audio record curve follows the NAB Standards curve (see Fig. 7-26 for details of the NAB curve).

The biasing systems, although quite different in appearances in Fig. 9-2*a* and *b*, are really very similar in operation. Where the instrumentation version uses the output of a crystal-controlled oscillator, binary-divided down to a 250-kHz square wave as a source of power, the video version has its own oscillator. This is a 200-kHz emitter-coupled multivibrator, the output of which is divided down by a flip-flop to a 100-kHz square wave. In both cases, the square waves are used di-

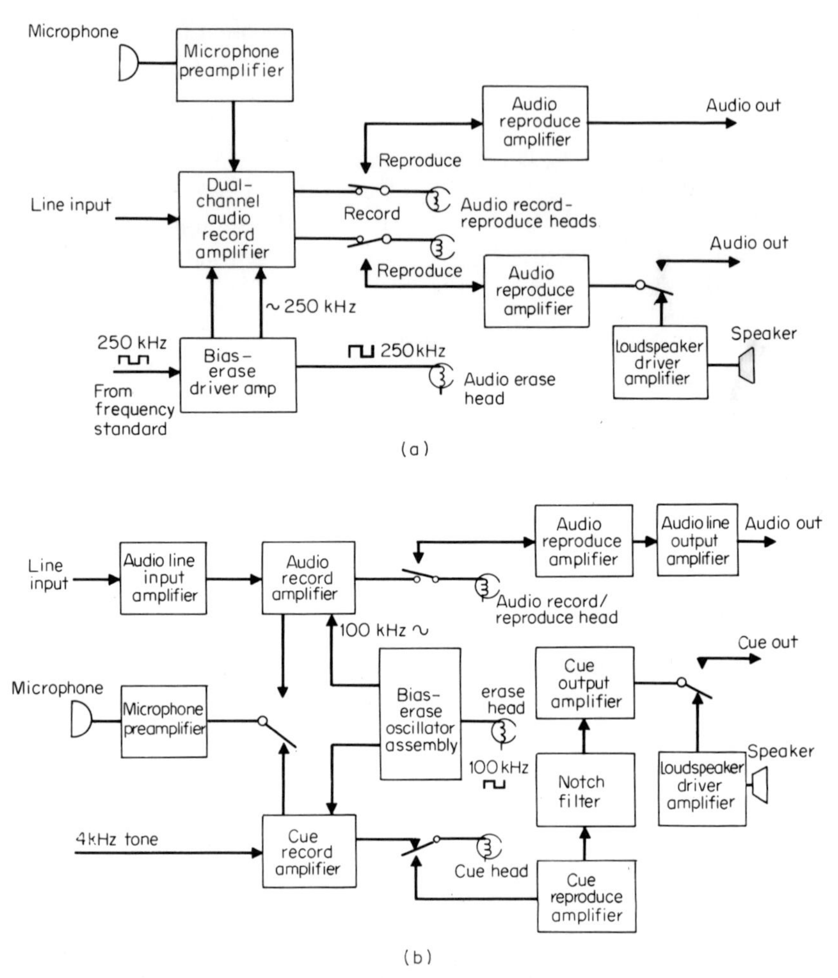

Fig. 9-2 Audio and cue block diagrams. (*a*) Rotary instrumentation and audio block diagram; (*b*) rotary video audio and cue block diagram.

rectly as the erase current. For signal bias current, some form of fundamental filter is used to extract the sine wave from the square wave. In the instrumentation version this is usually a series-resonant filter, and in the video circuits where transformer coupling is common, a tuned secondary winding is used to extract the 100-kHz sine wave. This is essentially a resonant filter.

LOW-BAND REPRODUCE AMPLIFIERS

The reproduce (playback) amplifier for video and instrumentation low-band systems are somewhat different from each other. This is not surprising if one considers that they are performing to different specifica-

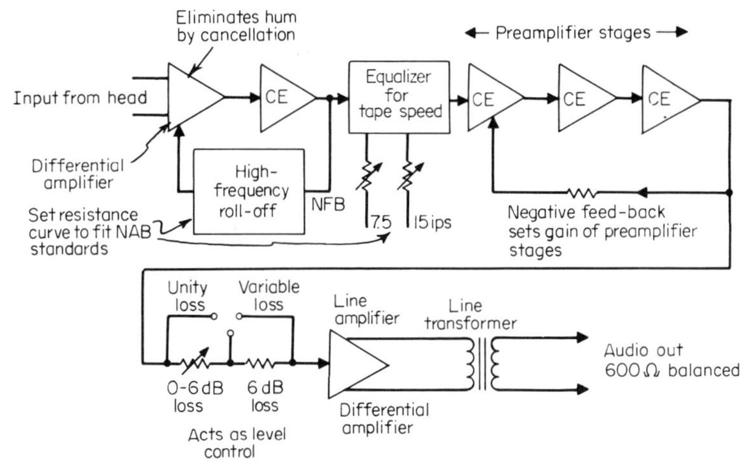

Fig. 9-3 Audio-reproduce and line amplifiers for video systems.

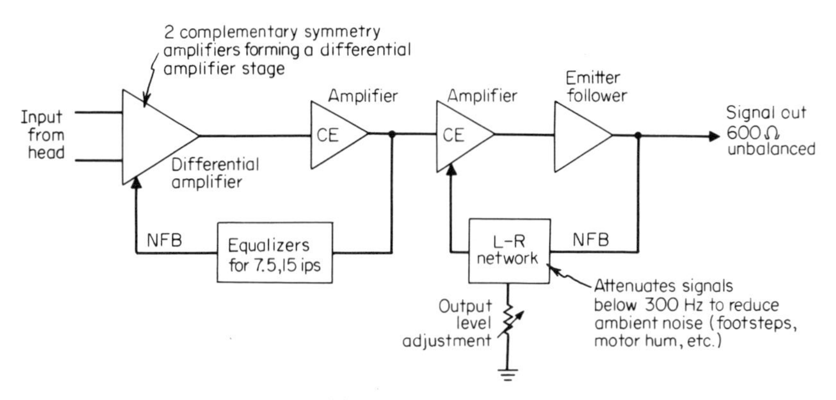

Fig. 9-4 Audio-reproduce amplifier for instrumentation.

tions. Where the instrumentation specifications call for a flat response (±3 dB) over a bandwidth of 300 Hz to 30 kHz and a tape speed of 12.5 ips, or a bandwidth of 300 Hz to 60 kHz at a tape speed of 25 ips, the video systems call for curves to fit the NAB Standards. The ranges of frequency are 50 Hz to 12 kHz at tape speeds of 15 ips and 50 Hz to 10 kHz at tape speeds of 7.5 ips. Figure 9-3 illustrates a typical video system, and Fig. 9-4 shows one for instrumentation.

INTERMEDIATE-BAND DIRECT RECORD/REPRODUCE ELECTRONICS— TO 600 kHz

The intermediate band is specified as having a frequency response of 300 Hz to 500 kHz at a tape speed of 120 ips (IRIG Standards), and

100 Hz to 7.5 kHz at a speed of 1⅞ ips. The industry has supplied a wider bandwidth than that (300 Hz to 600 kHz at 120 ips) with an excellent signal-to-noise ratio by improving the quality of the tape and heads and designing a very stable transport.

Figure 9-5 illustrates a typical intermediate-band-system block diagram. In this figure there is one bias oscillator for each track of record electronics. As a result, if the component values in each bias oscillator are not exactly the same, each oscillator will have a different output frequency. Such differences will cause beat notes that result in a high noise level in the system. To prevent this, the secondaries of the bias transformers are tied together so that they synchronize each other, and a common bias frequency is formed for all the tracks.

The data signal is fed via the record-level potentiometer to the first stage, usually an emitter follower. This stage provides the input impedance for the record amplifier. The permissible value of the incoming data varies quite widely with the different systems. It tends, however, to fit into one of the following categories:

0.1 to 25 V rms 0.7 to 10 V rms 0.2 to 10 V rms

The minimum-maximum ranges are primarily dependent on the value of the record-level-adjustment potentiometer. The nominal value is typically 1 V rms.

The output of the emitter follower is fed to the data signal amplifier. This normally consists of a common-emitter amplifier, followed by a phase splitter which drives a push-pull power-amplifier stage. Negative feedback from the output of the push-pull stage to the base of the phase splitter is used to stabilize the circuit. The output of the push-pull stage is coupled to the head via a parallel-resonant bias trap. This blocks the bias frequency from entering the data amplifier and causing distortion. The bias oscillator is a self-starting, push-pull sinusoidal oscillator. Master oscillators are often Butler oscillators or similar types. The bias is added to the data signal at a solder connection which may be considered as a summing node. Intermediate-band levels are most commonly 18 mA of bias to 1 mA of data. These levels are

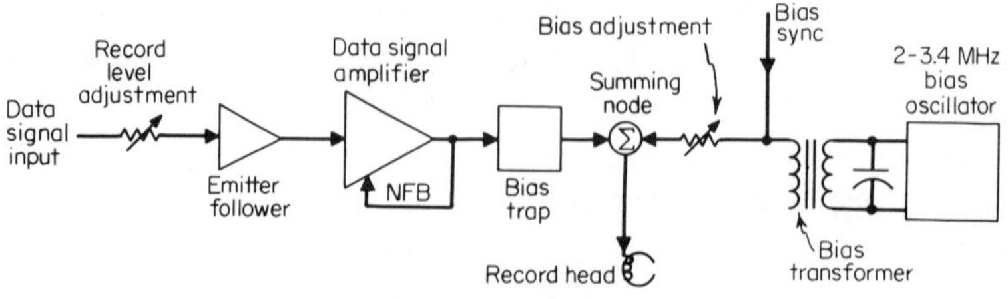

Fig. 9-5 Block diagram of single-speed direct record system.

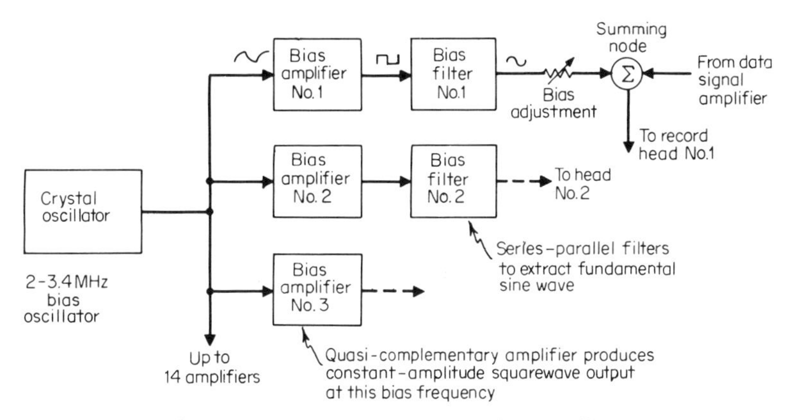

Fig. 9-6 Bias distribution system using master bias oscillator.

measured as a voltage drop across a precision resistor, located in the head-to-ground lead, and converted to current through Ohm's law mathematics.

Where a master bias oscillator is used (Fig. 9-6), the output of the master oscillator is fed to a bias amplifier and filter in each record card. There is an individual bias adjustment for each channel of electronics. Bias current for either of these systems is between 18 and 23 mA, and the data signal level, 1 and 1.5 mA.

Single-speed Reproduce Electronics and the Process of Equalization

Figure 9-7 shows a simplified diagram of a single-speed direct reproduce amplifier of the intermediate-band class. The quasi-differential input (one input to base and a second to emitter of the same common-emitter transistor) reduces the ground loops that may be formed with the preamplifier and coaxial cable feed to the reproduce amplifier. Transistor *Q*1 is a common-emitter amplifier, with its collector load formed by transistor *Q*3. *Q*3 provides low dc resistance and high ac impedance. Feedback from the emitter of transistor *Q*2 via resistor *R*12 to base of *Q*1 provides dc stability to the circuit.

The output of transistor *Q*2 is fed to a summing node Σ via resistor *R*3 and to the amplitude and phase equalizers. These equalizer circuits provide compensation for the response curves of the head and tape and the reactance of the internal *L*-*C* circuits. Data should have a linear amplitude response and a correct phase linearity by the time it has reached the base to transistor *Q*5.

Amplitude Equalization Referring to Figs. 9-8 and 9-9, the action of the amplitude equalizers may be explained as follows. This assembly

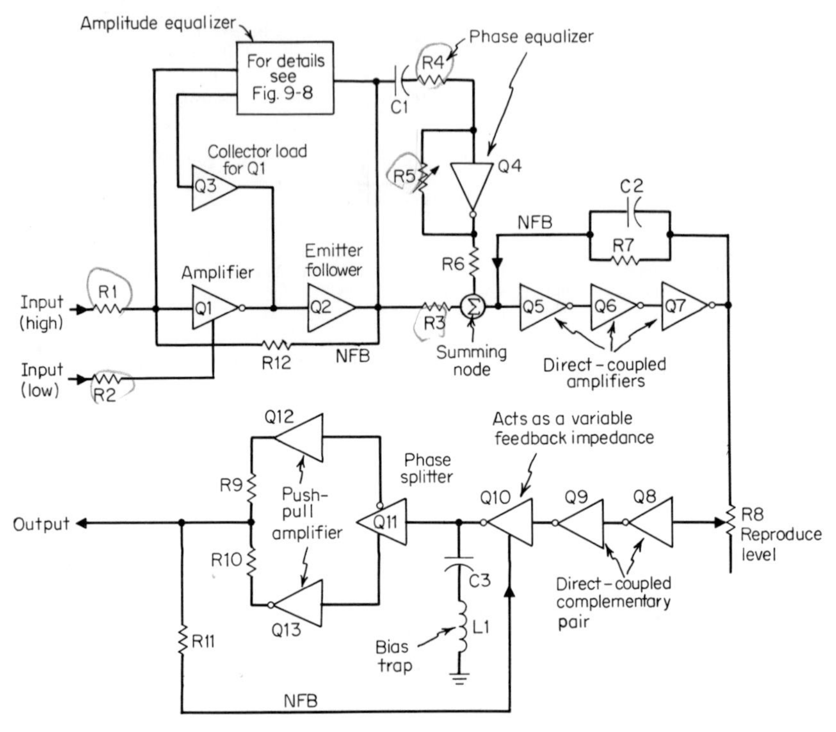

Fig. 9-7 Simplified diagram of single-speed direct reproduce amplifier.

provides the proper compensation for the reproduce heads and tape. The curve of frequency response is essentially the reciprocal of the head output. It is adjustable in two areas, as shown in Fig. 9-9, but the relationship of the other components, fixed in value by the designers, is a matter of interest:

1. Area *A* is controlled by the value of *R*1.
2. Area *B* is controlled by the value of *R*2 (the depth of midband attenuation).
3. The shape of the curve in area *C* is controlled by the ratio of *R*2 to *R*4. *R*4 is variable. Within area *C*, a decrease in the resistance of *R*4 will result in increased amplitude.
4. The peaking frequency in area *D* is determined by the values of *C*3 and *L*1.
5. The elevation of the peaking frequency in area *E* is controlled by the value of the variable damping resistor *R*5. Increasing the value of the resistor will increase the elevation of the peak.

Phase Equalization Phase linearity is required to permit the transmission of square waves with a minimum of overshoot and without

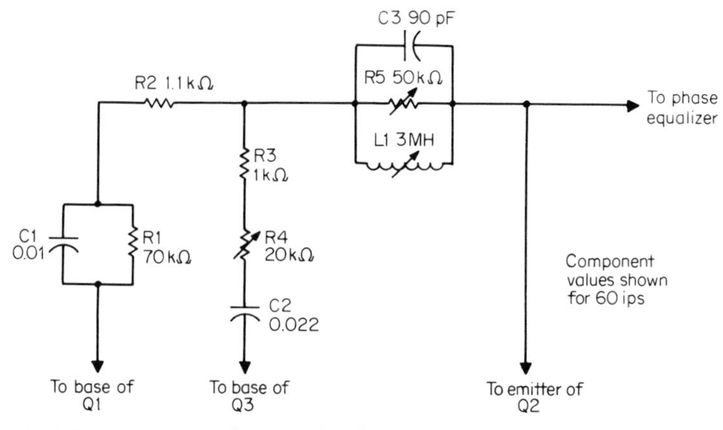

Fig. 9-8 Direct reproduce amplitude-equalizer schematic diagram.

significant degradation of amplitude response over a specific bandwidth. To achieve this, it is necessary to phase-equalize or phase-correct the incoming nonsinusoidal signal. The method by which this is done may be explained using Figs. 9-7 and 9-10 as references.

A square wave consists of the fundamental frequency and all the odd harmonics. Displaying only the fundamental and the third harmonic for simplicity, Fig. 9-10*a* illustrates the phase relationship of the voltage E and current I vectors if there is no phase error. Since any magnetic head has inductance, following the formula $X_L = 2\pi fL$, then the harmonics will be delayed by progressively larger amounts as the frequencies increase. Therefore the first and third harmonics have a vector relationship, not as in Fig. 9-10*a*, but rather as shown in Fig. 9-10*b*. A common vector will be formed by OB (E_1) and OD_1

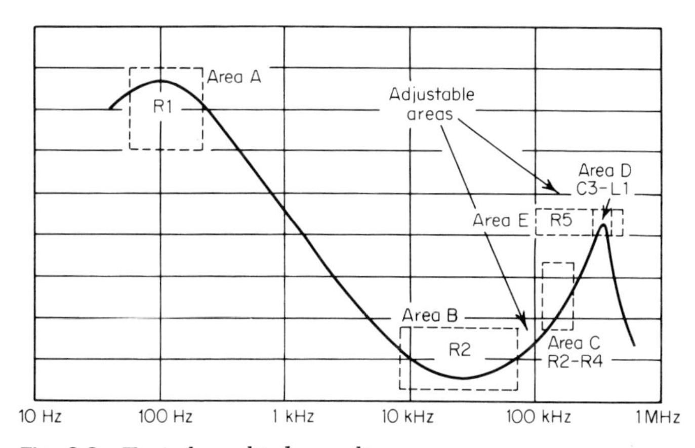

Fig. 9-9 Typical amplitude-equalizer curve.

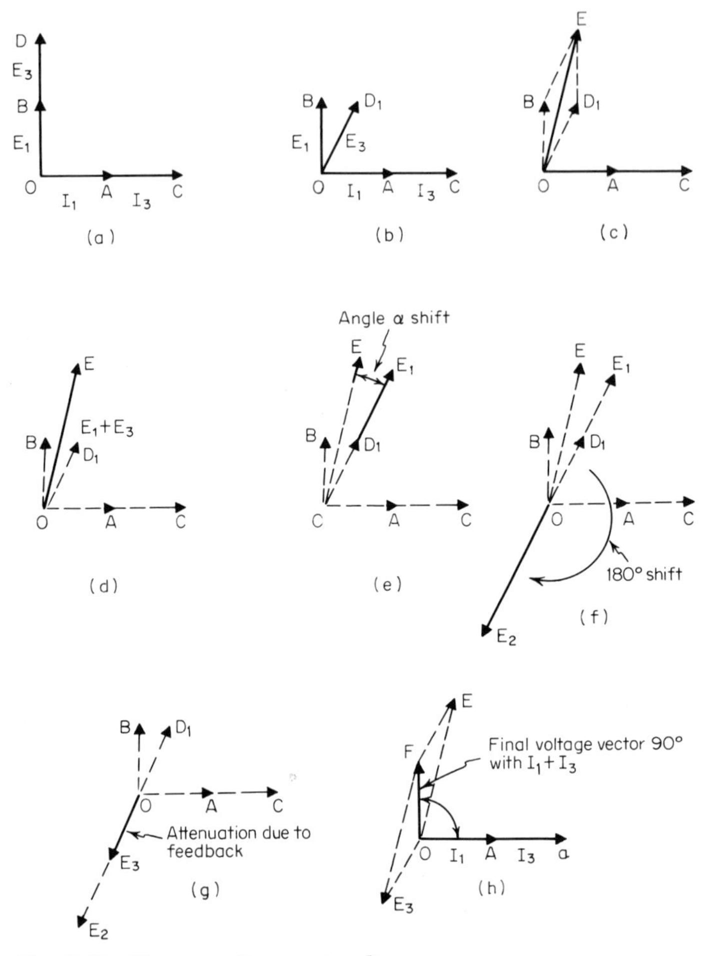

Fig. 9-10 Phase-equalizer vector diagrams.

(E_3), that is, OE. The relationship between the reproduced voltage vector OE and the original record current vectors OAC is not 90° (Fig. 9-10c) as it should be, and the reproduced square wave will be severely distorted. To correct for this, the data signal (voltage vector OE, Fig. 9-10d) is passed to the summing node Σ via $R3$ and to the phase-equalization network. This network consists of capacitor $C1$, resistor $R4$, transistor $Q4$, and potentiometer $R5$. As the data signal passes through $C1$ and $R4$, the phase of the combined vector OE is shifted, as shown in Fig. 9-10e, to OE_1, due to the capacitive reactance of $C1$. This shift of phase is called, in this book, the angle of alpha shift (α), for convenience. The vector is now shifted by 180° as it passes through transistor $Q4$, base to collector. The vector OE_2 of Fig. 9-10f

is now 180° out of phase with the original vector OD_1 that caused the problem, but is much larger in amplitude. The feedback via potentiometer *R*5 from the collector to base of *Q*4 attenuates the vector so that it is equal in amplitude to OD_1 (Fig. 9-10*g*). This vector OE_3 is fed via resistor *R*6 to the summing node, where it will be combined with *OE* (Fig. 9-10*d*) to form the new vector *OF* (Fig. 9-10*h*). It will be noted that the new vector *OF* has a 90° phase relationship with the record current vector *OAC*. Thus phase error has been eliminated.

Reproduce Circuitry after Equalization After the signal has been equalized, it is fed to a three-stage voltage amplifier consisting of *Q*5, *Q*6, and *Q*7. A compensated feedback loop via *C*2 and *R*7 provides signal and dc feedback around the voltage amplifier. Output from the voltage amplifier is taken from the wiper of potentiometer *R*8 and fed to the output amplifier (Fig. 9-7).

The output amplifier, consisting of transistors *Q*8 to *Q*13, is designed to provide power to drive low-impedance lines. Q8 and *Q*9 are a directly coupled, complementary pair. The output of transistor *Q*9 is capacity-coupled to transistor *Q*10, which acts as a variable-feedback impedance between the output and the phase splitter *Q*11. The output of *Q*11 is directly coupled to the push-pull-operated transistors *Q*12 and *Q*13. Resistors *R*9 and *R*10 provide current limiting to protect transistors *Q*12 and *Q*13, in the event of a short circuit across the output. *C*3 and *L*1 form a series-resonant bias trap to eliminate any carry-over of bias frequency to the output of the reproduce amplifier.

Multispeed Reproduce Electronics

When multispeed reproduce signal electronics is used, the direct reproduce amplitude and phase equalizers are automatically selected as speed is changed. Usually, a ground-seeking switch motor is used to select one of the six sets of equalizer pairs. Multispeed units come in units of four or six speeds. This usually doubles the width of the module physically, and reduces the number of channels of electronics that may be fitted into the bay, or rack. Despite this handicap, multispeed units are widely used because they increase the versatility of the electronics and rule out the need for changing the equalizer plug-ins each time the tape speed is changed.

1.5-WIDEBAND DIRECT RECORD/REPRODUCE ELECTRONICS—TO 1.5 MHz

The 1.5-wideband response is specified as being from 500 Hz to 1.5 MHz at a tape speed of 120 ips (IRIG Standards). The industry has provided an even better range, that is, 400 Hz to 1.5 MHz at tape speed of 120 ips and a signal-to-noise of 30 dB. The input impedance is selectable over a range of 75, 1,000, or 20,000 Ω. Input levels vary with the type of electronics, but for the most part fall into one of two categories: 0.25 to 25 V rms or 0.25 to 10 V rms. Output levels are normally quoted as 1 V rms nominal across 75 Ω ± 5 percent.

Block Diagram of a Typical 1.5-MHz System

Figure 9-11 illustrates a typical 1.5-MHz wideband direct record/reproduce system. During recording the data are fed to the direct record amplifier via a record-level-adjustment potentiometer. This is usually a 1,000-Ω potentiometer, located on the front of the bay, file card, or module for easy access by the operator. This adjustment is normally set for 1 percent third-harmonic distortion, measured at the output of the reproduce amplifier. The record amplifier is a feedback amplifier, or similar type, to produce a very constant gain over the entire bandwidth. It drives a head driver (which is a separate card or module) and a monitor amplifier. The monitor will be used to indicate the state of the input signal.

The head-driver card is used for several purposes:

1. To amplify and filter the 7-MHz bias, which is a square wave when it arrives from the bias oscillator, so that it can be linearly mixed with the data signal as a controlled amount of 7-MHz sine wave
2. To amplify the data signal without inversion, and pass it to the summing node Σ for mixing with the bias
3. To provide a bias trap to prevent the bias from feeding back into the data amplifier and causing distortion
4. To provide a means of adding dc voltage to the summing node to offset the second-harmonic distortion, which magnetizes the heads
5. To provide a means of monitoring the bias current

The master bias oscillator is designed to provide enough current for a full 14-channel system. It is usually a modified Colpitts, since the crystal makes it relatively stable. The oscillator is followed by a class C amplifier and transformer for impedance matching to the record-head drivers.

On reproduction, the signal from the reproduce head is applied to a 40-dB four-stage transistorized preamplifier. It will accept signals varying from 0.01 to 5.0 mV rms. Its output is passed to the reproduce amplifier, which is usually located some distance away from the head area (in the bay or card file).

The input signal from the preamplifier is fed to a voltage amplifier. This amplifier will have enough gain to offset the losses that occur in the subsequent equalizer stages. After amplification the signal is phase-

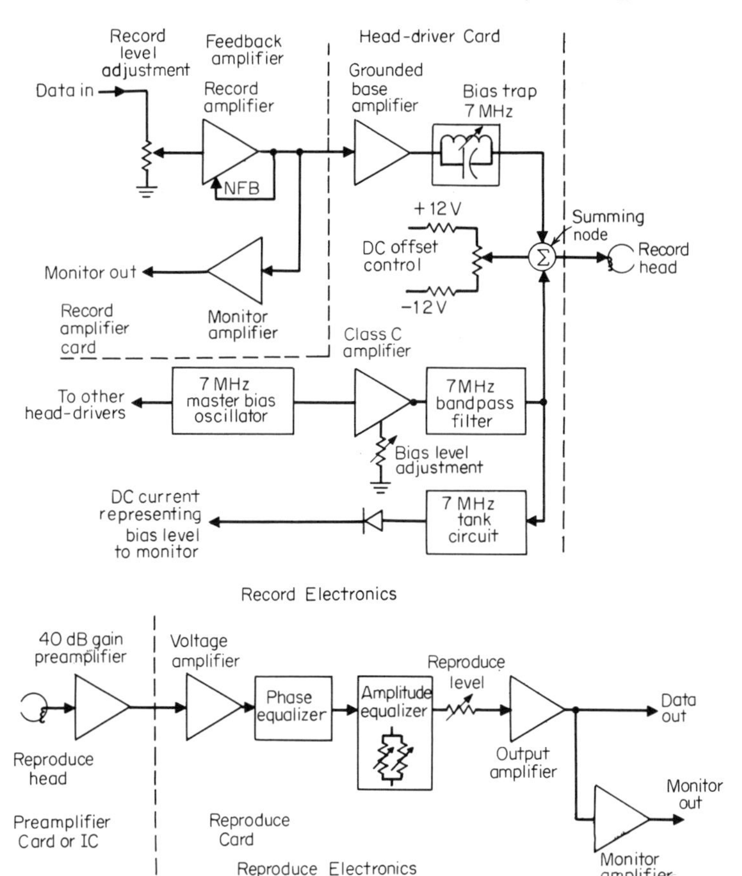

Fig. 9-11 Block diagram of a typical 1.5-MHz-wideband record/reproduce system.

and amplitude-equalized in exactly the same fashion as already discussed in the intermediate-amplifier section of this chapter. From the equalizer stages, the data signal is passed through the reproduce-level adjustment to the output amplifier. This level is normally set at 1 V rms. The output amplifier is a three-stage feedback amplifier with enough power to drive 100 ft of terminated 75-Ω line at 1.5 V rms. Additionally, an adjustable output for monitoring purposes is usually supplied.

2.0-WIDEBAND DIRECT RECORD/REPRODUCE ELECTRONICS—TO 2 MHz

The bandwidth of the 2.0-wideband direct record/reproduce system is from 400 Hz to 2.0 MHz at a tape speed of 120 ips down to 400 Hz to 31 kHz at $1\frac{7}{8}$ ips. It is normal with this type of signal electronics that there is a wide choice of input impedance and voltage ranges. For example:

Voltage ranges	*Impedance ranges*
0.25 to 1.5 V rms	75 Ω*
1.0 to 5 V rms	1 kΩ
3.5 to 20 V rms	20 kΩ

* Maximum of 10 V with this impedance.

There are two distinct types of 2.0-wideband direct signal electronics available. One, a multispeed, electrically selectable type that can be used for either 2.0- or 1.5-wideband systems, or intermediate-band systems, by changing the plug-ins. The other is used for 2.0-wideband systems only.

MULTIBAND WIDEBAND DIRECT SYSTEMS

The multiband group follows the same general pattern already established for the 1.5-MHz wideband electronics (Figs. 9-11 and 9-12). The first stages of the record amplifier contain the voltage- and impedance-range components. These are selectable via jumper plugs. Data from the record amplifier are fed to the record-head drivers, which will be located as close to the head area as the mechanical design permits, to reduce radio-frequency interference. The record-head drivers will provide power amplification to the data signal, linearly mix it with the bias, and drive the record heads.

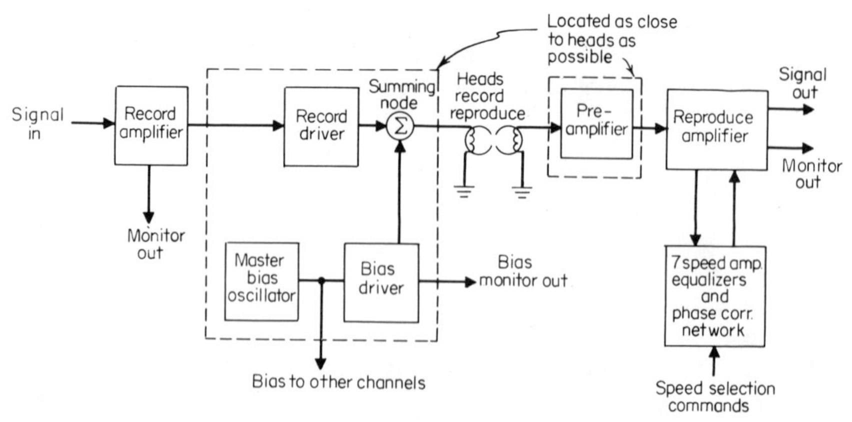

Fig. 9-12 Multiband 2.0-MHz-wideband direct system block diagram.

The bias driver is used to supply the necessary power amplification of the bias signal while maintaining minimum even-order harmonics. This stage was called the class C bias amplifier in the 1.5-MHz system. The output of the bias driver is mixed with the data signal at the summing node Σ.

The bias oscillator, like that of the 1.5-MHz system, is a modified Colpitts oscillator, designed to operate at 7.7 MHz. It will produce a 7-V output across a low impedance.

The reproduce preamplifiers may be either integrated circuits or transistor versions. Their function is to amplify the small data signal from the reproduce head by 39 dB while providing a broadband frequency response of 50 Hz to 2.4 MHz ± 0.5 dB.

The reproduce amplifier consists of a mother card and seven plug-in, shunt, amplitude equalizers and phase-correction networks. These operate in a similar fashion to the 1.5-MHz system. Selection of the correct equalizer for the tape speed in use is accomplished via speed-bus switching. The normal output of the reproduce electronics is 1 V into 75 Ω load. A monitor output is also provided.

2.0-WIDEBAND DIRECT SYSTEMS—SINGLE-BAND

The major areas of difference between the multiband and the single-band direct signal electronics is in the bias-driver stage of the record and in the equalizer stage of the reproduce system. A comparison of Figs. 9-12 and 9-13 will acquaint the reader with the most obvious of these.

Record System

Referring to Fig. 9-13, the incoming signal is passed first to an input attenuator and impedance-matching network so that selection of the proper input impedance and adjustment of the signal level can be made. From there it is fed to the record amplifier, which has a gain of 30 dB set by negative feedback.

The record driver performs the function of amplifying the data and passing them via a bias-reject filter (series- and parallel-tuned *L-C* network) and summing node to the record heads, in much the same fashion as previously described for the multiband version. Negative feedback is used to control the gain of the amplifier, and the record level is set by the record-level adjustment at the input to the amplifier.

A sudden change in the supply voltage or the loading of the master bias oscillator could change its output. Quite naturally, this will be reflected in the feed to the bias driver and be passed on to the summing

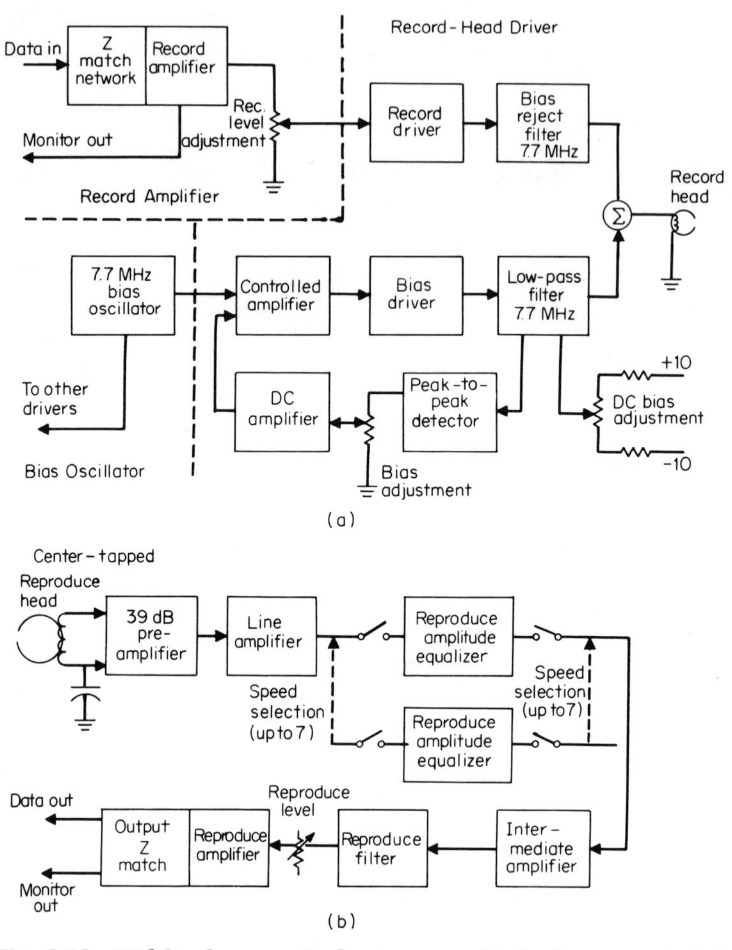

Fig. 9-13 Wideband group II direct system block diagrams—single-band. (*a*) Direct record system; (*b*) direct reproduce system.

node if correction was not made. To maintain bias stability, an automatic gain control (AGC) circuit is used. It consists of a peak-to-peak detector, a dc amplifier, and a controlled amplifier. The amount of bias fed to the head is sensed by the peak-to-peak detector (taken from the low-pass filter), rectified, and passed to the dc amplifier. This dc signal will be used to control the gain of the controlled amplifier. Thus, if the output of the master bias oscillator changes, the AGC circuit will reestablish the bias level as originally set by the bias-adjust potentiometer.

The balance of the record system acts the same as or similarly to the multiband electronics.

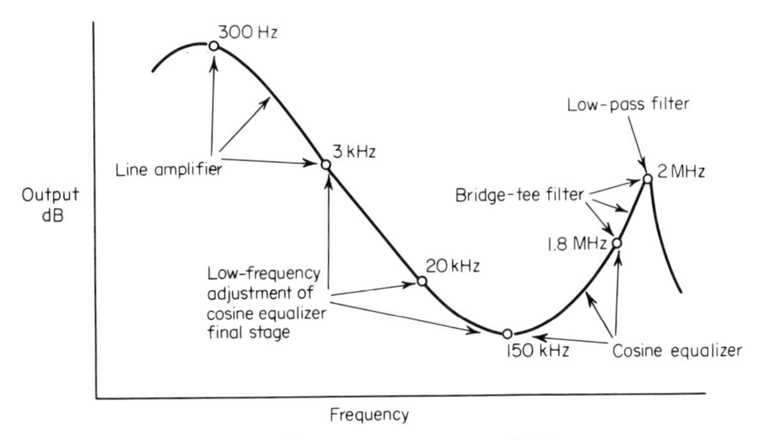

Fig. 9-14 Reproduce equalization curve—areas of influence.

Reproduce System

The signal coming from the reproduce heads (Fig. 9-13*b*) will be provided with a 39-dB gain by the preamplifier (an integrated circuit). The first stage of the IC is a differential amplifier which, with the center-tapped heads, provides cancellation of any phasing errors that occur due to the inductive reactance of the head. The output of the preamplifier is passed to a line amplifier which not only amplifies the data, but provides partial amplitude equalization (300 Hz to 3 kHz), as shown in Fig. 9-14. The data signal is then fed to the amplitude equalizer. Generally, there are equalizers for six speeds with large laboratory recorders, whereas portable or airborne versions have only two or four sets.

Cosine Equalizer—Aperture Corrector Figure 9-15 represents an amplitude- and delay-compensated equalizer. The first four stages consist

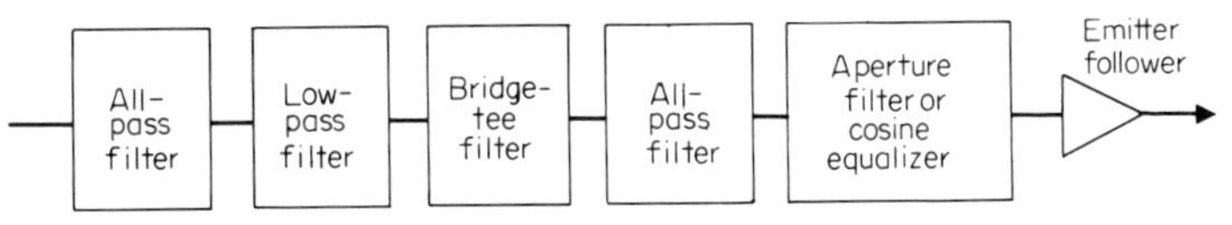

Fig. 9-15 Block diagram of reproduce amplitude and delay compensation equalizer.

of two all-pass filters, a low-pass filter (cuts off at 2 MHz), and a bridge-tee filter which is used to adjust the peaking value at 2 MHz. The bridge-tee filter affects the high end of the equalization curve, between the frequencies of 1.8 and 2 MHz. Equalization in the mid-frequency band (150 kHz to 1.8 MHz) is accomplished by an aperture filter, or cosine equalizer (it has been called both). This equalizer consists of a delay line and a differential amplifier. It must adjust the amplitude of the midband-frequency signals to compensate for head and tape response, without introducing phase shift (time delay) into the signal path.

The delay line (Fig. 9-16) acts like 120 ft of coaxial cable. When it is not terminated in its characteristic impedance (200 Ω), reflections will be set up similar to those occurring in unterminated coaxial cable. When it is terminated, no further reflections will take place. The delay line resonates at approximately 6.6 MHz, and therefore a wave passing down the delay line will be phase-shifted in accordance with its wavelength relationship to the resonant frequency of the delay line; i.e.,

3/8 wavelength = 2.4 MHz = 135°
1/4 wavelength = 1.6 MHz = 90°
1/8 wavelength = 833 kHz = 45°
1/16 wavelength = 417 kHz = 22.5°
1/32 wavelength = 208 kHz = 11.25°

Referring to Fig. 9-16, it can be seen that a signal at *A* coming from the all-pass filter will be shifted in phase through the delay line by an amount proportional to the frequency of the signal. This phase-shifted signal is shown as waveform *B*. Since the *Y* end of the delay

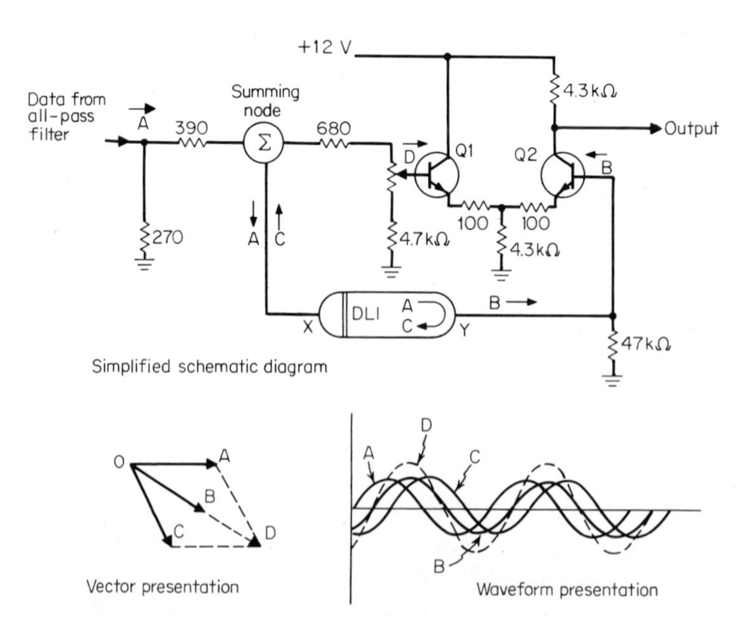

Fig. 9-16 Cosine equalizer—aperture-corrector action.

line is not terminated, the signal at B will be reflected back through the delay line and receive additional phase shift, as indicated by vector OC and waveform C. The reflected signal C will be added to the input signal A and appear as signal D, the sum of the vectors $OA + OC = OD$. It should be noted that the resultant signal D is in phase with the signal B but different in amplitude. If the phase rotation is the same (resultant signal D to signal B), the total amount of envelope delay (timing) will be the same through the equalizer network, and no envelope-delay distortion will have been introduced.

Signal B is fed to the base of the output stage of the differential amplifier $Q2$, and the vector sum OD is fed via the base-emitter junction of transistor $Q1$ to the emitter of $Q2$. The output of the final stage of the differential amplifier will vary, as by a cosine curve.

Low-frequency compensation (between 3 and 150 kHz) is accomplished by an R-C attenuation network, part of the final stage of the equalizer. An intermediate amplifier is used to provide 30 dB of gain to compensate for the losses that occurred in the equalizer stages. Refer to Fig. 9-13.

The reproduce filter that follows (a series of high- and low-pass filters) is used to reject any bias that may have passed through the equalizer networks. The final stage of the reproduce system is a 30-dB-gain amplifier and output impedance-matching network. There are two outputs from this network, one the data out and the other available for monitoring purposes. Where the data output impedance may be selected, the output impedance of the monitor is usually fixed.

LIMITATIONS AND ADVANTAGES OF THE DIRECT RECORD/REPRODUCE SYSTEM

Many of the advantages and disadvantages of the direct record/reproduce system have already been mentioned in earlier sections of this book. It might be well, however, to reiterate some of the major points:

1. As the frequency decreases, the output voltage from the reproduce head reduces, until it approaches the inherent noise level of the system. Thus, with near-zero frequencies (near dc), there would be no usable output. This is one of the principal limitations of the direct recording process.
2. There is a specific dynamic range for each recording system. The dynamic range is the ratio of the maximum signal which can be recorded with a given level of distortion to the minimum signal that can be recorded (determined by the inherent noise level of the system). As the recording level is increased, the distortion increases correspondingly.

3. Another limitation of the direct record process is amplitude instability. This is normally caused by head-to-tape separation. (Refer to Fig. 7-18 for details of separation loss.) These losses of signal (dropouts) are relatively unimportant in audio recording, because the human ear tends to integrate the various signal levels and is relatively insensitive to short periods of signal loss. This is not true with instrumentation, since the dropout might occur precisely at the instant that some important transient phenomenon was taking place.

In spite of these limitations, there are many advantages to using the direct recording process, such as:

1. It has the widest frequency bandwidth vs. tape speed.
2. It has a wide dynamic range and can handle moderate overloads without sudden or drastic increases in distortion.
3. It may be used for frequency-multiplexing a number of signals simultaneously on one track.

REFERENCES

1. Haynes, N. M.: "Elements of Magnetic Tape Recording," Prentice-Hall, Inc., Englewood Cliffs, N.J., 1957.
2. Spratt, H. G. M.: "Magnetic Tape Recording," The Macmillan Company, New York, 1958.
3. Stewart, W. E.: "Magnetic Recording Techniques," McGraw-Hill Book Company, New York, 1958.
4. Davies, G. L.: "Magnetic Tape Instrumentation," McGraw-Hill Book Company, New York, 1961.
5. Inter-Range Instrumentation Group.: Telemetry Standards (Revised January 1971), Document 106-71, Secretariat, Range Commanders Council, White Sands Missile Range, N. Mex., 1971.
6. Lowman, C. E., and G. J. Angerbauer: "General Magnetic Recording Theory," Ampex Corporation, Redwood City, Calif., 1963.
7. Lowman, C. E.: "The Magnetic Tape Recorder/Reproducer and the Concept of Systems Used for Recording and Reproducing FM Analog Test Data," vol. 1 of "Fundamentals of Aerospace Instrumentation," Instrument Society of America, Pittsburgh, Pa., 1968.
8. Athey, S. W.: "Magnetic Tape Recording," National Aeronautics and Space Administration, Technology Utilization Division, Washington, D.C., 1966.
9. Weber, P. J.: "The Tape Recorder as an Instrumentation Device," Ampex Corporation, Redwood City, Calif., 1967.
10. Stiltz, H. L.: "Aerospace Telemetry," Prentice-Hall, Inc., Englewood Cliffs, N.J., 1961.
11. Grob, B.: "Basic Television," 3d ed., McGraw-Hill Book Company, New York, 1964.

CHAPTER TEN

Frequency-modulation Record and Reproduce Electronics

GENERAL

Although FM record parameters have been established for low band, intermediate band, wideband group I, and wideband group II in the instrumentation field and single-sideband specifications for video and instrumentation rotating head and helical scan, this chapter will be devoted to the intermediate and wideband groups I and II only, for the reason that most low-band systems have been retired in favor of the intermediate band, because the same frequency range may be obtained with half the recording speed. The single-sideband systems for video and instrumentation rotating head and helical scan are discussed in Chap. 11. See Table 10-1 for instrumentation parameters.

Originally, the FM carriers were recorded on tape without the use of bias. This was particularly true with low band. Now, however, with the development of multispeed-, multiband-instrumentation FM signal electronics, bias is used with increasing regularity. The reasons are threefold:

1. Most modern multichannel recording systems permit the use of any one of a number of modulation techniques, singly or mixed together,

TABLE 10-1 Single-carrier and Wideband FM Record Parameters

Band	Tape speed, ips	Center-carrier frequency, kHz	Carrier plus deviation, kHz	Carrier minus deviation, kHz	Modulation frequency
Low	60	54	75.6	32.4	dc to 10 kHz
Intermediate	120	216	302.4	129.6	dc to 40 kHz
	60	108	151.2	64.8	dc to 20 kHz
Wideband group I	120	432	604.8	259.2	dc to 80 kHz
Wideband group II	120	900	1170.0	630.0	dc to 400 kHz

NOTE: For details at other speeds the reader is referred to IRIG Telemetry Standards, Document 106-71.

i.e., direct, FM, FSM, PDM, etc. In those systems that use direct electronics where the bias oscillator is not part of the module or card, i.e., where a master bias oscillator is used and the final stage of recording is a head driver, it is the practice to feed bias to every record channel. This means that when FM or any other modulation system is used that includes the head drivers as the final stage, the signal will be mixed with ac bias.

2. In intermediate- and wideband systems, due to the position of the carrier and its associated sidebands on the nonlinear portions of the tape's B_r-H curve, bias is added during recording, and amplitude and phase equalization in reproduce to improve the signal-to-noise ratio.

3. The operator does not have to change bias connections when various channels are not fitted with direct record electronics. Nor does he have to remove the master bias oscillator every time the type of modulation is changed.

INTERMEDIATE-BAND FM RECORD/REPRODUCE ELECTRONICS

A number of systems are used for the intermediate-band FM electronics. These systems may be broken down into the following groups:

1. Single speed. Where manual replacement of the frequency-sensitive elements (frequency-determining units or filters) must take place with each tape-speed change. These replaceable units are normally plug-in units.

2. Two-, four-, or six-speed. Where two, four, or six sets of frequency-sensitive elements are included as part of the modulator or demodulator card or module. Selection of the correct element (correspond-

ing to the requirements of the tape speed) is made via relay, transistor switch, or diode-switching matrix. These may be controlled by the tape-speed selection knob located on the transport.

3. Multispeed systems. Where two distinct ranges of modulator or demodulator frequency elements (binary or decade) are available by manual jumpering of frequency-sensitive elements; i.e., a binary pattern is used for tape speeds of 120, 60, etc., with carrier frequencies of 216 and 108 kHz, etc., and a decade pattern where the tape speeds are one-tenth of normal, as are the carrier frequencies, that is, 12 ips and carrier of 21.6 kHz.

Modulator Types

A number of modulator types are used for the intermediate-band FM electronics. One of the most popular is shown in Fig. 10-1. A second one will be found in the section describing multispeed, multiband wide-band group II systems. A third uses a relaxation oscillator as the basic modulator, and there is one where a voltage-controlled oscillator is used. The latter is similar in operation to a free-running multivibrator. Its output is a series of symmetrical square waves at the frequency required by the tape speed. The VCO frequency is controlled by varying the resistance of the discharge path of the cross-coupling capacitors. For the sake of brevity, only the monostable multivibrator of Fig. 10-1 will be described in detail.

Monostable Multivibrator Type of Modulator Referring to Fig. 10-1, the type of modulator used in this system may be called the monostable multivibrator. The output of the multivibrator will be a series of asymmetrical pulses, two times the frequency required for the tape speed. The pulses must be put through a binary stage for shaping and frequency division. To provide for four-speed operation, the binary chain is extended to four, and the selection of the precise carrier is made through the use of a diode-switching matrix. The final stage is a head driver, where the carrier is amplified to furnish the necessary current to drive the heads.

Multivibrator Action—No Data Input: The multivibrator acts initially as a free-running multivibrator at approximately 400 kHz. During its first few cycles, the multivibrator output will partially charge up and discharge a set of precision capacitors, which are part of a discriminator network. During the discharge phase, a negative-going pulse will be fed via the summing node Σ to the ramp capacitor. This will generate a negative-going ramp that will be amplified in the wideband amplifier and fed to the multivibrator. The first stage of the multivibrator contains a —6.2-V zener. When the negative-ramp size is sufficient to overcome the zener, the free-running multivibrator will be changed to a

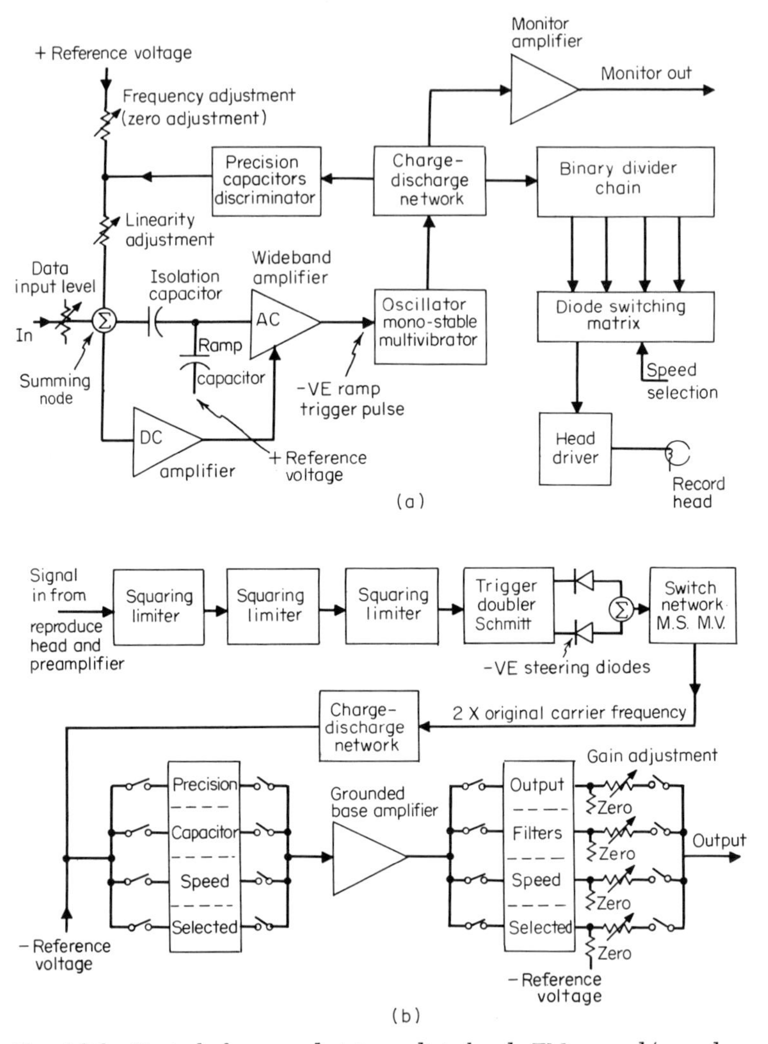

Fig. 10-1 Typical four-speed intermediate-band FM record/reproduce electronics. (*a*) Record electronics; (*b*) reproduce electronics.

triggered monostable multivibrator, with a period of 1.5 μsec. The frequency of the monostable multivibrator (better called the pulse repetition rate) will be set in the following manner:

1. Since the period of the multivibrator is 1.5 μsec, it will flip and charge the precision capacitors for that length of time each time it is triggered. The charge/discharge source is controlled by the use of a precision reference voltage, charge/discharge switching transistors, and level-setting diodes to 10.8 V. Thus the total charge fed to the

precision capacitors will be $E_{\text{constant}} \times t_{\text{constant}} \times C_{\text{constant}}$ (precision capacitors) $= Q_{\text{constant}}$.

2. When a constant-size discharge pulse is fed to the ramp capacitor, already charged to a positive level (reference voltage level), the ramp capacitor will discharge to some value and create a negative-going ramp, particularly if the *R-C* time of the ramp capacitor is long. As soon as the discharge pulse disappears, the ramp capacitor will charge up again to the level of the reference voltage source. See Fig. 10-2*a*.

3. If the discharge pulse were increased in size (less positive direct current mixed with it from the frequency-adjust potentiometer), the slope of the negative ramp would become sharper and the zener breakdown point would be reached sooner. See Fig. 10-2*b*.

4. If the discharge pulse were decreased in size (more positive direct current fed from the frequency-adjust potentiometer), the slope of the negative ramp would decrease and the zener breakdown point would be reached later. See Fig. 10-2*c*.

Since time is changing, so is frequency. Thus the PRR (pulse repetition rate) of the multivibrator can be changed by the adjustment of the frequency-adjustment potentiometer. This adjustment is more often than not called the zero adjust, and is used to set the basic carrier frequency, which will be two times the carrier required by the highest tape speed.

Multivibrator Action—Data Input: A positive or negative data signal going into the summing node Σ will be added to the ramp in such

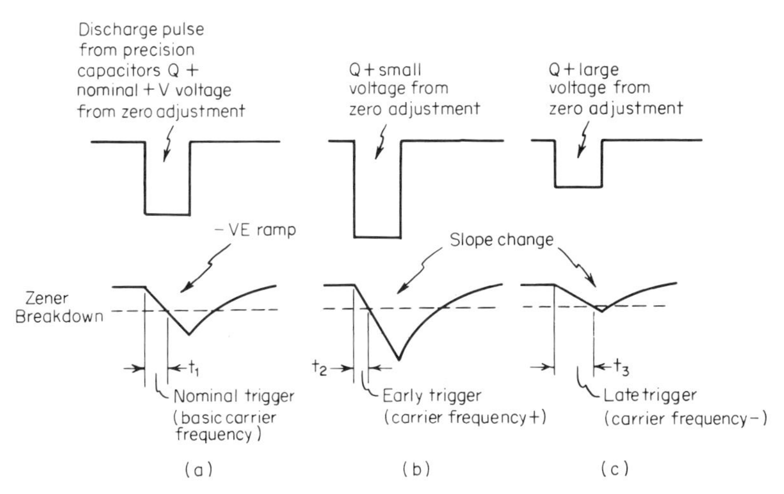

Fig. 10-2 Discharge pulse from precision capacitors and the resultant ramp pulse.

a way that the positive signal will increase the ramp size and thus sharpen the slope, and a negative input will decrease the slope. The inversion is due to the action of the dc and wideband amplifiers. Thus a positive input will increase the basic carrier frequency (earlier) and a negative input will decrease it (later). An ac signal can pass through the isolation capacitor and be added directly to the ramp signal. The dc signals and all frequencies below 10 Hz are considered dc by the system and are forced to go through a dc chopper amplifier to the wideband amplifier, effectively bypassing the isolation capacitor. The size of the data signal, and thus the amount of shift of frequency of the carrier (deviation), is controlled by the input-level potentiometer. The linearity of the system (same amount of increase or decrease in carrier frequency for equal amounts of positive or negative change in the input level) is controlled by changing the effective dc bias at the summing node via the linearity control.

Demodulator Types

A number of demodulator types are in use with intermediate-band FM. The one shown in Fig. 10-1*b* makes use of a monostable multivibrator and selectable precision capacitors and output filters for its operation. A second type uses a bistable multivibrator, a ramp generator, and a blocking oscillator (Fig. 10-5). A third type will be found described in the section on wideband group II modulators. But they all have the same function, that of changing the recorded carrier frequencies back to their respective dc or ac values. The demodulator of Fig. 10-1 will be used for explanation purposes.

Typical Intermediate-band Demodulator Referring to Fig. 10-1*b*, the basic principles of operation of this demodulator may be explained as follows:

1. OBJECT

   To produce a series of pulses of constant charge Q at a frequency equal to two times the deviated carrier and feed these to a low-pass filter so that the filters may extract the modulating frequency f_m originally recorded (data).

2. METHOD

   a. Amplify and limit the signal off tape until the square-wave resultant is a true representative of the crossover points of the sine-wave FM signal originally recorded. Additional squaring, using Schmitt triggers, provides fast rise and fall times to the square waves.

   b. Differentiate the two outputs from the Schmitt trigger, Q, $\bar{Q}$, and feed them through the negative-steering diodes to the sum-

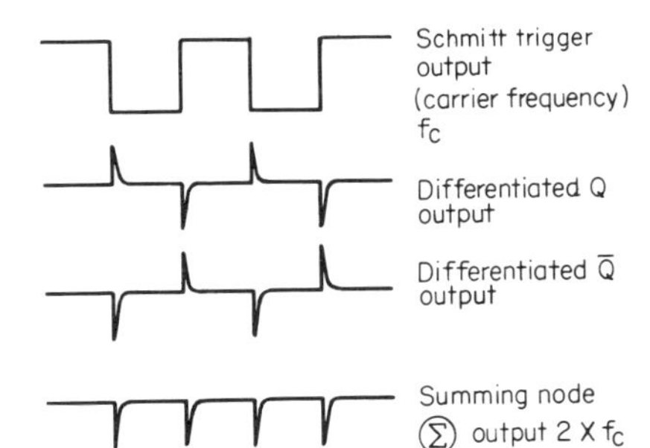

Fig. 10-3 Action of trigger-doubler circuit.

ming node Σ. There the number of sharp negative-going trigger pulses will be twice the carrier frequency. See Fig. 10-3.

c. The switch network (monostable multivibrator) has a period of 1.5 μsec and will be triggered by each negative trigger pulse. Similarly to the action of the modulator circuit, the multivibrator will deliver a charge to the precision capacitors each time it flips. The charge/discharge source is controlled by the use of a precision reference voltage, charge/discharge switching transistors, and the level-setting diodes to 10.8 V. See Fig. 10-4.

d. If a number of charges Q are fed to a series of low-pass filters upon the discharge of the precision capacitor, the output of the filter will reflect any change. Since the only change is f, the output of the filter is a true function of the modulation frequency.

3. SPECIAL CONSIDERATIONS

a. Since the intermediate-band carrier frequency for a typical tape speed of 60 ips is 108 kHz, and the signal bandpass (modulating frequency f_m) is dc to 20 kHz, it is difficult to extract the modulation frequencies from the lower deviation frequencies and still preserve a reasonable signal-to-noise ratio; that is, 20 kHz extracted from −40 percent deviated 108 kHz = 20 kHz from 64.8

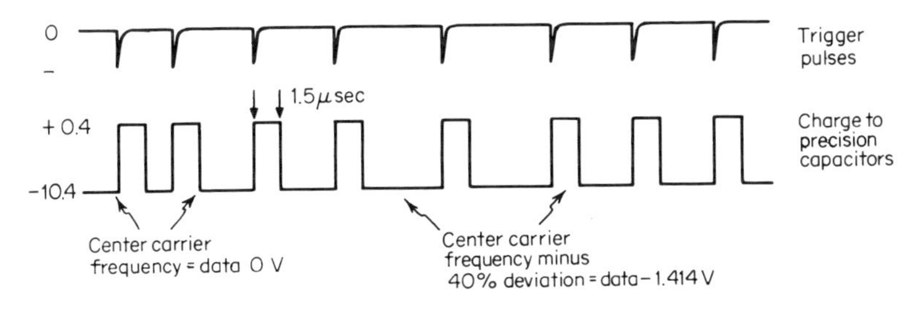

Fig. 10-4 Pulses produced in demodulator precision capacitor circuits.

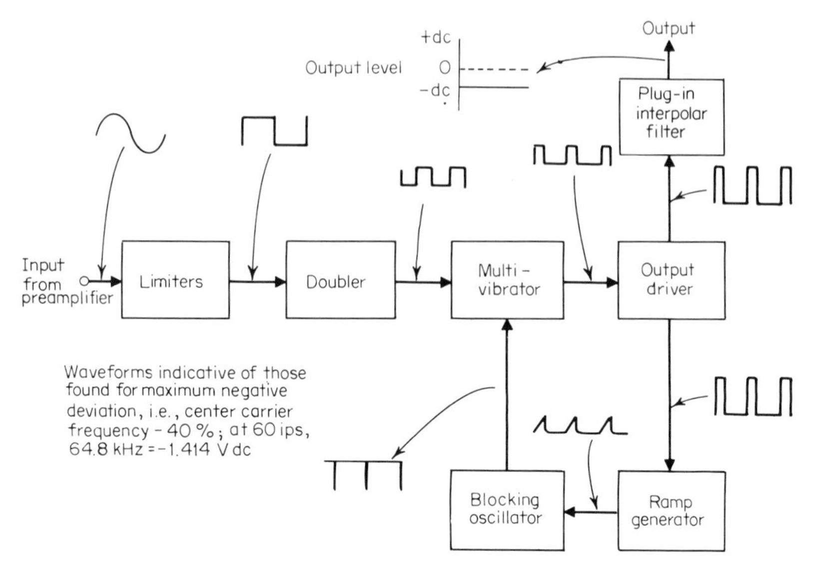

Fig. 10-5 Block diagram of type 2 FM demodulator.

kHz = extraction ratio of 1:3. It is the practice, therefore, to double the number of pulses fed to the output filters; that is, $64.8 \times 2 = 129.6$ kHz and extract the modulation frequency from that. Since the filters now have an extraction ratio of 1:6 to contend with, the signal-to-noise ratio is considerably improved.

b. There will always be a slight dc offset to the output of the filters due to the polarity of the discharge pulse from the precision capacitors. Therefore, when the center-carrier frequency only is present [no data or data at zero volts (crossover point)], the output will be slightly positive. To correct for this, a small amount of the opposite-polarity direct current is fed through the zero-adjust potentiometer to cancel out the offset.

WIDEBAND GROUP I FM RECORD/REPRODUCE ELECTRONICS

Wideband group I FM record and reproduce electronics comes in both single and multispeed versions. The single-speed units require manual replacement of the frequency-sensitive elements, such as the frequency-determining units and filters, etc., each time the tape speed is changed. These elements are normally potted plug-in units or, particularly with the later versions, etched cards which plug into the mother board. Many of the multispeed versions use triggered multivibrators followed by a series of binary dividers as the frequency-determining element (voltage-controlled oscillator, VCO). Choice of the carrier frequency is made by selecting the correct binary-divider output. The demodulator uses either a triggered one-shot multivibrator for producing a constant-sized pulse that will be fed to a series of low-pass filters or a triggered, free-running multivibrator that is used to recirculate a charge to a binary countdown circuit. This latter system will be explained under the heading of wideband group II demodulators. Since the multispeed units form the bulk of the systems in use with modern tape recorders, this type is selected to be used as an example of wideband group I modulators.

Multispeed Modulator

Figure 10-6 illustrates the block diagram of a typical multispeed FM modulator. It should be noted that the heart of the system (the voltage-controlled oscillator) operates on the principle of an operational amplifier. The reader might better understand it if the explanation of an operational amplifier preceded any discussion of the VCO.

Referring to Fig. 10-7, the major requirement for proper operation of the amplifier is that the summing node Σ be kept at zero volts (virtual ground). Assuming that an incoming signal develops a current across R_{in} of 1 mA in the direction shown, in order to satisfy the zero-volt-potential requirements of the summing node, the feedback current across R_{fb} must be equal and opposite in polarity to that of R_{in}. The amplifier must draw extremely small amounts of current (in the nature of microamperes) from the summing node. Its gain must be high, and it must have a phase inversion of 180°. In Fig. 10-7, to maintain a zero-volt level at the summing node, the output of the operational amplifier will have to be approximately −10 V (the gain of the amplifier being the ratio of the feedback resistor R_{fb} to the input resistor R_{in}).

In Fig. 10-6, all the components within the dotted line have been substituted for the single high-gain amplifier just discussed. The controlled multivibrator (CMV) will be triggered on for half a cycle each time the dc amplifier enables the NOR gate. The output of the CMV will be a pulse of precision width (a function of the free-running frequency of the multivibrator). A level stabilizer is used to control the amplitude of the pulse. The combined output of the CMV and the level stabilizer will be a pulse of precision width and precision amplitude. This is fed to the charge/discharge circuit that contains a number of precision capacitors. Each time a pulse arrives, the charge/discharge capacitors will receive a precision quantity of charge $-Q$. When the end of one-half cycle of the CMV operation occurs, the charge in the charge/discharge circuit will be dumped into the summing node

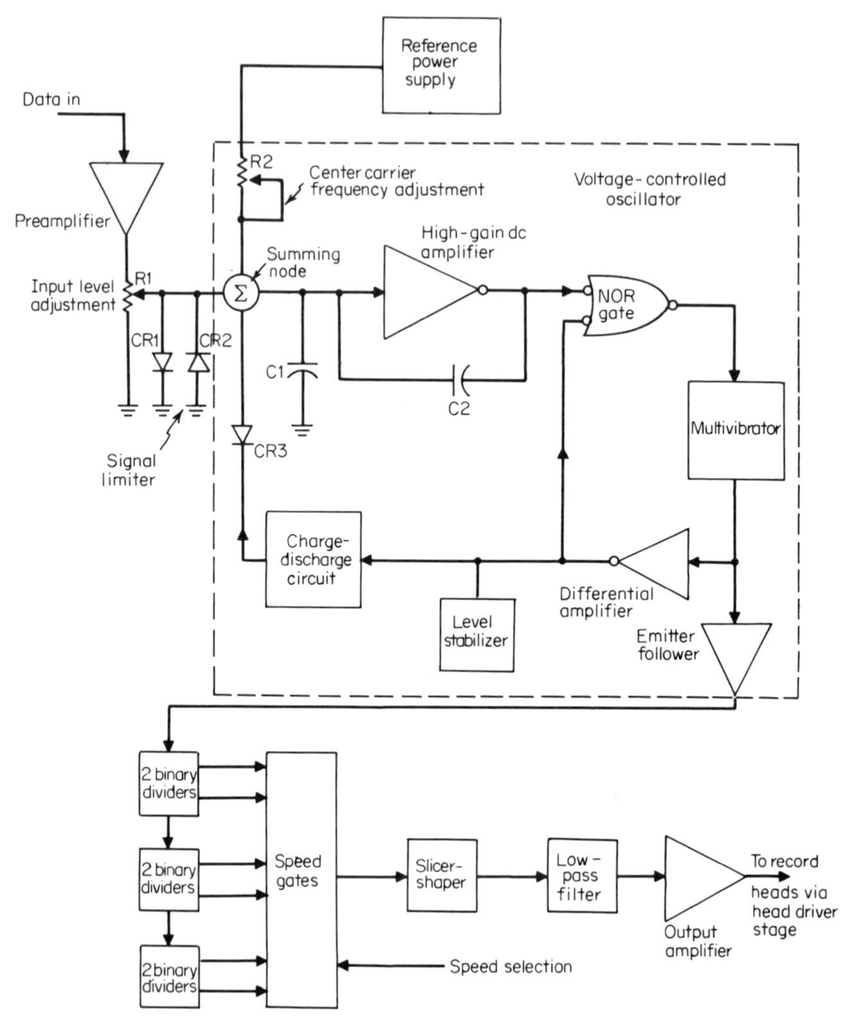

Fig. 10-6 Block diagram of FM wideband group I modulator.

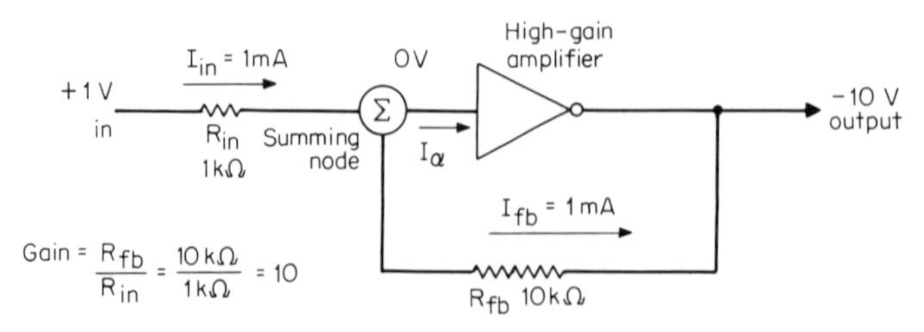

Fig. 10-7 Operational-amplifier block diagram.

Σ. This precise negative charge will offset the current flowing from the reference power supply and data input. Capacitor $C1$ will take on a slight negative charge due to the discharge of the $-Q$-dump circuit (Fig. 10-8), and the high-gain dc amplifier output will switch to a positive value. At the same time the output of the CMV via the differential amplifier will be positive and the NOR gate will be inhibited. This action ensures that the CMV is not triggered until the output of the high-gain amplifier once more returns to a fixed negative value. As soon as pulse $-Q$ has been dumped into the summing node, capacitor $C1$ will start to charge in a positive direction until the output of the high-gain amplifier becomes negative enough to trigger the NOR gate, and the whole process repeats itself.

The triggering time of the CMV is a function of the current being fed into the summing node from the reference power supply and the data input. If there are zero data in, the slope of charge of $C1$ will be nominal and the triggering time will be reached in nominal time. If the polarity of the incoming data is the same as the reference power supply, a greater amount of current will be available and the charging slope of the capacitor $C1$ will be sharper. The triggering time of the

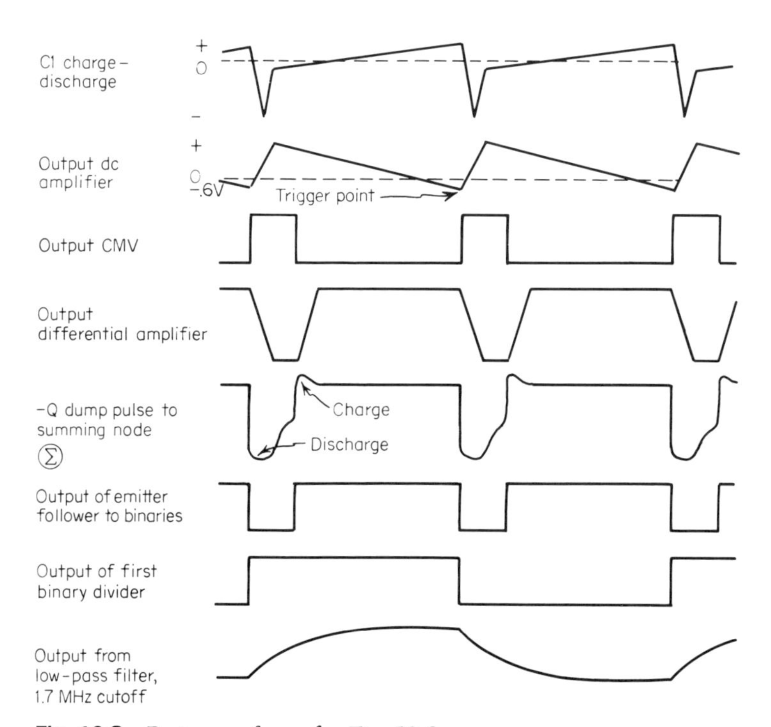

Fig. 10-8 Basic waveforms for Fig. 10-6.

CMV will be reached sooner. If the polarity of the data is such that it detracts from the reference-power-supply current, the slope of the capacitor charge will be longer and the triggering time of the CMV will become greater. Thus, as the triggering time of the CMV changes, so does the frequency of the system. See the waveforms of Fig. 10-8. With a zero data input, the time required to trigger the CMV will be equivalent to two times the required center-carrier frequency. This asymmetrical output will be shaped to a symmetrical waveform and divided by 2 in the first binary-divider stage which follows the VCO.

A series of binary dividers is provided so that the center-carrier frequencies for a number of tape speeds may be obtained from the same VCO. In Fig. 10-6 six binary dividers provide the center-carrier frequencies of 432, 216, 108, 54, 27, and 13.5 kHz for tape speeds of 120, 60, 30, 15, 7½, and 3¾ ips, respectively. Quite naturally, if there has been a data input, these carriers will be deviated by some amount, deviation being dependent upon the amplitude of the incoming signal. For FM wideband group I systems, IRIG Standards specify 40 percent as maximum deviation. Therefore, with a center carrier of 432 kHz and full positive deviation, the frequency will be 604.8 kHz, and with full negative deviation, 259.2 kHz. The range of modulation frequencies for such a carrier is from dc to 80 kHz.

After binary division, the carrier and deviation are passed to a center slicer and shaper. This stage is used to select the center line of the square wave as reference to ensure waveform symmetry. After the slicer/shaper stage, the data are usually fed to a 1.7-MHz low-pass filter. In tape recorders used for wideband group I, the final stage before the record head is the head driver. Since this stage is used for all modulation techniques, it is normal practice to introduce ac bias at this point. Unfortunately, unless otherwise prevented, the ac bias and the FM carrier frequencies will generate spurious cross-modulation products. As a result, the low-pass-filter stage is designed to have a response that is almost flat throughout the FM carrier passband and a high-end roll-off that will attenuate the fifth and higher harmonics of the carrier. This gentle roll-off introduces no group delay distortion of the FM carrier signal and prevents the generation of the cross-modulation products when the bias is added to the carrier. The final stage of the FM modulator is the output amplifier. Its job is to provide sufficient drive for the head driver.

Multispeed Demodulator

Figure 10-9 illustrates the block diagram of a typical modern multispeed FM demodulator. It should be noted that automatic selection of equalizers, timing components, and filters is provided.

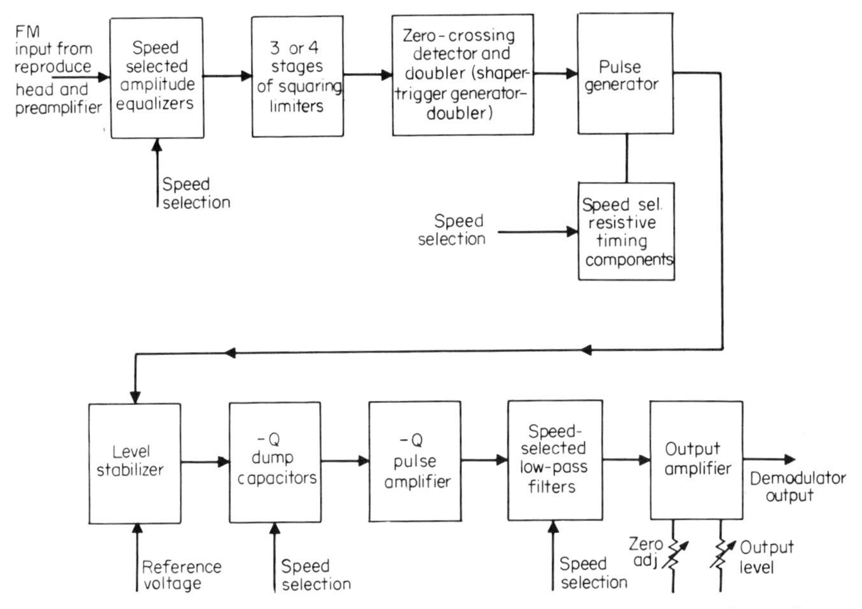

Fig. 10-9 Block diagram of FM demodulator—wideband group I multispeed.

In the first stage of the demodulator, since the center carrier and upper and lower sidebands of the FM signal are obtained from a 6 dB/octave head curve, the amplitude of each carrier will not be the same. To correct for this nonlinearity, high- and low-frequency adjustments are provided in the equalizers.

The next stage of the demodulator consists of a number of squaring limiters. The gain of these stages varies from system to system, but averages at 20 dB per stage. The output of the limiters will be a square-wave signal that will be fed to a Schmitt trigger and doubler circuit. This circuit is often called a zero-crossing detector and frequency doubler. The square wave from the limiter will be used to produce a trigger pulse every time the FM input signal passes through zero. The trigger is developed by using the output of the Schmitt to feed a double-ended pulse transformer and two steering diodes. Each time the Schmitt changes state, a pulse is produced from one of the sections of the transformer. Thus two trigger pulses will be produced for each full cycle of Schmitt operation.

The pulse generator accepts the series of trigger pulses and, for each one of these, produces one positive- or negative-going pulse of fixed duration (polarity depends on system), the width of the pulse being established by the timing components of the one-shot-multivibrator pulse generator. Each time the tape speed is changed, a different set of timing components will be added to the one-shot circuit, and the width of the pulse will be changed. The slower the tape speed, the wider the pulse.

The output of the pulse generator is fed to the level stabilizer. This circuit operates in the same fashion as the level stabilizer of the modulator. Thus the combined action of the pulse generator and the stabilizer will be to provide one pulse of a fixed width and a fixed amplitude to the frequency-to-dc converter each time a trigger pulse is generated in the zero-crossing detector and frequency doubler.

The precision-sized pulse will be used to charge up one of a series of precision capacitors (—Q-dump capacitors). The particular —Q-dump circuit is selected by relays with tape speed. At the end of each precision pulse, the precision capacitors dump their charge via an amplifier to a relay-selected low-pass filter, the function of these filters being to extract the data information from the carrier frequencies. The data at this point will be varying about a dc reference level. As a result, it is necessary to feed the signal from the low-pass filter to a restoration network (zero adjust). A reference voltage and a zero-set potentiometer are used to cancel out any offset dc level by adding a correction voltage to the data. Such a system provides cancellation of the offset, and hence a zero-volt output for a center-carrier input.

Amplification of the low-level signal from the low-pass filters is provided in a final amplifier stage. It is not uncommon to use a pair of cascade-connected differential amplifiers for this purpose, followed by a complementary pair of emitter followers. The gain of such an amplifier is usually controlled by series feedback.

In addition to the circuits just described, there are several other systems, very similar to those used with FM wideband group II systems, differing only in the carrier frequency, equalizer ranges, and low-pass-filter passbands. These circuits will be discussed under the heading of wideband group II.

WIDEBAND GROUP II RECORD/REPRODUCE ELECTRONICS

Wideband group II FM eelectronics is available in both single- and multispeed versions, similar to those of the wideband group I. The single-speed versions require manual replacement of the frequency-sensitive elements each time the tape speed is changed. The multispeed groups use triggered multivibrators followed by a series of binary dividers, as did the modulators of wideband group I, or a variation of this theme. The demodulators are based on a system of standard charge dispensers or that of a recirculating charge dispenser. Like wideband group I, the correct center-carrier frequency, output filters, and/or count-down circuits are selected with tape speed.

It has been the trend in the past few years to build one set of wide-band FM electronics that will not only satisfy the wideband group II requirements, but also fit those of the wideband I, intermediate band, and low band as well. Thus provisions are generally built into the boards or modules so that, with a series of jumpers or manually selectable switches, a change of range from one frequency band to another can be made. It is true that the complexity of the electronics increases and that a great deal more care must be taken by the operating staff to place the jumpers or switches in their correct position. Nevertheless, the fact that only one set of electronics plus a number of plug-ins can be used for all four bands of FM (wideband groups I and II and intermediate and low bands) reduces the overall cost and provides a convenient means of setting up various data systems.

Multispeed, Multiband Modulators

Figure 10-10 illustrates a simplified diagram of a typical multispeed-multiband FM modulator. It should be noted, comparing this diagram with the block diagram of the wideband group I system (Fig. 10-6), that there are a great number of similarities, the major changes occurring in those areas used to make band changes possible, i.e., input-stage attenuators and impedance networks, center-carrier frequencies, binary-divider stages, and filters.

It is normal in the multispeed, multiband versions that the input attenuator have three or more ranges of input. Each input range may be selected by either a jumper or a switch. This arrangement materially assists in setting up procedures for the system. After attenuation and impedance matching, the data are fed to a preamplifier which acts as a buffer between the data source and the voltage-controlled oscillator. The operation of the VCO is the same as that previously discussed under wideband group I, with the following exception: the center-carrier frequency adjustment has a wider range of adjustment so that it can be used to provide a center-carrier frequency of either 1.8 or 1.728 MHz. The 1.8 MHz will be used as the center-carrier frequency for wideband group II (with no data input), and the 1.728 MHz will be used for the wideband group I and intermediate and low bands. The center carrier is presented to two sets of binary dividers, where the selection of the various group frequencies is made. In the case of wide-band group II, the 1.8 MHz will be divided by 2. For wideband group I, 1.728 MHz will be divided by 4, for intermediate band, the 1.728 is divided by 8, and for the low band, the 1.728 is divided by 16. These first binary outputs represent the frequencies required for a tape speed

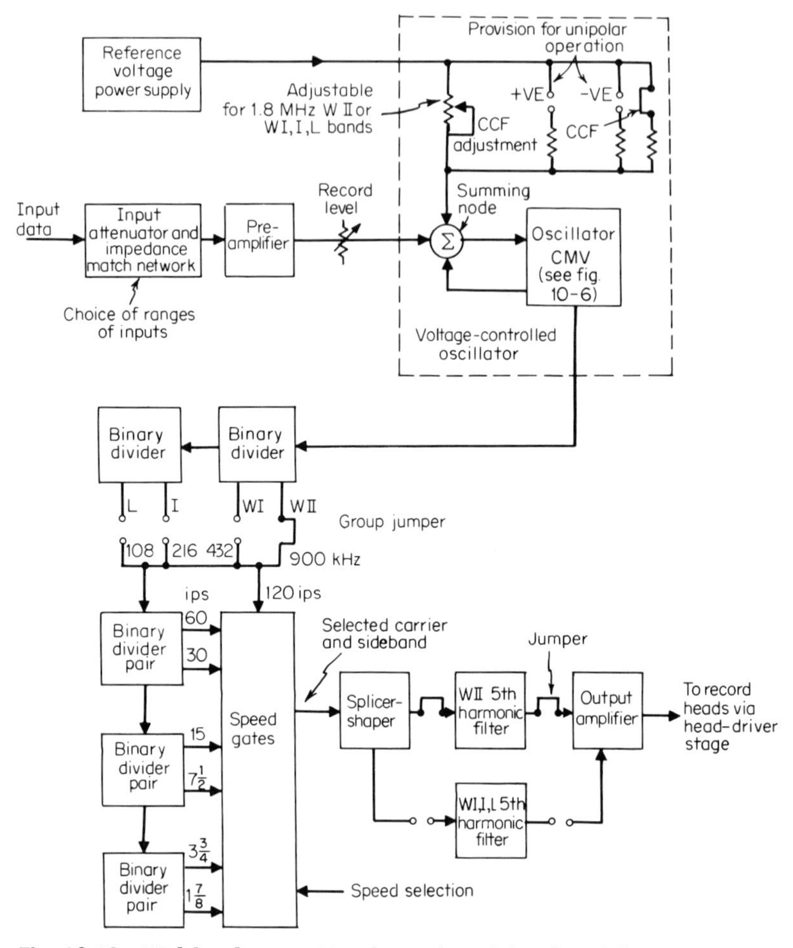

Fig. 10-10 Wideband group II multispeed, multiband modulator.

of 120 ips. Carrier frequencies for the other tape speeds are selected via the balance of the binary-divider stages.

After binary division, the carrier and its deviation frequencies are passed to a center slicer and shaper for shaping and for feeding to the fifth-harmonic filter. This filter is selected to fit the FM group desired. The fifth and higher harmonics are removed from the carrier at this point. The filter provides a gentle roll-off beyond the fifth harmonic so that no delay distortion will be introduced. This prevents the generation of cross-modulation products when ac bias is added to the carrier in the head-driver stage. The final stage of the modulator is the output amplifier. Its job, as in the wideband group I system, is to provide sufficient drive to the head driver.

Provision may also be made for unipolar operation in this type of system (Fig. 10-10, VCO area). Additional resistors can be placed into the circuit that will adjust the amount of reference voltage fed to the summing node. In this way the deviation will start from the maximum positive or maximum negative deviation point and go in one direction toward the opposite deviation value.

Multispeed, Multiband Demodulators—Recirculating-charge-dispenser Type

Figure 10-11 illustrates a simplified diagram of a typical multispeed, multiband FM demodulator, of the recirculating-charge-dispenser type. Since it is not possible to mount the electronics for all the FM bands on the mother board at one time, it will be necessary to make physical changes to the plug-in components. Those items that must be replaced when changing from wideband to another band are:

1. The input amplifier and equalizer assembly
2. The data filter assemblies
3. The output filters (carrier filters)

Data from the reproduce head and preamplifier are first fed to the input amplifier and equalizer stage. This circuit contains a bias-rejec-

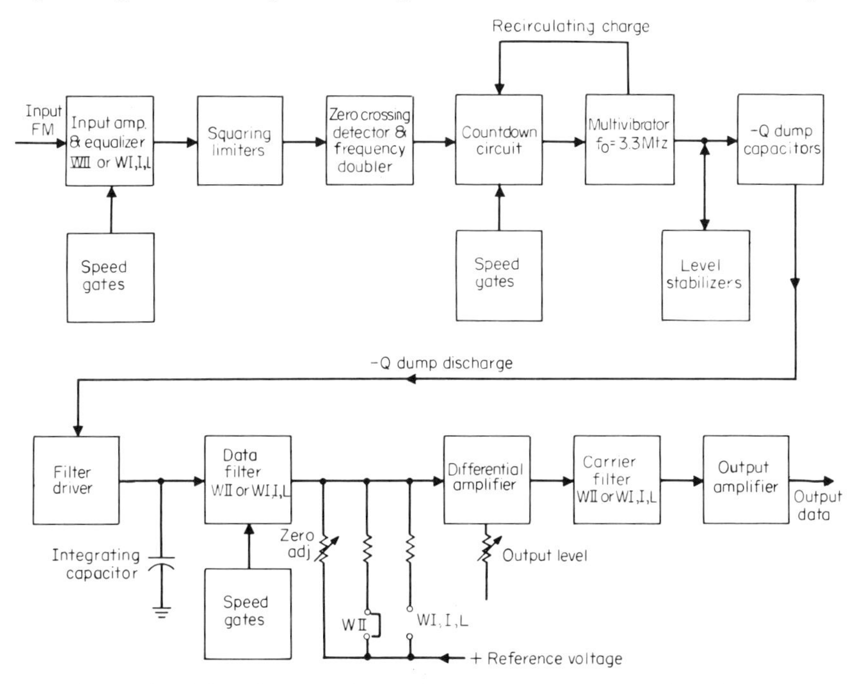

Fig. 10-11 Simplified diagram of a multispeed, multiband FM demodulator—recirculating-charge-dispenser type.

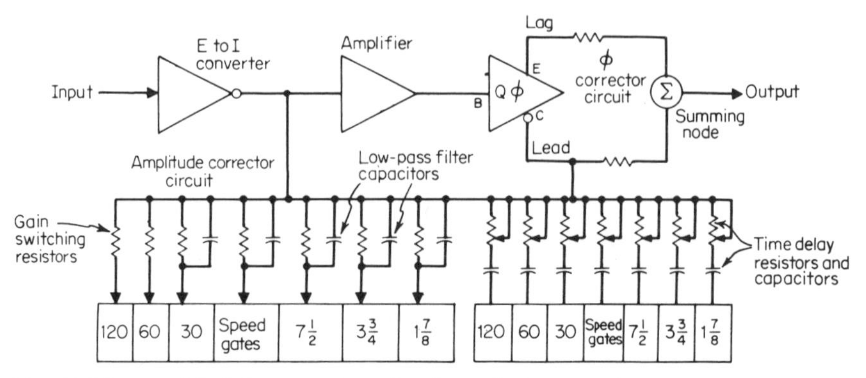

Fig. 10-12 Simplified diagram of amplitude- and phase-equalizer stages.

tion stage, an *E*-to-*I* converter, and amplitude- and phase-equalization networks. The bias-reject circuit is used to reduce any bias feed-through to a reasonable level. The *E*-to-*I* stage acts as the input amplifier and buffer. Amplitude equalization is achieved by altering the gain of the stage with speed-selected load resistors. This compensates for the differences in the amplitudes of the FM carrier arriving from the reproduce heads when the tape speed is changed. When tape speed is high, no gain is added. When the tape speed is low, up to +25 dB of gain is added. See Fig. 10-12.

Following the amplitude equalization, phase correction is provided by a first-order linear time-delay network. This circuit consists of a gain-stabilized amplifier and a set of speed-selected *R-C* networks (Fig. 10-12). Since the FM carriers see a varying impedance at the reproduce heads ($X_L = 2\pi fL$), the midband and upper band of frequencies will be slightly lagging in phase with respect to the lower carrier frequencies. For this reason a system of cancellation correction is employed. The carrier fed to transistor $Q\phi$ base is taken from its emitter and fed to a summing node without inversion. This signal will be lagging in phase. The carrier signal is also taken from the collector (leading in phase by a nominal 180°) and fed to the summing node. An adjustable amount of phase lead is added to this signal via the linear-phase and time-delay networks. The lead cancels the lag, and the phase-corrected signal at the summing node will be fed to the limiter stage.

Referring once more to Fig. 10-11, the limiter stages amplify and square the signal so that the output is a series of square waves representing the crossover points of the original FM carriers. This square wave will be fed to a Schmitt trigger and doubler stage, which, like that of the wideband group I multispeed demodulator, is often called a zero-crossing detector and frequency doubler. The output of this stage will be a series of trigger pulses at two times the original carrier frequency. Each of these pulses will be of short duration (approximately 100 nsec) and negative-going. Its operation is similar to that of the wideband group I circuits.

The next stage of the demodulator is called the recirculating charge dispenser. At a tape speed of 120 ips, the circuit will produce a precision pulse each time it is triggered. The timing between the pulses will be in direct proportion to the timing between the trigger pulses. At one-half the top speed (60 ips), two pulses will be produced each time a trigger pulse arrives. These pulses are amplified and used to charge a series of precision capacitors (—Q-dump). With a tape speed of 30 ips, 4 pulses will be produced with each trigger pulse; with 15 ips, 8 pulses; with 7½ ips, 16 pulses; etc. (Fig. 10-13). At the end of each charging pulse, the —Q-dump capacitors discharge into an integrating capacitor via a filter-driver-amplifier stage. A negative-going ramp is produced for each discharge pulse, and a positive-going restoration level is reached with each charge pulse. The number of negative ramps per group of pulses will depend on the control time of the multivibrator via the countdown circuit, and thus on speed selection.

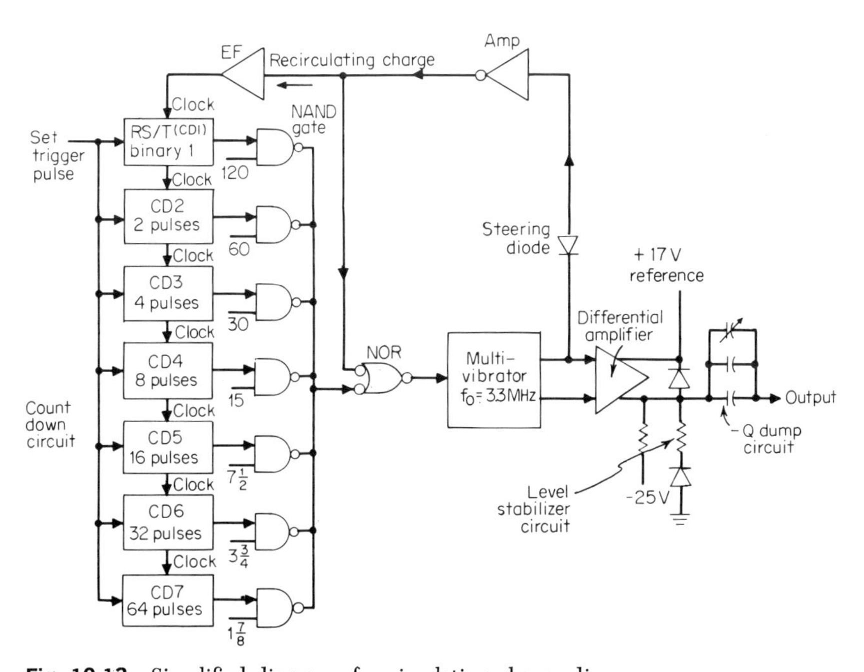

Fig. 10-13 Simplified diagram of recirculating charge dispenser.

The output of the integrating capacitors drives the data filters, whose purpose it is to extract the data information from the carrier. There will be one data filter for each speed of the wideband group II system, and another set for each speed and frequency range of the wideband group I, intermediate-band, and low-band systems. The particular filter is selected by means of a diode-switching network or by relay. The output from the filters will vary around a negative dc reference level. Therefore a positive dc offset voltage (zero adjust) must be added to the data signal at this point. A combination of a fixed and a variable resistor feeding a positive voltage from a reference power supply is used for this purpose. The signal is then passed to an amplifier which contains the output-level adjustment, and from there to a carrier filter. This low-pass filter eliminates the second and higher harmonics of the carrier from the data signal and improves the signal-to-noise ratio. The output amplifier usually contains a pair of complementary emitter followers which provide the power amplification to drive the output load.

ADVANTAGES, LIMITATIONS, AND MAJOR APPLICATIONS FOR THE FM RECORDING PROCESS

The advantages of the frequency-modulation recording process are:

1. The ability to record low frequencies down to direct current
2. Freedom from the effects of tape dropouts
3. Excellent phase shift vs. frequency characteristics and the attendant ability to preserve accurately the waveform of the recorded signal

The disadvantages of the FM process are:

1. Less efficient utilization of the tape, in that at least four times tape speed is required to give the same upper-frequency limit as the direct recording method
2. Additional complexity of the electronic circuitry required for the modulator, demodulator, and low-pass filters

The major applications for the FM recording process are:

1. The recording of low-frequency-signal information such as vibration, noise, underwater sound, etc.
2. The recording of transient phenomena, such as shock, blast, and ignition, where accuracy of waveshape is important
3. Changes in time base, permitting the speeding up or slowing down of a given event
4. Video (TV) recording (covered in Chap. 11).

REFERENCES

1. Athey, S. W.: "Magnetic Tape Recording," National Aeronautics and Space Administration, Technology Utilization Division, Washington, D.C., 1966.
2. Inter-Range Instrumentation Group: Telemetry Standards (Revised January 1971), Document 106-71, Secretariat, Range Commanders Council, White Sands Missile Range, N. Mex.
3. Lowman, C. E.: "The Magnetic Tape Recorder/Reproducer and the Concept of Systems Used for Recording and Reproducing FM Analog Test Data," vol. 1 of "Fundamentals of Aerospace Instrumentation," Instrument Society of America, Pittsburgh, Pa., 1968.
4. Lowman, C. E., and G. J. Angerbauer: "General Magnetic Recording Theory," Ampex Corporation, Redwood City, Calif., 1963.
5. Stiltz, H. L.: "Aerospace Telemetry," Prentice-Hall, Inc., Englewood Cliffs, N.J., 1961.

CHAPTER ELEVEN

The Television Recorder

INTRODUCTION TO THE TELEVISION RECORDER

The invention of the video tape recorder was based on the requirements of the U.S. Television Standards, i.e., a 4.2-MHz bandwidth for a 525-line 30-frame system (Table 11-1). The problem was one of designing a magnetic recording system capable of handling signals that went from the television frame rate to the limit of resolution of a good camera, such as the 4½-in.-image orthicon. Thus the designers were faced with putting together a magnetic tape recorder that was able to record (with a flat response) a band of frequencies from 30 Hz to 4.2 MHz, and do it at a time (1952) when the best longitudinal tape recorder was capable of recording only up to 100 kHz and 60 ips.

For reasons already discussed in Chap. 8, the rotary-head technique was selected, and a head gap of 100 μin. The original video heads used platinum as gap-spacing material. This had to be hand-lapped to the required 100 μin. With this gap, the optimum wavelength that could be recorded was 200 μin. During the development stage it was decided that an FM system would be used to record and play back the video signal. The maximum or optimum frequency that would be required would be 7.5 MHz. Although this was not the precise fre-

TABLE 11-1 World Television Standards

Number of lines	Frame rate	Bandwidth, MHz	Countries or areas using Standard
525	30	4.2	United States of America, Japan, Latin America, except Argentina. Referred to as the U.S. Standard.†
625	25	5.5	Rest of the world (see above), except those listed as using separate standards. Referred to as the CCIR,* or EBU,* Standard.†
405	25	3.0	In some parts of the United Kingdom; being rapidly phased out. Referred to as the United Kingdom Old Standard.
819	25	8.0	France and Belgium black-and-white. They use 625/25 standards for color.

* CCIR, Consultative Committee for International Radio; EBU, European Broadcasters Union.

† By 1975 these two should be the World Standards.

quency that was finally chosen, nevertheless it was the design objective toward which the development team worked.

Using an FM system having a maximum carrier frequency of 7.5 MHz and multiplying it by a wavelength of 200 μin. gives a required head-to-tape speed of 1,500 ips ($v = f\lambda$). There are a number of ways of achieving a head-to-tape speed of 1,500 ips (most of them have been discussed in Chap. 8), but in the earlier stages of the design of the video recorder, longitudinal tape speeds of nearly 800 ips were tried. Experiments conducted by the British Decca Corporation and RCA achieved almost simultaneously in 1950 a working video tape recorder. These used ½-mil tape, stationary heads, a tape speed of 800 ips, and head gaps of approximately 60 μin. Both companies achieved pictures; in fact, RCA produced color pictures that were quite good. Such longitudinal tape speeds were impractical because of the very short recording time per reel of tape; obviously, that particular format had no commercial application. Thus it was that the rotating-head principle was chosen as a basis of television magnetic recording.

THE QUADRUPLEX VIDEO RECORDER

The Quadruplex Head

In the quadruplex system, the four heads are mounted on the periphery of the drum 90° apart. Referring to Fig. 11-1, the tape is moving toward the reader with a forward speed of 15 ips. The tape normally used is 2-in.-wide 1-mil polyester, with a coating thickness of 0.4 mil (Chap.

6). The head drum is 2 in. in diameter; therefore its circumference, or periphery, is $\pi \times 2$ in. = 6.28 in. Using the U.S. Standard, the head was made to rotate four times the vertical sync rate, or $4 \times 60 = 240$ rps. For the CCIR Standard the rotational rate is five times the vertical sync, or $5 \times 50 = 250$ rps. With these drum sizes and rps speeds, the head-to-tape contact speed for the U.S. Standard is 1,507 ips and the CCIR Standard, 1,570 ips. For details of modern high-band heads, the reader is referred to Fig. 5-7.

Television vertical sync pulses were chosen to synchronize the rotation of the head because TV signals, by their very nature, are repetitive. The period of a line is fixed at so many microseconds. There are so many lines in a television field. Two television fields make a television frame. These are things that we can rely on. They are there, they remain there, and they will always be this way. See the glossary of television terms in Appendix B.

Tip Width Originally, a head-tip width of 10 mils was chosen to provide a wide enough track on tape to align enough domains to give a reasonable signal-to-noise ratio. Now, however, with the improvement in the technologies for the manufacturing of both tape and heads, the width can be reduced to 5 mils. Today, therefore, there are two standards of head sizes, 10 mils for longitudinal tape speed of 15 ips and 5 mils for tape speeds of 7½ ips.

Female Guide In Fig. 11-1 we see that the tape is formed by the female guide to fit the periphery of the head drum. A vacuum system

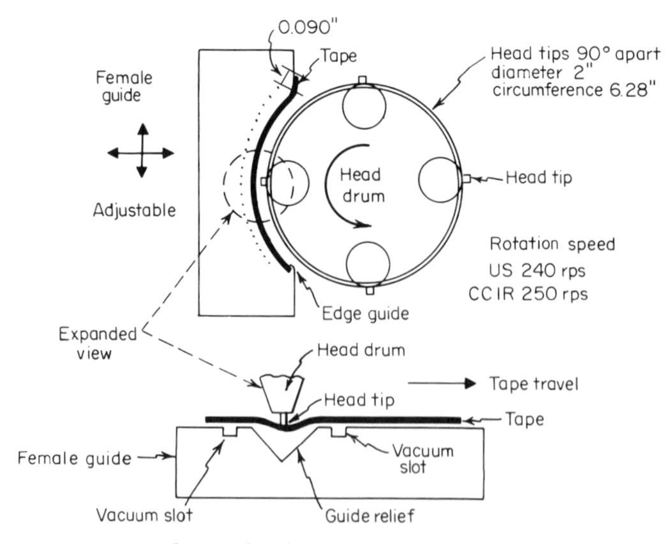

Fig. 11-1 Quadruplex-head- and tape-guiding details.

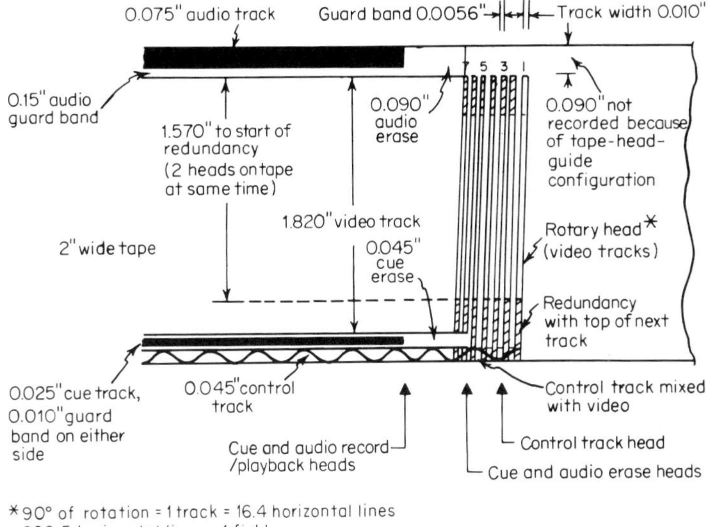

Fig. 11-2 Tracks recorded by high-band video recorder.

is used to hold the tape to the guide. The guide height must be set so that the center lines of the tape and of the drum are coincident. Additionally, the guide must be positioned so that the penetration of the head tip is correct. The tape is edge-guided by a small tit which is part of the female guide. If the guide height is set correctly, in accordance with the standards established by industry agreement, the video head should contact the tape approximately 90 mils down from the upper edge. It should record a track 10 mils wide across the tape, as shown in Fig. 11-2. One-quarter of the circumference of the head drum will be 1.57 in. This means that after a given head has gone 1.57 in., the next head picks up contact with the tape and there is redundancy for approximately 340 mils. Since the tape is being moved longitudinally at the same time that the head is moving, tracks will be laid down across the tape at a slight angle (an angle of 33 minutes of arc with respect to the perpendicular). The tracks are tilted due to the forward tape motion. This ensures, when the next tip comes along, that the next track will be laid downstream with respect to the first one. The space between the two tracks is called the guard band, and the specific values that have been quoted in Fig. 11-2 apply to the U.S. Standard. The ratio of track widths to guard band has always been approximately 2:1. Obviously, if the guard band were made smaller and the tape speed reduced, tape economy would go up, but tape tracking would become an almost impossible problem. For this reason the industry has established the 2:1 ratio as a means to ensure that the design of the capstan servo is reasonably simple and practical.

It will be noted in Fig. 11-2 that the information on the bottom of the track, which has been arbitrarily called track 1, has also been recorded on top of track 2. This is redundant information, since in the record mode all the four heads are being driven continuously by the same FM signal. There is no attempt at gating the four heads in sequence during the record mode. The question has been asked, "Why not?" The answer is readily apparent in Fig. 11-3. Let us assume that the guide height is set too high, as in Fig. 11-3*a*. If the gate is electronic, its operation is based on shaft information to turn it on. Since the heads would not be touching the tape at the start of the

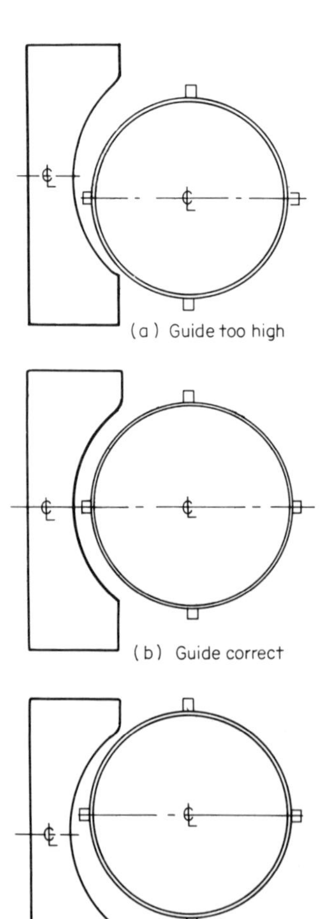

Fig. 11-3 Relationship of female guide to head-drum-guide height.

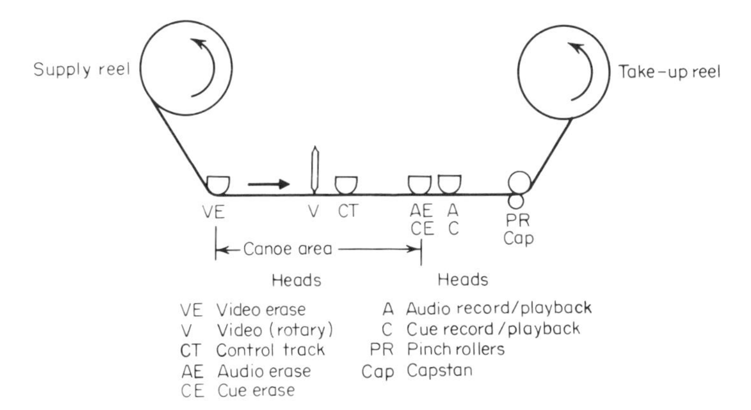

Fig. 11-4 The tape path of the quadruplex video recorders.

turn, there would be missing information. If the gate is to be turned off at a fixed time, and the guide is too low, as in Fig. 11-3*c*, there would be missing information at the bottom of the tape track. Thus, to use a gating system, the assumption must be made that the guide height is perfect at all times as in Fig. 11-3*b*—a dangerous assumption under the best of circumstances. The approach of the design engineers was that if more information was recorded than was needed, the redundancy could be eliminated by suitable electronic switching during the playback mode.

Tape Path Figure 11-4 illustrates the path that the tape takes from the supply reel to the take-up reel. Modern video recorders are fitted with 2-in.-wide video erase heads immediately following the supply reel. This degausses the tape and prepares it for the recording of the video and control tracks to follow. The older machines had no erase head, but were fitted with an idler. The erase-head assembly or the idler is used to determine the angle at which the tape will enter the tape-path area called the canoe. Its shape is critical, and has been standardized by industry agreement. From the erase head or the guide, the tape then goes to the rotary-head assembly.

Control-track Head The control-track head is mounted ¾ in. downstream from the center of the head drum. The 45-mil-wide control track is recorded on the bottom edge of the tape and will be used to define the speed of the longitudinal tape movement in playback (Fig. 11-2). Control-track information will be passed through a filter during playback to remove any remnant of the video information that is mixed with it.

Audio and Cue Erase Heads After the rotating and control-track-head assemblies, the tape passes on to a set of audio and cue erase heads. The audio track is at the top, and the cue track is at the bottom of the tape (Fig. 11-2). The audio track is used for recording the audio information that accompanies the video, and the cue track is used for editing purposes. This particular feature was not available until about 1959. The audio and cue record and playback heads are located farther downstream. Generally, these are combined heads (used both for playback and record). To provide the means of replacing the audio information at any time (dub in the voices to fit the pictures, etc.), the audio erase head erases the top 90 mils of the tape, even though the video tracks do not start for 90 mils from the edge of the tape. The audio record head lays down a 75-mil track and leaves 15 mils for a guard band between it and the video tracks. The cue erase head erases a 45-mil portion of the video tracks directly above the control-track slot. Its record head lays down a 25-mil track in the center of this erased band. Thus, referring to Fig. 11-2, it should be noted that the audio, video, cue, and control tracks are separated from each other by guard bands, as are the individual tracks of video. It might be remembered that, of the 1.82 in. of video, only 1.57 in. is nonredundant.

Tape-speed Control

One of the design considerations for a video tape recorder, be it a quadruplex or helical-scan system, is that it must pretty well attempt to duplicate what a film system would do. With film, due to the sprocket holes, a single frame of information is spaced a fixed number of sprocket holes apart. When reproducing that film, the criterion is merely to make sure that the speed of the motor moving the film is the same as it was when the film was exposed. With magnetic tape, the problem is to ensure that the video information coming from a camera being triggered by a very stable crystal-controlled sync generator is transferred to tracks on tape. The transfer must be as close to perfect as possible. Mechanical motion that departs from the ideal by even a small amount causes errors; therefore the design intention of a video tape recorder was to make sure that the departure from ideal motion was within acceptable limits. The specifications were set up by the Federal Communications Commission (FCC) in the United States and by the European Broadcasters Union (EBU) or other government agencies in the various countries abroad. In video recording/playback systems we are concerned with keeping the velocity of the rotating heads as close to absolute constant velocity as possible. Any difference between the velocity in record and playback will give rise to time-base errors. It is possible, even if the head-drum speed was sloppy during record, that by accident it could be made to wander at exactly the same rate in playback. Under these conditions, the time-base error would be zero. However, the more stable the head-drum motor is kept during the record mode with respect to the signal coming into the machine, the simpler the problems of control become in playback.

Timing Rings and Tachometer Systems To control the timing of the head drum in the earlier machines, a timing ring was used. One-half of it was silver and reflective, and the other half, black and nonreflective. As shown in Fig. 11-5*a*, an exciter lamp was used in conjunction with a photo diode to produce timing pulses. Whenever the light hit the silver surface, it was reflected to the diode, and a large output was obtained. When the light shone on the black surface, there was no output. This system became known as the PEC (photoelectric cell system). It had a definite disadvantage because the lamp was always on whether the head was rotating or not, and like any other lamp, these lamps burned out. When that happened, the whole system failed, because everything in the machine was predicated on an output from the photocell. This output was used to control the head and the longitudinal tape speeds in record and playback, and it was used for editing, etc. With the advent of the high-band recorders, the PEC system was replaced by a tachometer system like that shown in Fig. 11-5*b*. There, a tachometer disc is fitted to the head-drum motor shaft, at the opposite end from the head drum. The tachometer disc itself is nonmagnetic, but there are a series of magnetic inserts, two of them on one side of the disc and a single one 180° away. As the head rotates, a double-

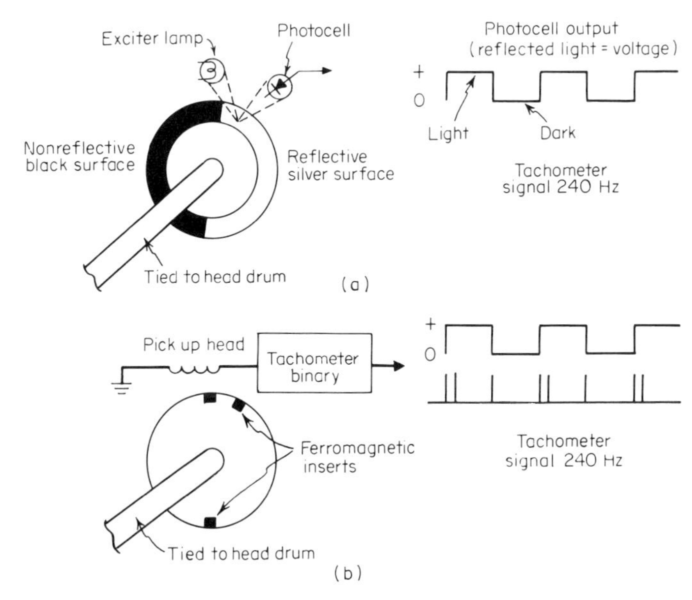

Fig. 11-5 Timing devices used to register head-drum speed.

pulse train is introduced in a magnetic pickup head when the double insert goes by, and a single pulse is introduced when the single insert passes. These pulses are fed to a tachometer binary system which produces a series of square waves. The frequency of the square waves will be an exact replica of the speed of rotation of the head drum (240 rps). The second pulse in the double-pulse train is used as a steering pulse to ensure that the logic of the tachometer binaries always locks up correctly. The advantage of this system is that nothing will wear out. In conformity with the redesign, the name was changed from PEC to TACH. The tach signal is used extensively in the control systems.

THE THREE SERVO SYSTEMS OF VIDEO RECORDERS

There are three distinct servo systems in most video recorders:

1. The head-drum servo, which is used to control the absolute position and velocity of the head in record and playback
2. The capstan servo, which ensures that the tape speed is correct in record and playback with respect to the head speed
3. The tape-guide servo, which ensures that the engagement of the head with respect to the tape is the same in playback as it was in record

Capstan and Head-drum Systems

Capstan and head-drum servo systems may be classed in two groups, original and modern, the original acting pretty much as already described in Chap. 8, whereby, in the record mode, the head was locked to the incoming video signal, and in playback the head was timed by the ac line. Unfortunately, the ac line bore no relationship to the local sync generator. This meant that, in early recording systems, the video tape recorder was a maverick in the entire system. Its signal would lock up neither vertically nor horizontally with respect to what was happening in the studio. As a result, whenever the operators switched from live camera to a tape signal, the effect on the home receiver was that the picture would roll. Another problem was that is was impossible to mix a local camera signal with a tape signal, since they bore no relationship to each other in terms of vertical coincidence, and no relationship in terms of horizontal. These facts led to the development of the modern servo system, one of the forms of which is that used by the Ampex Corporation, called intersync. As the name implies, it was intended to intersynchronize the playback of the tape recorder with respect to the rest of the studio. It is a playback device. It permits vertical lock of the tape playback signal with respect to the local sync to within ±150 nsec. In the record mode, intersync acts as a head-drum servo.

Head-drum Timing To restate, the requirements of a video-tape-recorder servo system in the record mode is:

1. A head-drum drive timed to the video input
2. A capstan drive locked to the head drum
3. A control-track signal recorded on tape

The rotating video head drum must always be locked to the video input so that if the video input should change, the head drum would change also, and lock up to the new signal. On the other hand, the capstan must be locked to the head drum because, if the head should speed up and the longitudinal speed of the tape were not increased, two tracks could be recorded on top of one another. If the head drum should slow down, the track would be too widely separated. Thus, to ensure constant spacing of tracks on tape, the capstan is locked to the head-drum speed. It may be remembered from earlier discussions that the heads for speeds of 7.5 ips are 5 mils wide and those used for speeds of 15 ips are 10 mils; the guard bands between tracks are 2.8 and 5 mils, respectively.

From a study of Fig. 11-6 one could almost conclude that, since the head drum is locked to the video input, which in turn is derived from a crystal standard, the system should be reasonably stable. Unfortunately, this is not so, the reason being that the head-drum comparator is comparing 60-Hz vertical information from the video input with the signal coming from the shaft of the head drum, which occurs at a 240-Hz rate. This means that every fourth spin of the head drum provides a pulse that operates the servo—a rather inefficient method of control. Due to this combination, the record stability of modern recorders, even those with air-bearing heads, is no better than ±1 μsec, with a rate change of 4 to 8 Hz.

Playback systems without intersync use the same circuit for playback as for record. Thus the servo stability is typically ±1 μsec with a rate change of 4 to 8 Hz (instead of ±150 nsec with intersync). The rate of change is one of the reasons why tape signals from the older systems, or systems without intersync, could not be mixed with local studio signals. The local signal has no rate change because it is the clock for the system. Therefore, with no vertical coincidence between the tape recorder and local sync, when signals from the two are mixed together, a windshield-wiper effect is produced in the video. However, it may be said of any video tape recorder, be it rotating-head or helical-scan type, that as long as the rate of change of head or scanner speed

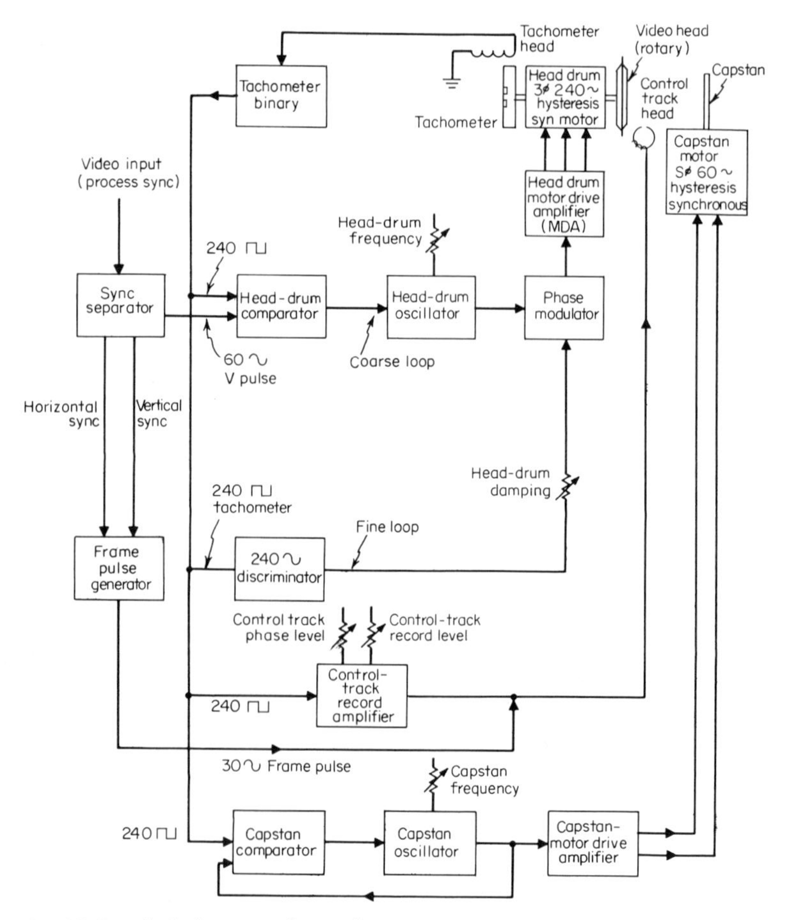

Fig. 11-6 Block diagram of record servo system.

is relatively low (4 to 8 Hz), the home television receivers can easily track the changes, and the picture is stable.

Record Timing The term record timing should be defined at this point; refer to Fig. 11-7. Record timing is correct when in the record mode the tachometer signal has its positive edge coincident with the leading edge of the third serration of the vertical sync. Referring to Fig. 11-7, the vertical sync is the broad pulse, three horizontal lines long, that is serrated, or broken up, every 32 μsec at a half-line rate. This provides the horizontal information to the horizontal oscillator so that it does not get out of sync during the vertical period. The third serration is exactly on the center line of the tape. The choice of this position was made to increase the ease of making mechanical or electronic tape splices.

As previously discussed, the tachometer disc has three magnetic inserts in it, two on one side and one on the other, at 180° (Figs. 11-5 and 11-6). The first of the double inserts is in line with head 4. The logic of the system is such that the leading edge of the 240-Hz square wave produced always starts with the first of the double inserts. The second of the two inserts acts as an inhibit or steering device, so that the logic of the circuit cannot lock up backward. When the positive edge of the 240-Hz square wave is received, the operator knows that head 4 is in the middle of the tape. The head-drum servo system takes the 240-Hz pulse and uses it to generate a ramp in the head-drum comparator. The vertical signal, also being fed into the head-drum comparator, is used to generate a sample pulse. By sampling the ramp, a 60-Hz stepped dc signal is produced. This will indicate the difference in timing (error). The dc signal will be used to change the frequency of the head-drum oscillator. See Fig. 11-8.

During the initial setup of the servo system, the drum frequency is set at 240 Hz with the head not spinning. This will mean that when the servo turns on, it will lock up in the minimum time. Many video tape recorders have a head override or ready switch so that the servo can be locked or ready before the playback or record buttons are pressed and the female guide brought in. This is called the fast start.

Coarse and Fine Head-drum Controls The coarse control might also be called speed, or frequency, control; the fine control, the velocity, or phase, control. To understand the coarse-loop system, it is necessary to realize that in the record mode the system is also in EE (electronics to electronics, a system whereby the record modulator output is fed to the record heads and also directly to the reproduce electronics). In this fashion, the signal recorded can be monitored. Additionally, since the signal contains composite video and sync, sync information can be extracted from the signal and used to sync the recorder. This type of sync is used in preference to that coming from the sync generator, since (unless used in closed-circuit work) the sync generator might not bear any relationship to the signal being picked off the air. The

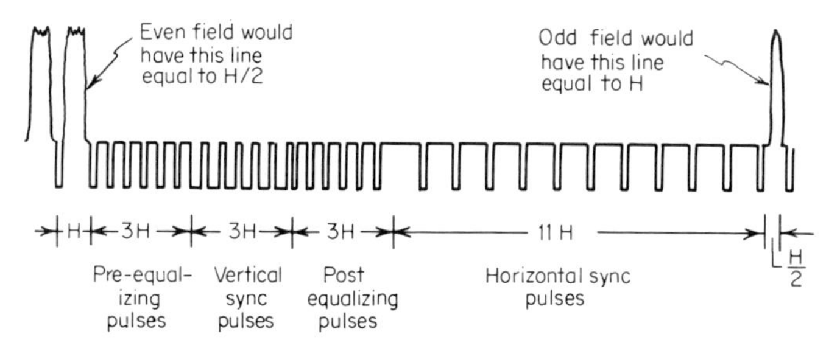

Fig. 11-7 A typical television vertical blanking pulse following an odd field.

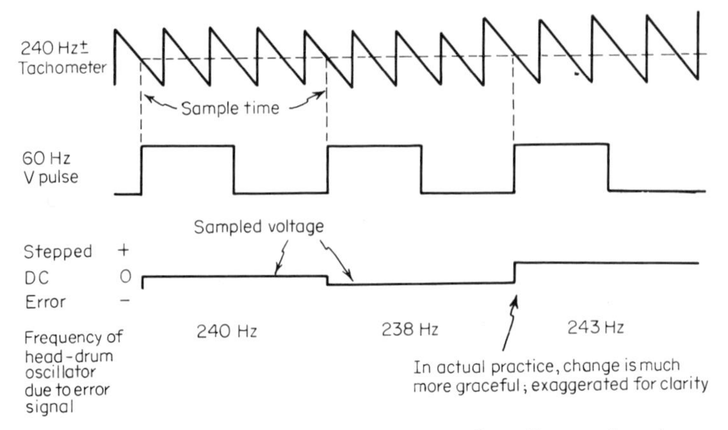

Fig. 11-8 Action of head-drum comparator and oscillator with tachometer output varying.

sync (proc sync, Fig. 11-6) is separated in the sync separator into its vertical and horizontal components, and the vertical pulse, which is a 60-Hz pulse, goes to the head-drum comparator. This pulse, when separated, corresponds to the first serration. Since it is desired to have the timing correspond to the third serration, a delay circuit is added. As shown in Fig. 11-8, the 240-Hz signal from the tachometer is compared with a 60-Hz vertical pulse, to find out if, in fact, the drum oscillator is synchronous with the input. If it is not, a correction voltage is generated. Should the video input be changed, by switching to a nonsynchronous input, or an input that varies from the local one, the head would have to reposition itself. This ensures that the new signal has vertical sync in the middle, where it was until the time of the nonsynchronized switching. The signal from the comparator is a dc error voltage (stepped) used for controlling the oscillator frequency (Fig. 11-6). The output of the oscillator passes through a phase modulator, which in the coarse loop is essentially a monostable multivibrator, to the head-drum power amplifier or motor-drive amplifier. There the low-level single-phase signal is changed to a high-level three-phase signal to drive the three-phase hysteresis-synchronous motor.

The head drum will be revolving at a 240-Hz rate, whereas the vertical sync is occurring at a 60-Hz rate. This means that sampling will take place only every fourth rotation of the head. During the intervening three rotations there is no vertical pulse. Thus, due to loading, etc., the head drum could go wild and speed up or slow down out of the correction range of the servo. To prevent this, an adapting loop, consisting of a one-cycle memory, is used to stop any change when there was no vertical pulse. In the fine loop, the output of the tachometer is compared on a rotation-by-rotation basis in the 240-Hz discriminator. If there are no differences between two pulses, there is no velocity change. If there is a difference, that difference will be sensed, and the heads will be speeded up or slowed down to counter the error. The output of the discriminator goes through the head-drum damping potentiometer to the phase modulator. In the modulator the phase of the output of the head-drum oscillator is retarded or advanced as required to slow down or speed up the head drum to maintain a constant speed. The amount of drive to the phase modulator is determined by the head-drum damping control. This loop is vital for correct operation of the machine, and its setting is extremely critical. For example, if a 2° slowdown of the head were sensed and the error voltage from the drum damping fed to the phase modulator advanced the phase by 2°, there would be critical damping. On the other hand, if a 2° slowdown were sensed and the head-drum damping were adjusted for 4° advance, the system would be overdamped. With a 1° advance the system would be underdamped. Of all the controls on the servo, drum damping is the one control that affects record stability.

Capstan Servo—Record Mode

In the record and playback modes the capstan is locked to the head drum for effective control. Referring to Fig. 11-6, the 240-Hz tachometer signal is compared with the output of the capstan oscillator. Any variation between the two signals will create an error voltage (stepped in a similar fashion to that of the head-drum comparator). This voltage will be used to change the output frequency of the capstan oscillator. Anything that happens to the tachometer signal causes a proportional change in the capstan frequency, and the tape speeds up or slows down. Such a system effectively ignores momentary changes in head speeds but takes into account the long-term changes. This type of action provides a smooth control.

Control-track Signal in Record

A control track is required because no two machines are ever completely identical. Even though the head drums are being driven within 1 μsec or better of the video signal, which is relatively constant, and though the capstan motor is being driven with the same stability, this does not mean that the tape will be moved longitudinally at the same speed. There can be slippage in the capstan drive system; i.e., belts could be incorrectly torqued, the capstan shaft could be too smooth, pinch-roller pressure could be incorrect, etc. In addition, tape will expand and contract with temperature and humidity change; thus it is necessary to put down some form of signal on the tape which will act as a memory of the longitudinal movement. In the record mode, the 240-Hz signal from the tachometer is laid down on the bottom edge of the tape (Fig. 11-2). The original square wave from the tachometer is changed to a sine wave in the control-track record amplifier for a quadruplex machine and into a non-return-to-zero (NRZ) type of signal for a helical-scan recorder. Additionally, a 30-Hz frame pulse is added (Fig. 11-9), to aid the operator in making mechanical splices. It is used to tell the tape editor where to cut the tape. The frame pulse is generated from the video signal itself. It is always generated by the field preceded by the half line. It is desirable to mix the 30-Hz pulse with the 240-Hz sine wave at 90°. For this reason, the control-track record amplifier is provided with both phase and amplitude adjustments. The 240-Hz sine wave is adjusted in phase so that the negative-going edge of the frame pulse is at 90° with the positive peak (it will become a positive edge on playback). The level of this control-track signal is important because no bias is added. The level in record is set so that on playback the control track has just reached the knee of saturation, and the signal looks as shown in Fig. 11-10. The reason that this level is critical is that there is no control over the level of the frame pulse, and if the 240-Hz sine wave is overdriven, the 30-Hz frame pulse will

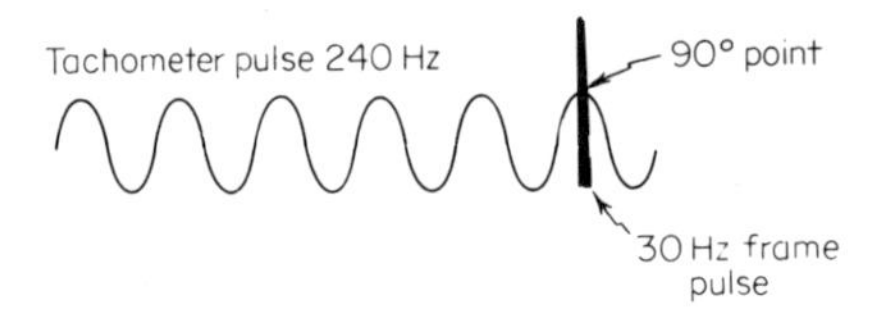

Fig. 11-9 Control track with frame pulse added.

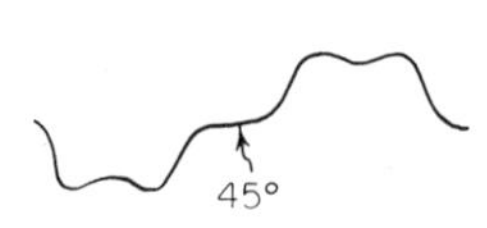

Fig. 11-10 Appearance of control-track signal on playback when the record level is set correctly.

be hidden, and there will be no way of extracting it. With proper setting the 30-Hz pulse will always be twice the amplitude of the 240-Hz control-track sine wave.

Summary of Capstan and Head-drum Servo Requirements in the Record Mode

The major requirements of the servo system in record are:

1. The head drum is locked to the video signal so that there is control over the absolute position of the video head with respect to the signal itself.

2. A damping loop keeps the head speed constant during the three rotations when there are no vertical pulses.

3. A capstan is locked to the head drum.

4. A control-track signal remembers the tape slippage so that correction can be made in playback.

Capstan and Head-drum Servo Action in Playback—Discussion Delayed

The action of these servos require information that has been recorded and played back. Thus discussion of their action will be delayed until after the record and playback processes have been covered.

HEAD-TIP PENETRATION—FEMALE-GUIDE HORIZONTAL-SYNC-PULSE RELATIONSHIP

The U.S. Standard group specified that the record timing be within 10 μsec, but the video-recorder manufacturers usually specify that there is no timing error. They have built test points and scope displays into their equipment to permit proper setting. Record timing is defined as the relationship that exists between the tachometer on the head wheel, or head drum, and the video signal coming in. The difference between the two signals is the record stability. Therefore requirements of any servo system maintaining record timing are such as to permit the head to be positioned and held in position with respect to the input signal.

Although the position of the head is kept as close to perfect as possible by making sure that the drum is perfectly balanced on the shaft of the drum motor, and so forth, there are problems in maintaining constant contact between the tape and head. The tape is held to the female guide by vacuum pressure.

Venetian-blind Effect

Positioning the female guide positions the tape, and the head is permitted to push against the tape surface. With the proper setting of the guide, the head does in fact penetrate into and stretch the tape by an amount almost equal to the height of the head tips. If the amount of penetration during the playback mode is not exactly the same as that during the record mode, the amount of tape stretch will be different. When this happens, the horizontal frequency will be increased or decreased. If the frequency is lower, the picture will show a bend, or skewing, to the right. This gives a venetian-blind effect, caused by the sync coming in later. If the frequency is increased, the skewing will be to the left.

Tip engagement with the tape is a very important consideration in video recording, for the reasons just discussed and because of separation loss. On the surface one would assume that the deepest penetration possible would be the best, since this would give reasonable immunity to imperfections in the tape and less separation loss. This is not true, however, since the farther the head penetrates into the tape, the greater the head wear will be. The best possible compromise between the reasonable head life and dropout immunity was taken into account by the industry when they set up the standards for head penetration. To ensure proper head engagement, the three manufacturers of broadcast video tape recorders (Ampex Corporation, RCA, and the German firm Fernseh) have come out with a standard alignment tape. It was intended to define two things: (1) the correct penetration and (2) the correct guide height.

To summarize, if the head-tip penetration on the playback machine is the same as the record machine, the vertical information will be reproduced vertically. If it differs, there will be a skewing effect to the left or to the right, depending upon whether the engagement is lesser or greater than it was when it was recorded. To define what the penetration should be in record, there is a standard alignment tape which *must* be used before every recording. There are *no* exceptions to this rule.

Scalloping

Referring to Fig. 11-3*b*, it can be seen that the guide-height center line and the drum center line are coincident. If the guide is pulled directly away from the head but still bears the same relationship between the female guide and the tape held to it and the heads, there will be a linear change. When the guide is changed upward (Fig. 11-3*a*), a differential change will be established between the tape and the heads. This will cause a scalloping type of error in the picture on playback. Scalloping is the result of a differential error rather than a linear one. If the guide height on the playback machine is adjusted to get rid of the scalloping effect, all that can be said is that the guide-height relationships have been matched between record and playback. It does not necessarily mean that there are correct guide heights. That can be proved only by the use of the alignment tape.

Referring once more to Fig. 11-3*a*, there the tape did not touch the head at the beginning of the pass, and it almost stopped the head at the end of the pass. If the guide is too low, the head will be almost stopped at the beginning of the pass and will not be touching tape at the end of the pass. This will create nonuniform velocity of head travel across the tape. Constant head velocity is achieved when the center lines of the guide and head drum are coincident.

Velocity errors are particularly hard to live with in color, since the effect of the head changing speed as it makes a pass across the tape will cause the colors to shift their hue. Pure red can be shifted to its purple or orange tone, depending upon whether the guide is too high or too low. Correction of this problem on modern recorders involves the use of expensive accessories such as Amtec,[1] Colortec,[1] and velocity compensation.

Thus it may be said that two of the problems that occur in the quadruplex system, or in a system where the tape is being guided by a vacuum guide, are skewing and scalloping. Skewing is caused by the difference in head penetration (playback vs. record), and scalloping by guide-height differences between playback and record. The correct relationship can be established by using the standard-alignment tape provided with each machine. That tape is made under laboratory conditions and is intended to be used as a standard, agreed upon by the industry.

Quadrature Error

There is a third error, an error of geometry, caused by the use of four heads, called quadrature error. This effect has nothing to do with the vacuum guide but rather the angular relationship that exists between adjacent heads on the drum (Fig. 11-11). Assume that 90° of head rotation corresponds to 16.4 horizontal lines. In a system that is limited in bandwidth to 4.2 MHz, one complete cycle is 240 nsec, and a half cycle is 120 nsec. By definition, one picture element can be seen, and is equal to a half-cycle. Therefore, in one cycle, the picture could go from one element to another, or directly from black to white. Now

[1] Registered trademark of the Ampex Corporation.

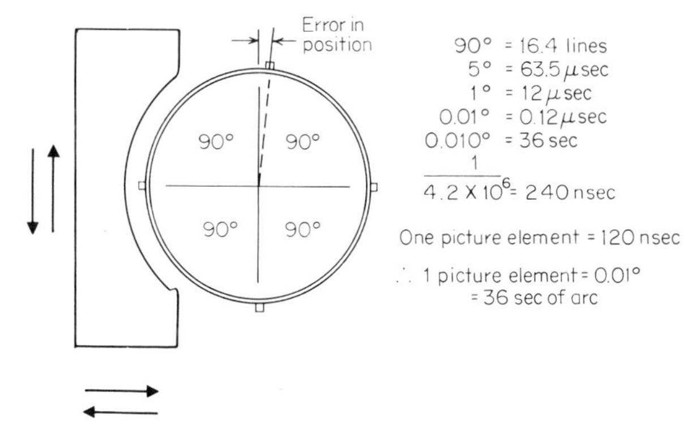

Fig. 11-11 Quadrature error.

if 90° of rotation is 16.4 lines, 5° of rotation is 63.5 μsec, 1° is 12 μsec, and 0.01° is 0.12 μsec, or 120 nsec. Therefore 0.010° is equal to one picture element. Since 0.01° is equal to 36 seconds of arc, this means that if the head departs in the 90° delay relationship by more than 36 seconds, a problem will show up on interchange. For example, if one head were 89.99° in relationship to another, there would be an angular error equal to one picture element. This would show up as a displacement in time of the channel due to the switching logic in playback.

An alignment tape is not used to check for quadrature, but rather the procedure for quadrature check and setup is as follows. Since there are four heads on the transport, there will be four possible combinations on playback: head 1 plays back the same track as it recorded (home track), or it plays back track 2 or 3 or 4. Home track is defined in video usage as a situation that exists when a given head plays back the track it recorded. Quadrature error is impossible on the home track because the early tip is playing back the early track and the late tip is playing back the late track. However, if all four possible positions of tracking are gone through, band displacement can be seen if there is quadrature error. All modern video tape recorders have a vernier tracking control which has a range of ±45° of adjustment (the capstan control). Additionally, they have a coarse shifting device which permits switching heads a full 90°. Once tracking is optimized on the home-track position, the track is slipped by 90° of delay and any band displacement will show up. This allows identification of the transducer (head) that is responsible for the quadrature error. The transport is then turned off, and a small set screw is adjusted to move the head earlier or later, as the case may be, to apply correction. The adjustment is made until the error is beyond the point of visibility. Video record heads coming from the factory have a quadrature set to within 0.02 μsec, which is 6 seconds of arc, or 1/6 of the error just discussed.

RECORD PROCESS

The input to the signal system, composite video, is fed to the modulator, where it is transformed into an FM signal within the particular range of the system. Refer to Fig. 11-12 and Table 11-2 for the block diagram and bandwidths involved. Before the video is applied to the modulator, it is preemphasized and the gain is preset for the different amounts of deviation required by the various standards the modulator is required to work on. A preset amount of direct current is also applied to the modulator to provide the major frequency shift among the standards. The video input gain control can be thought of as a deviation control

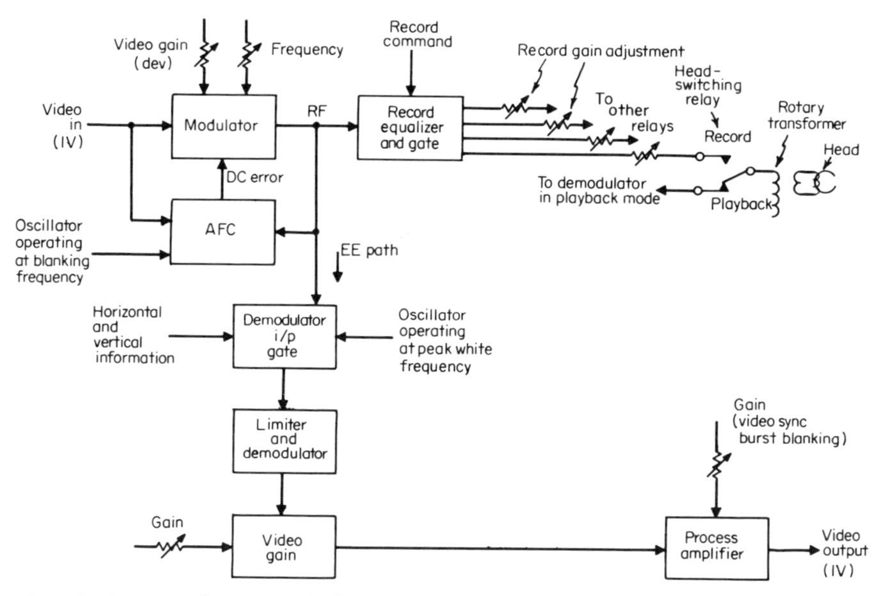

Fig. 11-12 Record-process block diagram.

for the modulator. The frequency control is a vernier adjust to put the modulator on the correct frequency. Once these two controls have been set for one standard, they should track for all standards involved.

Modulators

There are several ways in which the correct carrier frequency may be generated. Two oscillators could be used, one at 100 MHz and the other at 108 MHz. Feeding both outputs to a mixer and taking the difference frequency will provide the necessary 8 MHz. This is a simple standard heterodyne modulator. There are disadvantages to this type of scheme, however, since a large amount of second-order harmonics will be generated. A better system would modulate two oscillators that

TABLE 11-2 Modulated Frequencies of the Television Standards

Modulated frequencies	525/30 low-band mono	525/30 low-band color	525/30 high-band mono/color	625/30 low-band	625/25 high-band
Tip of sync, MHz	4.28	5.5	7.06	5.0	7.06
Carrier frequency (blanking), MHz	5.0	5.76	7.9	5.5	7.8
Peak white, MHz	6.8	6.5	10.0	6.8	9.3

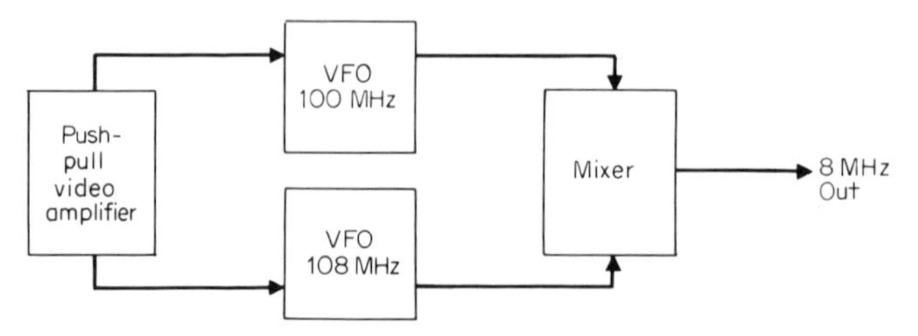

Fig. 11-13 Double-heterodyne modulator.

have a difference frequency of 8 MHz and cause one of them to go up in frequency at a video rate, the other to go down (Fig. 11-13). The advantage of the double-heterodyne, or push-pull, system over the single-heterodyne is that the second harmonics are eliminated and linearity is much better. Both systems are in use, however, as is a third method. This method is a form of astable multivibrator. In this type of system the video signal is fed straight to the multivibrator and generates the fundamental frequency directly, instead of using the technique of a difference frequency of the heterodyne system. The advantage of such a system is its simplicity. But it is much more difficult to design a stable system of this type than the double-heterodyne system.

All the modulator systems described make excellent recordings, both in monochrome and color. Which shall be used depends upon the choice of the design engineer. To keep the modulator on correct frequency, an automatic frequency control (AFC) system is used. It will take a look at the output of the modulator during blanking and see whether its frequency is correct. If it is not, a dc correction voltage is fed to the modulator to modify its frequency and make it correct.

Record-process Block Diagram

Referring to Fig. 11-12, the FM signal from the modulator is passed to the record gate, record equalizer, and record drivers. The FM signal is also fed to the demodulator input gate. This forms part of the EE path, where the FM will be demodulated back into video as a check of the signal path. The record gate is turned off, and no FM passes this point until a record command is given. The function of the record equalizer is to shape the voltage characteristics of the signal so that a constant current is fed to the head, since head impedance changes as the frequency changes. The equalization is done before the signal is split into four identical parts to drive the individual heads. Having adjusted the voltage response of the signal, the individual levels are now controlled so that each head receives optimum drive. There will be a slight difference in the amount of drive because no two heads are absolutely identical. This process is called optimizing. In most recorders, microphones and current meters are built into the tape transports for the purpose of optimizing. The amount of current chosen is the one which will give the maximum signal off tape in reproduce. To define the optimum setting in another way, the amount of record drive for each head will be minimum to give maximum RF return in playback. The amount of current will be such as to magnetize the tape slightly below saturation point. From the final record amplifiers, the signals are fed to relays on the head assembly that will switch the head between record and playback. In record, the relays will be energized and the FM will be fed to the heads via rotary transformers. Although the rotary transformers were developed for high-band recorders, they are now available for all video quadruplex machines. In the very early series of video recorders, commutators and slip rings were used. These required a tremendous amount of maintenance. The brushes often went bad, and more important, the capacitance of the commutator itself added to the head resonance in playback and caused that resonance to be at a higher level than desired. With the rotary transformer, because there is usually a ratio of 10:1 in the windings, any head resonance in record is completely swamped out by the extremely low impedance presented to the transformer. In playback the transformer acts as a step-up device and provides a 10:1 amplification of the voltage from the playback head.

EE Signal

Using the heads for both record and playback presents problems. There is no means of monitoring the signal during recording. Although there is a record current meter in the final record amplifier to indicate that the heads are being driven, until such times as the tape is played back, there is no certainty that there is anything on tape. Looking at the input signal will not in itself indicate that any recording is taking place. On the other hand, if after this signal is changed to FM a part of it is fed to the demodulator, where it is changed back to video, the signal viewed at this point would indicate that the signal electronics are working correctly. In other words, if the quality of the signals at the input to the video tape recorder and the demodulator output is the same, we could rightly assume that the signal system is operating correctly. The portion of the FM signal that is fed to the demodulator is called the EE signal (electronics to electronics).

The demodulator input gate is inserted in the circuit before the demodulator. It selects either tape FM or EE FM to be fed to the demodulator. It is controlled by the machine mode. In the record mode, EE FM is fed to the gate along with a white calibration signal.

The latter is RF from a crystal oscillator corresponding to the peak-white frequency for the standard involved. This white RF is switched into the EE FM path during the vertical interval. On demodulation it will correspond to a peak-white signal and will be used as a reference to set the deviation control (at the input to the modulator). The function of the EE signal is to indicate that the signal system is operating correctly in record. Not until the signal is played back off the tape is there a certainty that it was recorded correctly on tape.

Record-system Accessories

In systems that include a processor as an accessory (these are usually called proc amp), the video output of the demodulator passes through it. The proc amp is used to extract the sync that will be used to control record timing in the servos of the video tape recorder. The question might be ased, "Why was this system designed this way? Why not, instead, use the sync from the recorder's own sync generator to determine the record timing?" There is an excellent reason why this was not done. Assume that the composite video input is coming from the television network via AT&T. Assume also that the local studio's sync generator is not locked up to this network, as is the custom of many broadcasters. Under these conditions the local sync pulses will bear no relationship to the sync of the signal coming in from the network. By taking the sync from the EE signal, the system is assured of correct synchronization to the network. The function of the proc amp in playback is to generate a brand-new blanking signal and replace the sync signal which may have noise and switching transients. The original tape blanking is eliminated and replaced by the electrically generated blanking that is correctly timed to the tape signal. This means that any noise in the blanking is completely erased. The proc amp also takes the sync from the tape signal and cleans it up and puts it back in again, or uses it to generate a brand-new sync pulse which is timed to tape sync. Thus the proc amp will take everything that is happening during the blanking period, clean it up, and either reinsert it into the signal or use it to generate brand-new sync pulses. This means that the only noise that is in the output of the video tape recorder is the noise that occurs during the video, and that is the data being recorded. The sync and blanking pulses are extremely clean.

PLAYBACK PROCESS

In the playback process, the object is to take the FM signal off tape and convert it back to video. The reconstructed video should suffer no degradation. The output off tape should be a perfect replica of what was put on tape. The four separate bursts of RF will be recombined into one continuous stream.

Preamplifier

Refer to Fig. 11-14. The tape signal is first amplified in a preamplifier that is located on the head assembly to minimize losses. Depending upon the type of preamplifier used, the head resonance could be fairly low, and fall into the bandpass of the video signal. In high-band recording, we are interested in recording and playing back an FM carrier plus the first order of upper sidebands. The carrier goes to 10 MHz at peak white, and the signal bandwidth is 4.2 MHz. This means that the system should be capable of reproducing perfectly anything up to 14.2 MHz off tape.

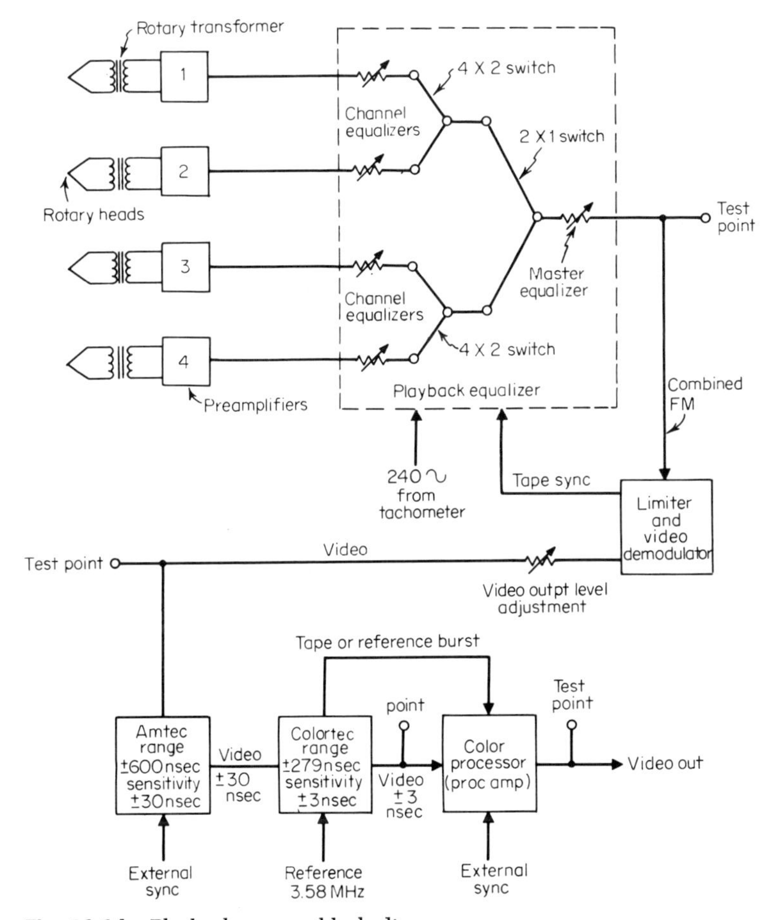

Fig. 11-14 Playback-process block diagram.

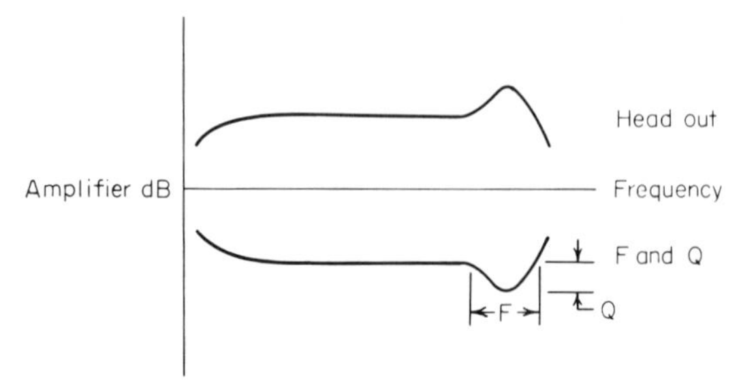

Fig. 11-15 Action of *F*-and-*Q* circuits.

When nuvisters are used in the preamplifier, the head resonance point may be within this bandpass, and distortion will occur. To overcome this resonance problem, an *F*-and-*Q* circuit is often installed between the preamplifiers and the playback equalizers. The *F* portion of the circuit will allow the operator to tune the resonance point of the head back and forth to provide a reciprocal characteristic curve of the resonance of the head output. The *Q* portion permits broadening or peaking of the amplitude. See Fig. 11-15.

Channel Switching and Playback Equalization

Following the preamplifiers, the RF is combined into one continuous stream. Ideally, the switch would take place every 90° of head rotation. This would correspond to 16.4 lines and would mean a head switch during the active picture portion. Before switching is done, some response shaping takes place in the form of equalization. There are individual equalizers to match the four heads so that they all have the same output characteristics. Following the switcher is an overall master equalizer. The types of equalizers used are those that produce no differential phase shift between frequencies.

The record track is 1.82 in. long and contains approximately 18 horizontal lines. Referring to Fig. 11-2, it is seen that only 1.57 in. of the recorded track is useful information (16.4 lines); the balance is redundant. The recorder should take advantage of this to switch at a time where there will be no noticeable disturbance in the picture. In the early stages of the playback electronics, the time-base error is typically in the order of ±150 nsec. In an FM system, if two signals are combined when they are not within 45° of each other, random reinforcement and cancellation will take place. Considering a 10-MHz signal, one cycle would be 100 nsec, and 45° would approximate 12.5 nsec. This figure is considerably smaller than the 150 nsec of typical time-base error. If the signals were combined during the active scanning portion of the picture, a visible switching transient would be produced. Switching signals must be developed to combine the signals at a time during the horizontal-synchronizing interval. This way, if a transient is developed, it will not be seen. The time chosen is on the front porch of horizontal sync of the television signal (Fig. 11-16). The fact that there is overrecording allows us to choose the time when we want to switch. The number of lines used in each head pass is 16 or 17. Only one head at a time is gating through information; the other three are gated off (Fig. 11-14).

To tell the channel switcher when to switch, a 240-Hz square wave, derived from the tachometer, is used to combine the RF into two channels in a 4 by 2 switch. The tachometer signal combines channels 1 and 2 and 3 and 4. The two combined streams of RF are then fed into a 2 by 1 switcher. A 480-Hz square wave, from an oscillator that is phase-locked to the tachometer, is used to operate the switching signal for the 2 by 1 switch. To perform the switching at the exact time, the 480-Hz signal will be timed from horizontal information so that the switch takes place exactly on the front porch of horizontal sync.

Since the 480-Hz square wave is locked to the tachometer and correctly phased, a half-cycle of the 480 Hz will correspond to approximately 90° of head rotation. Figure 11-14 shows a simple representation of the switches in the playback equalizer.

After combination, the four signals are passed to a master equalizer, where the overall response can be shaped to meet the specifications of the system. If the system had been optimized correctly, it should be possible to leave the equalizers in their flat position which defines a flat response. The master equalizer should be in the flat position also. The recording and playing back of a test signal should look identical under these conditions.

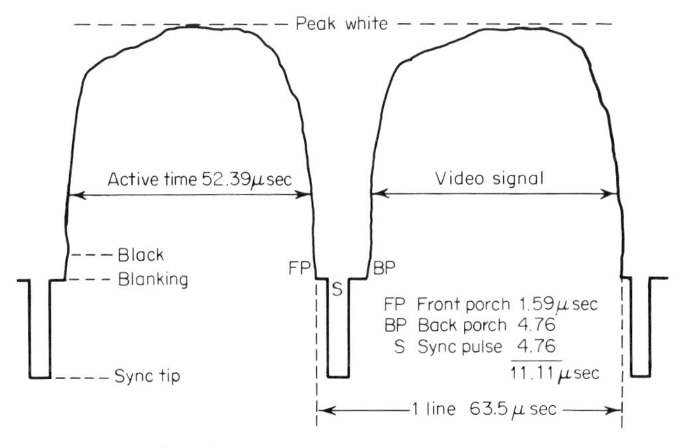

Fig. 11-16 The television signal—horizontal blanking.

Limiting before Demodulation

Referring once again to Fig. 11-14, the output of the playback equalizer is fed to a limiter. Its function is to limit the signal before demodulation. Limiting is required in any FM system, but particularly with rotating-head video recorders, because spacing losses generate objectionable amplitude modulation of the FM signal. These are the result of tip contour, head-to-tape contact, etc. Typically, about 70 dB of limiting is used.

Limiting may be defined in two ways: (1) static limiting, or the amount the input has to be attenuated before the output of the limiter shows a 3-dB drop; (2) dynamic limiting, or the reduction by a certain percentage of the amplitude modulation on the input signal. It is normal to use a 50-dB figure for dynamic and 70 dB for static limiting.

Demodulator

A number of different types of demodulators have been used. The most common in present use is the pulse-counter type. It involves delay lines or shorted pieces of coax. The limited FM signal is, by this stage in the signal chain, a very good square wave with excellent rise time. The edges of the square wave are used to generate precision pulses which will be counted per interval of time to generate the original video signal. A low-pass filter is used as an integrator. With many pulses fed to it, a large voltage output results. With fewer pulses the output is a lesser voltage. Thus frequency is converted to voltage and complete reversal of the record process.

The output of the demodulator will be a replica of the video signal within the limits of the filters that are used. Generally speaking, the filter values are laid down by the appropriate government bodies that authorize television transmission. American standards generally use 4.2 MHz for low-band and 4.5 MHz for high-band systems. The European broadcasting standards use 5 MHz for low-band and 5.5 MHz for high-band systems. When a system is being used in a closed-circuit application, the filter response is normally expanded to 6 MHz since no "on-the-air" transmission is being done. The filter numbers indicate the 3-dB point; for example, with a 4.5-MHz filter, the filter is flat to 4.2 MHz and down 3 dB at 4.5 MHz.

AUTO-CHROMA AND DROPOUT ACCESSORIES

Auto-chroma

Referring to Fig. 11-17, the modern video recorder may have a further device added to the signal system—auto-chroma. This could be con-

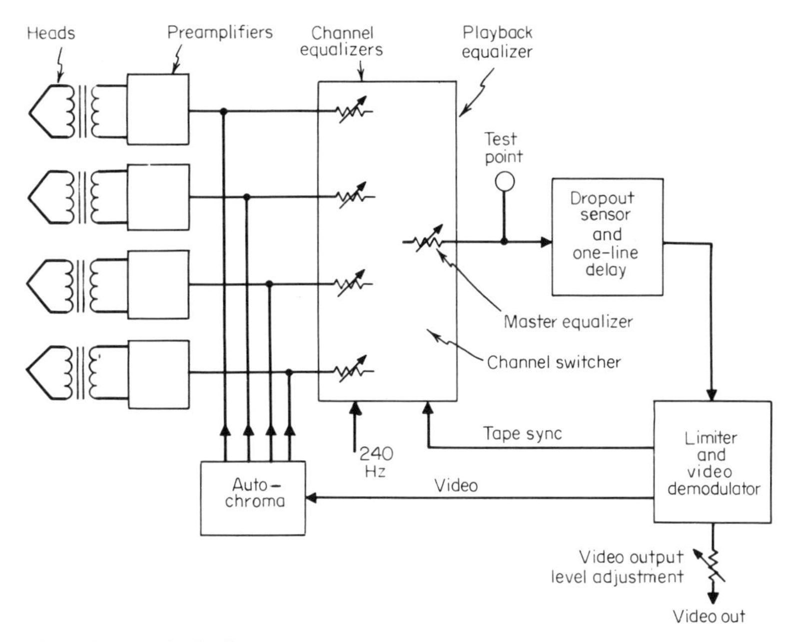

Fig. 11-17 Playback process with auto-chroma and dropout sensing.

sidered an automatic equalizer. Its function when installed is to look at the video coming out of the demodulator in the playback mode on a channel-by-channel basis to determine whether the color burst is 40 IEEE divisions. If it is not, then the auto-chroma will generate a voltage automatically and feed it to the equalizer circuits. This voltage will control the equalizers so that the burst will be increased or decreased as necessary to maintain a 40-IEEE division level. With this device the operator is no longer required to decide subjectively whether his equalizers are set correctly or not. This unit operates only with color and will automatically turn itself off with monochrome, making the manual controls available for operator adjustment. Using the auto-chroma device there is no reason why the operator has to adjust the four individual channel equalizers and the master equalizer to get the color bars correct. Thus there is no reason why there should be chroma bands on the home receiver.

Dropout Sensor

The RF coming off tape may contain dropouts caused by imperfections of tape, low tip projection, etc. Just before the demodulator, a dropout sensor and a one-line delay may be added to compensate for them. The dropout sensor senses a dropout as a predictable drop in level of the RF entering the limiter. Whenever a dropout takes place, the dropout sensor replaces the signal with reference black from the black oscillator, a part of the signal system. The black replacement is much less annoying to the viewer than a white streak or flash of color. When a one-line delay circuit is fitted (a 63.5-μsec quartz delay line), the system is effectively storing one complete line of RF. This will be recirculated in the system, and if a dropout occurs, it will replace the dropout. Referring to Fig. 11-18, the RF from the playback equalizer and channel switcher is fed to three places, the dropout sensor, the limiter, and the delay line. The dropout sensor has a sensitivity control that sets the point where the unit senses when a real dropout takes place, and not just a head-to-tape amplitude fluctuation. For example, the control could be set for a sensing level of 10 dB. When the sensor does operate, it will cause a tunnel diode to turn on and connect the signal from the quartz delay line (which was the signal that was available 63.5 μsec earlier) to the limiter. The RF stored in the one-line-delay line will be recirculated continuously until new information is available (the dropout is ended). Thus, if the head is clogged for 10 lines, the last of the RF before the dropout occurred is used to replace the 10 lines of missing data. When the dropout ends, the system goes back to the direct signal. Quite naturally, there will be some loss in the delay line, so that there is a little color change in the signal. However, even with this effect the viewer sees a much more satisfactory picture than with the original dropout or the replacement black of the standard dropout-sensor unit. Most dropouts are of short duration, rarely lasting a line.

An alternative system could provide improved correction. It would contain, in addition to the one-line-delay correction, a video dropout-correction network. This is added to the output of the demodulator. The video correction is downstream of the dropout sensor and one-line-delay unit; therefore it can be turned on a fraction of a second before the dropout occurs. This means, as far as the viewer is concerned,

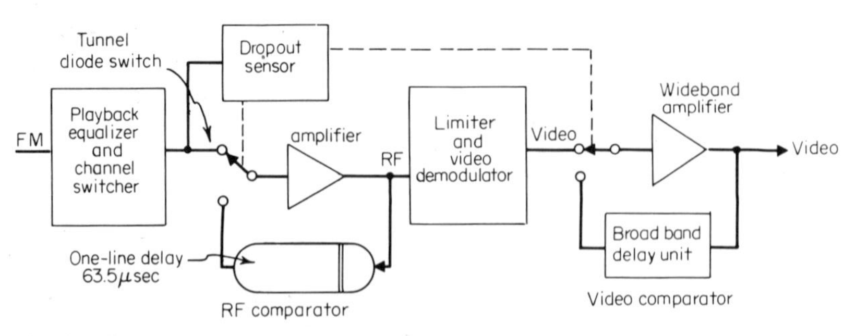

Fig. 11-18 Dropout correction using both RF and video correction.

that no dropout took place. Using this 18-octave broadband device, whatever correction was missed in the RF area will be caught farther downstream by the video dropout corrector (Fig. 11-18). When both of these devices are used, the stringent requirements for high-band video tape may be relaxed somewhat.

TIME-BASE ERROR

The output of the demodulator will contain time-base error even in the best possible case, where the recording was played back immediately on the same transport. In such a case the head geometry would be the same, and there would be no skewing, scalloping, or quadrature errors. This is assuming that the systems were perfectly set up and that the intersync servo system was in one of the tight control modes. Under these conditions the typical time-base error is ±150 nsec. Now, since one picture element in a 525-line TV standard is 125 nsec, a time-base error of 150 nsec will permit the error to be seen. Thus, when we take this signal and mix it with other local signals, the difference between the two will be visible. It is therefore necessary and desirable to reduce the time-base error substantially, much beyond one picture element, so that it cannot be seen. To do this, a series of correction devices have been developed, called Amtec, Colortec, and processor. Refer to Fig. 11-14.

Correction with Amtec

Initially, the job of Amtec was to correct for any and all geometric errors in the picture. Now, however, when used with intersync servo in either of the two precision modes (horizontal or automatic), it corrects the time-base error down to 30 nsec. It has a correction range of ±600 nsec and a sensitivity of 30 nsec. Since 30 nsec is one-quarter of a picture element, with this amount of correction no visible geometric error and no time change between the tape of one machine and another will be visible. It will be possible to mix a black-and-white signal from a video tape recorder with a local camera signal, and there will be no visible error. Therefore, with Amtec, a video tape recorder can be integrated into standard black-and-white studio operation.

The Amtec looks at the leading edge of the tape sync on a line-by-line basis to find out where it is with respect to where it should be. Based on its findings, it will generate a correction voltage that drives an electronic delay line consisting of a fixed inductance and 88 varicaps. As the voltage on the varicaps is changed, so also will the amount of delay be changed.

Referring to Fig. 11-19, we find that the Amtec gate bears an important

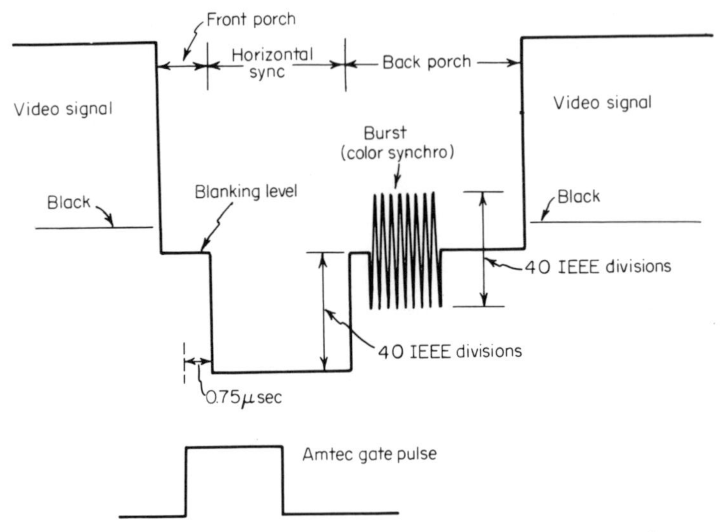

Fig. 11-19 Relationship of Amtec gate and edge of horizontal sync.

relationship to the sync. It must always start 0.75 μsec before the edge of sync, assuming that the sync is perfect and there is no timing error. Obviously, if there is jitter, the edge of the sync is at a random basis and is going to be early or late with respect to where it should be. Thus the Amtec looks at the edge of sync, finds out where it is with respect to where it should be, and generates an appropriate negative or positive correction voltage to increase or decrease the delay of the composite signal going through the delay line. This is a fixed dc voltage that is held for 63.5 μsec or until another sync pulse comes along and the Amtec ascertains whether the next line is the same as the previous one. This means that switching must always take place prior to 0.75 μsec before the edge of sync.

Unfortunately, a time-base stability for NTSC color is 5°. Another way of stating it is that if the color signal is changed by more than 5°, it will be visible to the average observer, and the flesh tones will shift toward blue or green. When dealing with a frequency of 3.579545 MHz, one cycle is 279 nsec. Therefore 5° is approximately 4 nsec. This means that to record color perfectly, time-base error at the output of the system must be less than 4 nsec. To achieve this, a second Amtec type of device is placed in series with the output of the Amtec, called Colortec (Fig. 11-14).

Correction with Colortec

This additional electronic delay line is designed to compare the bursts on the back porch of the signal coming off tape (if it is a color signal) with the local 3.579545-MHz signal generated by the color standard. This latter standard is not gated, but rather is a continuous signal. The logic of the Colortec is to gate through only the burst portion of the tape signal and compare it with the appropriate number of cycles of the reference 3.579545-MHz signal. The difference between the two becomes a correction voltage to a smaller delay line which has about one-half the range of the Amtec. This acts essentially as a vernier delay line with final correction down to ±3 nsec or better. Three nanoseconds corresponds to 4°. In most cases, the Colortec will correct so well that maintenance personnel cannot pick out any errors even on a vectorscope. Vectorscopes resolve down to 2°.

Summary of Correction Devices for Time-base Errors

This system has been called the direct color system. It is a combination of three different devices: the precision servo, which gets the signal within the capture range of the Amtec; the Amtec, which has an output beyond the specifications required for monochrome, but not quite to specifications for color; and finally, a precision Amtec called Colortec, which operates on the burst portion of the signal and corrects all the way down to 3 nsec or better. Such correction has exceeded the NTSC specifications, and as a result it is possible to eliminate the tape burst and put brand-new burst in place of it. If the error specifications have been exceeded, there should be no visible difference on color playback when new burst is used. There are a number of advantages in using new burst: The rise time is perfect. The number of cycles is perfect. Thus, when using intersync, it is possible to lock the video-tape playback vertically and horizontally with respect to the local sync. This will be an obvious requirement if the output of the tape recorder is to be mixed with live camera. Timing will have to be correct if such studio techniques as split screen, superimposition, mixing, and dissolving, etc., are to be used.

Color Processor—Proc Amp

Referring again to Fig. 11-14, the last unit in the chain is the color processor, commonly called the proc amp. It has several functions:

1. To generate a brand-new blanking signal (the primary use)
2. To clean up the sync or to generate a new sync pulse

When a new blanking signal is generated its timing must be perfectly coincident with that of the tape blanking. For this, a 31.5-kHz oscillator is used. This oscillator is always locked to the horizontal frequency of the signal coming in off tape. The output of the oscillator is fed through a delay line with many taps on it. These taps are used as start and stop pulses for flip-flops. The output of the flip-flops become sync, blanking, and gating signals—all the signals required for correct operation of the system.

Referring to Fig. 11-19, the proc amp generates brand-new blanking which will completely wipe out everything from the start of the front porch to the end of the back porch. Such action will destroy the color burst and sync. Therefore, when using proc amp, color burst must be sent over another path than that used for the original combined signal. In Fig. 11-14 the cable from the Colortec to the color processor labeled "tape or reference burst" has only burst on it. It will be either tape burst or a brand-new burst directly timed to the original tape burst and having the correct number of cycles on it. This means that after new blanking has been added to the original signal, new burst can be put back in again without introducing errors. Prior to the time that new blanking is added to the signal, however, it will be necessary to strip out and process the sync.

The proc amp operates in the following fashion: The composite video signal from the output of the Colortec is fed to two sections of the proc amp, the first section being a separator, where the color signal is separated into brightness components (luminance) and its color component (chrominance). The chrominance signal is set aside. New blanking will be added to the luminance after the establishment of the correct black-and-white levels; then the chrominance will be added, as well as a new or regenerated sync pulse. This sync pulse comes from the second section of the proc amp. There it had been extracted from the composite signal and cleaned up or used to generate a brand-new sync pulse. In modern systems, provisions are also made for adding external sync at the output of the proc amp. Such sync should be perfect since it is coming from a sync generator. The logic of operation of this portion of the proc amp is such that when external sync mode is selected and the composite signal is not locked up vertically and horizontally, the proc amp will not add external sync, but rather it will regenerate the sync internally.

CAPSTAN AND HEAD-DRUM SERVO ACTION IN PLAYBACK—DISCUSSION RESUMED

Capstan Servo—Playback Mode

In playback the servo system is considerably more sophisticated than in record. The tape signals are used to achieve vertical and horizontal lock with the local systems, in effect providing intersync or inter-synchronism of the tape recorder with respect to the sync generator or local or external sync. A number of modes are available, each having the degree of stability indicated in Table 11-3. It should be noted that in preset, normal, and vertical, the time-base error (TBE) is fairly large. As a result, these modes are not suitable for use with color. Horizontal and automatic are the only modes that can be truly called intersync modes. Since there is no vertical lock in the horizontal mode, the question arises as to why the mode is provided. In automatic, lockup is achieved sequentially, the vertical lock being provided first, followed by the horizontal. Should there be an interruption in the tape or reference sync during the playback, vertical lock must take place before horizontal lock can be reestablished. During this period of reframing vertically and relocking horizontally, the Amtec and Colortec cannot do their job. Due to the loss of color synchronization, there will be no color on the home receiver. In the horizontal mode, on the other hand, should there be interruption of the tape or reference sync immediately upon restoration of the sync, there is full horizontal lock. The home receiver will have reframed, taking care of the vertical lockup. Thus the advantage of the horizontal mode is that full color is restored immediately after restoration of sync; but it has disadvantages too. For example, it is not possible to superimpose the tape machine over a local video source. In addition, an external sync cannot be used in the proc amp. The automatic mode is recommended for several reasons, such as:

1. The external sync can be used with the proc amp, and in doing so the timing can be made absolutely perfect in terms of the breezeway (the spacing between the trailing edge of the sync pulse and the side of the burst).
2. It locks both vertically and horizontally.

If the horizontal and automatic modes are the only ones recommended, why the other three? These modes are essentially the same as those

TABLE 11-3 Lockup Conditions of Intersync Modes

Mode	Lockup	Stability
Preset	No vertical or horizontal lockup	± 1 μsec TBE
Normal	No vertical or horizontal lockup	± 1 μsec TBE
Vertical	Vertical lock but no horizontal lockup	± 1 μsec TBE
Horizontal	No vertical lock, has horizontal lockup	± 0.15 μsec TBE
Automatic	Both vertical lockup and horizontal lockup	± 0.15 μsec TBE

used in the older drum-and-capstan servo systems. In normal, the ac line is used to lock up the head drum. In preset, there is a choice of using the external sync as timing reference without establishing coincidence between the tape recorder and the sync that is being used for a stable drum reference. In vertical, only vertical lock is achieved. Thus these three modes are handy for test purposes. If there are problems in the intersync and it works correctly with these modes, the technician can use his time troubleshooting the circuitry that is unique to the horizontal or automatic modes.

Operation of Intersync—Preset and Normal Modes

A comparison of Figs. 11-6 and 11-20 illustrates the differences between record and preset and normal modes of reproduce. While the proc amp sync has been replaced by external sync or 60-Hz pulses from the ac line in the head-drum circuit, the major differences occur in the capstan servo circuit. In the record mode, it may be remembered, the capstan was locked to the head drum. In playback the composite control-track signal has been taken from the tape via the control-track head and used as a part of the control system for the capstan speed. The signal consists of 240 Hz, previously recorded as an indication of head-drum speed in combination with the longitudinal tape speed, and the 30-Hz frame pulse. The signal will be considerably distorted (Fig. 11-10) since no bias was used during the recording. The composite signal is fed to the control-track playback amplifier, where its level is adjusted to approximately 10 V peak to peak and the phase adjusted so that it is the same on playback as it was during record. After filtering, the control track (240 Hz) is fed to the capstan comparator, where it will be compared with the tachometer signal coming from the head-drum tach. Since the control track was the tachometer signal in record, in playback we are comparing the recorded tachometer signal against the actual tachometer signal. The same circuits are used in preset and normal modes as in the record, so that the servo does not know that it is not in record. Consequently, any difference between the 240-Hz signal from the tachometer and the 240-Hz signal from the control-track playback amplifier indicates a need for changing the tape speed. The capstan comparator will generate an error signal which will change the frequency of the capstan oscillator to speed up or slow down the capstan. If the head-drum speed had changed during recording, this would mean that the tape would speed up or slow down as well. Consequently, in playback, the control-track signal would have a different frequency or phase from the normal 240 Hz. The playback servo would have to correct for this speed change. In addition, if there were any

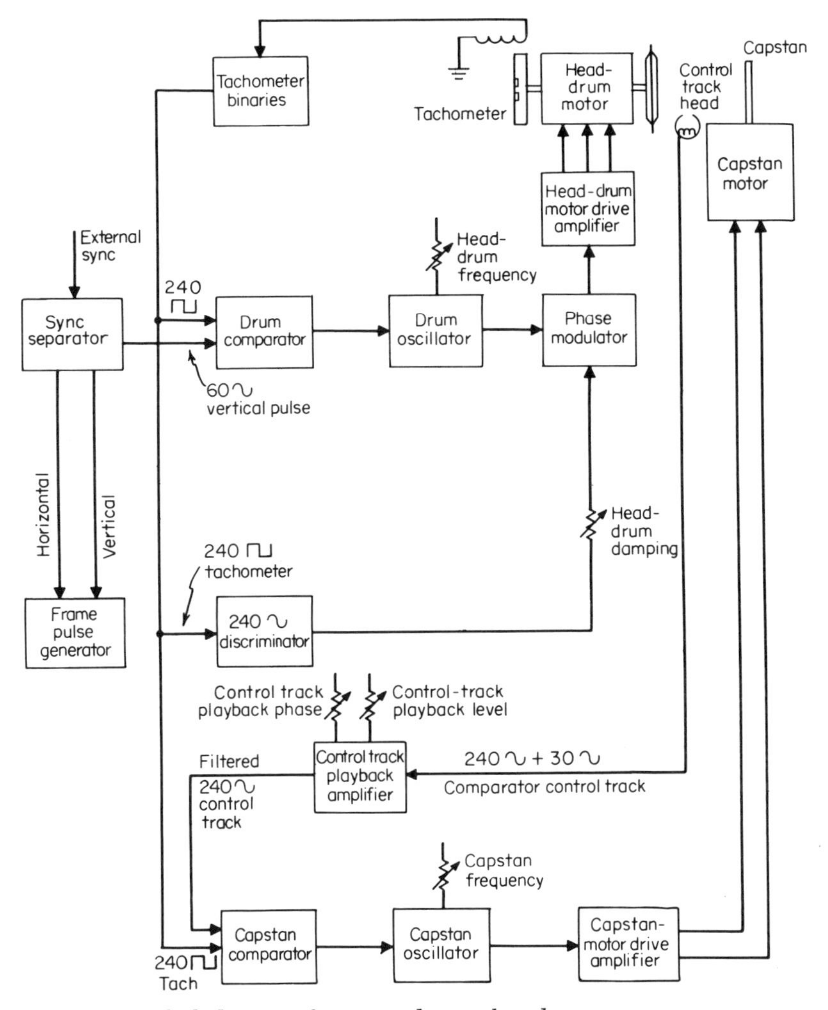

Fig. 11-20 Block diagram of preset and normal modes.

tape-speed change in record due to slippage, the capstan system would also have to correct for this in playback.

The playback phase control is located normally on the front panel of the recorder and is called the tracking control. When it is set correctly, the phase of the signal coming off the tape is exactly the same as the phase of the signal going on the tape.

In summary of the preset and normal modes, it might be said that the capstan and head-drum servos form simple and straightforward systems. The same circuits are used for head-drum control in playback as in record. The capstan comparator in playback is working as a true servo, and is comparing the 240-Hz control track with the 240-Hz tachometer pulses. Any difference between the two becomes a correction voltage for the capstan oscillator to speed up or slow down the capstan motor to keep the tape speed correct.

Operation of Intersync—Vertical, Horizontal, and Automatic

In the three modes of operation, vertical, horizontal, and automatic (Fig. 11-21), more than one step of capstan control is used. These three modes are referred to as the intersync modes. It should be remembered that in the vertical mode the time-base error is no better than it was

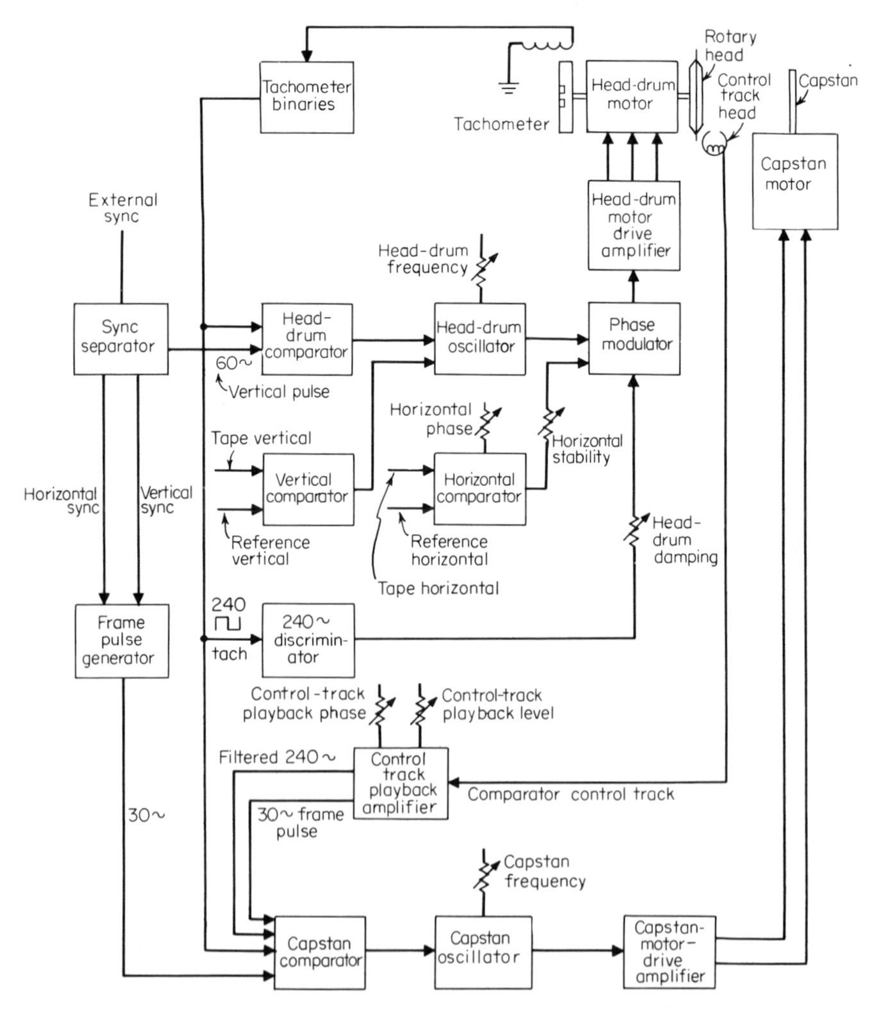

Fig. 11-21 Block diagram of reproduce intersync—vertical, horizontal, and automatic.

in record, because the speed of the head drum is controlled only every four rotations. The major difference in the vertical mode from normal and preset is that the 30-Hz pulse coming off tape (control-track pulse) is used to achieve vertical lock. At face value this can sound confusing, since in fact we are achieving vertical lock of the video output, not by controlling the head drum, but rather by controlling the tape speed (capstan). The means by which this is done is as follows: In the record mode a 30-cycle frame pulse was laid down on every eighth control-track pulse. This meant that when head 4 was recording the end of the first serration of the vertical sync pulse in the middle of the tape, at that same instant of time the frame pulse was being put down on the bottom edge of the tape. This means that if we can reproduce the control track so that the 30-Hz frame pulses are reproduced off tape at the same rate as they were recorded, and if the head can be controlled with the same reference as was used in record, then when the female guide comes in, the head must be on top of the vertical sync. In other words, there are eight cycles of the control track, only one of which is the right one to define where the vertical sync is in the body of the tape. To pick out this particular one, a reference frame pulse (generated in the same circuits as it was in record) is used. Using external sync, this pulse is compared with the 30-Hz pulse off tape. Any difference between the two will create an error signal which will be used to speed up or slow down the tape until there is no difference. This means that when the female guide comes in, if a normal start was being used (where the guide is programmed to be delayed 3 sec before it snaps in), during the 3-sec period the tape speed is made correct, so that when the guide does come in, the head is right on top of the vertical sync. If, for any reason, the 30-Hz signal is missing or of wrong amplitude, the logic of the intersync is such that as soon as the guide comes in, and there is tape sync available, a frame pulse is generated from the video instead of the control track. This generated frame pulse is used in the capstan circuits to achieve vertical lock. Once this is done, the system automatically goes to 240-Hz operation. Using 240 Hz, the sampling is done eight times more often than with 30 Hz, and so the degree of correction is that much better. This switching from 30- to 240-Hz operation is automatic in both the vertical and automatic modes. Obviously, if it did not happen automatically and there were no 30-Hz pulse from the tape and none generated from the video, the machine would carry on indefinitely, waiting for information to control the tape speed.

A second step of head-drum control is used to take care of a loss of vertical lock during playback. Comparing the tape vertical sync with external sync means that, rather than revert back to 30-Hz capstan operation, the head drum will be allowed to rephase itself slowly to achieve vertical lock, without any action on the part of the operator. If, however, there is any interruption of the control track, the logic resets itself so that as soon as the control track appears, the system starts out just as if it were going into an initial playback mode.

The only difference between the vertical and the automatic mode is that a third step is added. In the automatic mode, 30-Hz capstan operation is used to achieve vertical lock. Once vertical lock is reached, the system reverts to 240-Hz capstan control for the balance of the playback. During the period that the capstan is being framed at a 30-Hz rate, the system is using an external vertical sync versus 240-Hz tach, the same type of control used in record. This means that the head should be up to speed within 1 μsec of where it was in record. Once vertical lock is achieved, the system proceeds to step 2 to make sure that capstan did its job. Now the comparison is the tape vertical vs. the reference. The third step is to go to fine control of the head drum. The female guide has been in for at least 1 sec before this will occur. In this step the system takes a look at the tape horizontal pulse vs. the reference horizontal (Fig. 11-21). Now it is correcting 72 times more often than happened in record. In fact, the system is correcting every 5° of head rotation. It should be remembered that one horizontal line is 4.7° of head rotation. Initially, the correction was being done once every four rotations; now it is being done every 5°, or 72 times more often. Consequently, in the horizontal and automatic play modes, playback stability is plus and minus one picture element, or 0.15 μsec.

Color Reproduction For color reproduction, where the time-base stability should be 4 nsec or better, control must be done in three discrete steps, using either horizontal or automatic modes:

1. Tight control of the head drum by means of the intersync servo will get the system within the correction range of the Amtec.
2. Amtec has corrected down to its limits.
3. Colortec takes over and gives the final vernier correction.

Referring to Fig. 11-21, the 30-Hz frame pulse is taken off tape via the composite control-track signal and fed to the capstan comparator via the control-track amplifier. External sync is used to generate a 30-Hz frame pulse in the frame-pulse generator, which is in turn fed to the capstan comparator. Any difference between the reference 30-Hz pulse and the tape pulse will generate a correction voltage for the capstan circuit. Once vertical lock has been achieved, these circuits are disconnected, and 240-Hz circuits are connected for the balance of the playback mode. If, however, no 30-Hz pulse is available from the tape, the moment the female guide comes in, a frame pulse will be generated from the video signal. Once vertical lock has been achieved, the system is again locked into the 240-Hz circuit for the balance of the playback.

In the case of head-drum control, additional or vernier circuits are added; i.e., a vertical comparator, where 60-Hz signals from tape vs. 60-Hz reference are compared to provide a fine control for the head-drum oscillator, and a horizontal comparator, where horizontal pulses from tape are compared with reference horizontal, are used to give ultrafine control.

Both the automatic and horizontal modes use the horizontal comparator for fine control of the head-drum speed. By comparing the horizontal pulse from the tape with the reference horizontal, a correction signal is generated for every 5° of head rotation (at an approximate 15-kHz rate). Obviously, the 240-Hz oscillator cannot be controlled at such a rate, but the phase of the signal can be shifted every 5° through a phase modulator. The two controls connected to the horizontal comparator are adjusted so that there is minimum jitter between tape video and the local sync.

By changing the phase of the 240-Hz signal going to the phase modulator by 1°, a change of system timing of 12 μsec will be accomplished. This ensures that the signal at the head will be 12 μsec early, so that all the cumulative delays of the machine and signal electronics are compensated for. This is particularly important when working with color, where there are two prime timing requirements:

1. Horizontal phase. Adjusted to set the edge of the sync at the output of the tape recorder with respect to the local sync at the master switching point.
2. System phasing. Allows adjustment of the phase of the 3.58-MHz chroma coming off tape with respect to the 3.58 MHz at the switching point. This permits the operator to chroma-key titles off a local signal on top of the tape signal with no change in phase and no change in time.

Helical-scan Video Recorder—Location of Detail

The reader is referred to Chap. 8 for details of the single- and dual-head helical-scan video recorders, their drive and servo systems.

Video Cassette Recorders—Location of Detail

Due to the increased popularity of the cassette recorder, a separate chapter has been devoted to these devices. The reader is referred to Chap. 12 for details of their construction and operation.

REFERENCES

1. American Telephone and Telegraph Company, Long Lines Department: "Television Signal Analysis," 2d ed., rev. by Network Transmission Engineering Advisory Committee, April, 1963, New York, 1963.
2. Anderson, C. E.: "The PAL Color System," Ampex Corporation, Redwood City, Calif., Sept. 1, 1967.
3. Bretz, R.: "Techniques of Television Production," 2d ed., McGraw-Hill Book Company, New York, 1962.
4. Coleman, C. E.: "Off-tape Differential Phase Measurement in a Color VTR," Ampex Corporation, Redwood City, Calif., Jan. 1, 1968.
5. Coleman C. E., and C. E. Anderson: "Magnetic Tape Recording of the Composite PAL and SECAM Color Signals," Ampex Corporation, Redwood City, Calif., Dec. 1, 1967.
6. EIA Standard: RS 170—Electrical Performance Standard—Monochrome Television Studio Facilities, Electronic Industries Association, New York, 1957.
7. Fink, D.: "Television Engineering Handbook," McGraw-Hill Book Company, New York, 1957.
8. Grob, B.: "Basic Television," 3d ed., McGraw-Hill Book Company, New York, 1964.
9. Ginsburg, C. P.: Interchangeability of Video Tape Recorders, *J. SMPTE,* November, 1958.
10. Kietz, E.: Transient-free and Time-stable Signal Reproduction from Rotating Head Recorders, 1963 *Nat. Space Elec. Symp. Paper* 4.3, IEEE Professional Technical Group on Space Electronics and Telemetry, Miami Beach, Fla.
11. Kiver, M. S.: "Color Television Fundamentals," 2d ed., McGraw-Hill Book Company, New York, 1959.
12. National Television System Committee: Color Television Standards, McGraw-Hill Book Company, New York, 1961.
13. Wade, W. L.: "Television Tape Recording Systems: A Guide for School Administrators," Ampex Corporation, Redwood City, Calif., January, 1964.
14. Zettl, H.: "Television Production Handbook," 2d ed., Wadsworth Publishing Company, Inc., Belmont, Calif., 1970.

CHAPTER TWELVE

Cassette and Cartridge Systems

INTRODUCTION TO THE CASSETTE

Although modern reel-to-reel tape recorders have numerous advantages, they also have some unsatisfactory aspects, for example, the size of the reels, the necessity to thread the tape, the need to rewind or run the tape off one reel before removing it from the transport, and the risk of contamination of the tape when handling. These factors led to the development of the cassette and cartridge recorders.

The cassette recorder was introduced by the Philips Corporation of Holland in 1963. Referring to Fig. 12-1, the tape is contained on two reels in a small plastic box. Each end of the tape is attached to a reel by a section of clear-plastic leader. The cassette is ready for immediate use. It can be inserted or removed from the recorder regardless of the amount of tape that has been used. Recording and playback becomes a simple matter of inserting the cassette into the transport and pushing the appropriate mode button or buttons, since the necessary drive and playback devices engage with the insertion of the cassette.

Initially, Philips marketed the cassette recorder in Europe, using monaural format for voice recording. Their portable, battery-powered recorder gained widespread acceptance, first in Europe and then

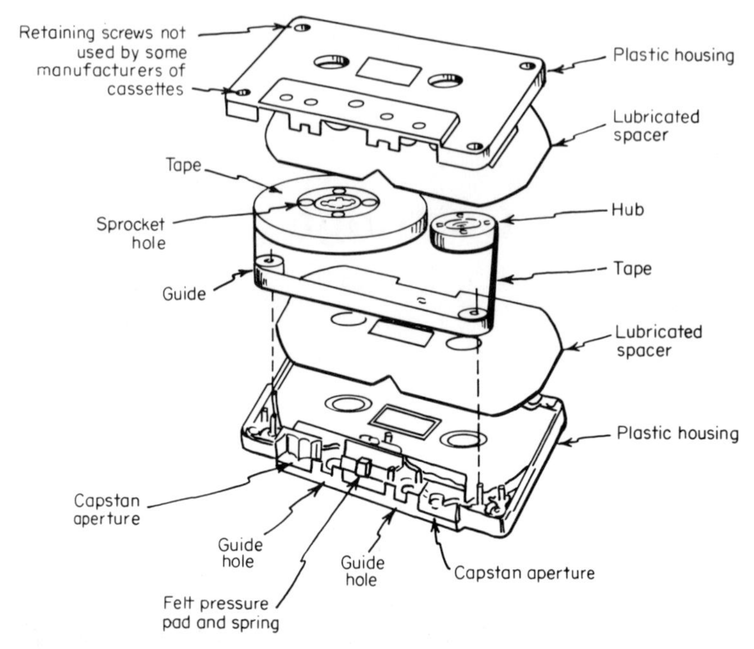

Fig. 12-1 The tape cassette.

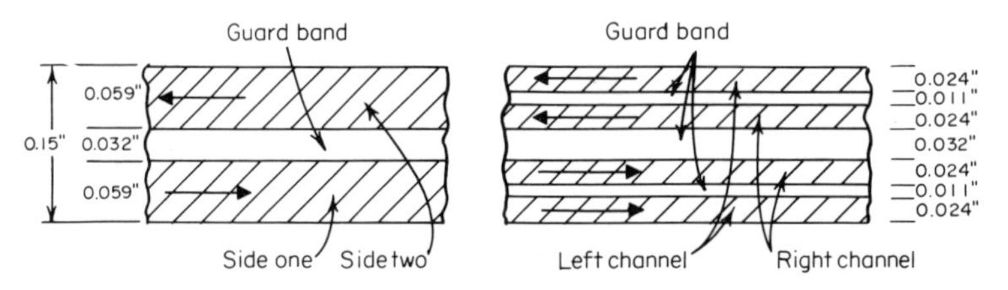

Fig. 12-2 Cassette recording format.

throughout the world, as Philips licensed other manufacturers to produce their device. In doing so, Philips was assured that there would be worldwide acceptance of their format as a standard. The recording formats and the cassette dimensions are shown in Figs. 12-2 and 12-3. The Philips Corporation introduced stereo cassettes in 1966.

INTRODUCTION TO THE CARTRIDGE

Oddly enough, the automobile industry can be given the credit for the popularity of the endless-loop cartridge system. When Learjet introduced the cartridge in 1965, the auto industry saw it as a means to provide stereo entertainment as an optional feature in their new cars.

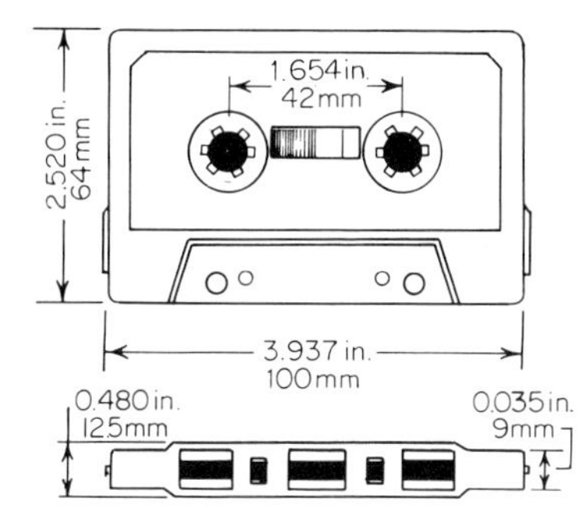

Fig. 12-3 The Philips cassette —dimensions.

The cartridge player (for it is essentially a playback system) was intensively promoted and became extremely popular for car use by 1967. It did not achieve the same popularity for home use because of the design of the cartridge, which made it difficult to make quality recordings. Its design does not allow for fast forward and rewind speeds, making search runs virtually impossible. It is approximately four times the size of the cassette, and therefore does not lend itself to the portability of the cassette recorder. Figure 12-4 shows the design of the cartridge with the side cover removed to illustrate the details of the loop.

USES OF THE CASSETTE RECORDER

The cassette principle is being used in virtually every magnetic recording application today. Cassette recorders are relatively easy to operate even

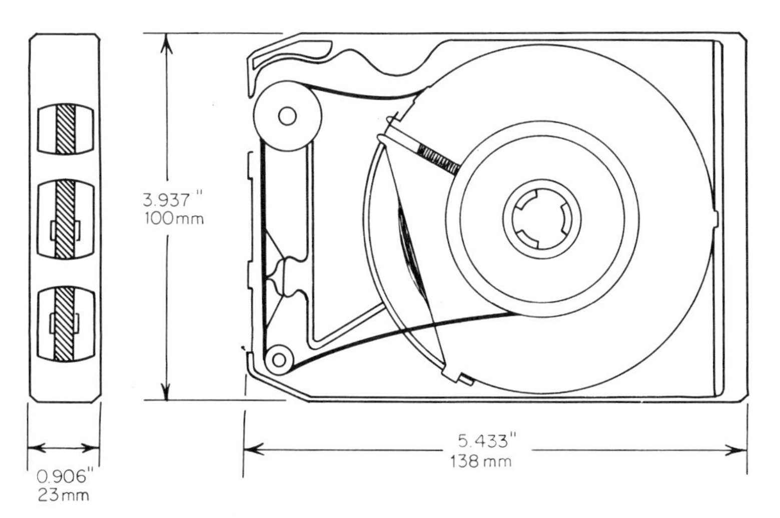

Fig. 12-4 The Learjet cartridge.

by the child or novice, and their manufacturing cost is fairly low. They are extensively used for audio entertainment in monaural, two- and four-track stereo formats. They have made large inroads into the dictating-machine business. They have great impact on the digital computer field as part of the minicomputers. They are becoming an important factor in the educational field where there is increasing emphasis on individualized instruction, particularly with lower-grade students. They are having a tremendous influence on the home video recorder market. They are discussed below in their three major classes: audio, digital, and video.

The Audio Cassette

There are more than 60 companies producing a total of more than 160 cassette recorders for audio use. Many more companies produce cassettes for specialized uses, such as digital or video. The audio cassette recorders are all based on the Philips format, with a tape speed of 1⅞ ips (4.75 cm/sec) and a cassette size, as shown in Fig. 12-3. The tape width is 0.15 in. (381 mm), and the tape length is such that the playing times are 120, 90, 60, 30, and 15 min. For example, the tape that gives 120 min of recording time is known as C120. It will provide two sides of 60 min of recording each. It has 60 min × 60 sec × 1⅞ ips + spare length of tape, or approximately 580 ft of tape in each cassette. The recording format is shown in Fig. 12-2.

For the most part, cassette recorders are battery-powered, using either 4.5, 6, 7.5, or 9 V dc as supplied by four or five C or D cells. A few also use 12 V dc with an optional adapter for use with 117 V ac. Such units are fitted with voltage dividers and power-rectifier circuits. (See Fig. 12-5.)

When the cassette is loaded into the cassette compartment, it automatically positions itself so that it is ready for recording or playback (Fig. 12-6). The tape is automatically placed between the capstan and pinch roller, and the hubs of the cassette reels over the supply and take-up turntable assemblies. When the playback or record mode is selected, the pinch roller (which is mounted on a movable plate linked to the mode buttons) is forced against the capstan, clamping the tape to the capstan. Both the record and playback buttons are normally pushed together to select the record mode. This requirement is a safety device to prevent unintentional recording over prerecorded or just recorded tapes.

When the playback or record mode is initiated, power is fed to the dc drive motor and the capstan is turned through a rubber drive belt-and-pulley system (Fig. 12-7). Since the tape is clamped to the capstan by the pinch roller, it will be moved at the speed of the rotating capstan.

Quite naturally, it is also necessary to take up the tape as it is fed from the capstan. Therefore a drive system, run from the capstan-drive belt, is used to turn the take-up turntable. The take-up drive is derived from a slip roller that is held against the rubber drive belt that goes to the capstan pulley. The other end of the slip pulley is held against the take-up turntable. See Figs. 12-7 and 12-8.

If a fast mode is desired, the pinch roller is held away from the capstan, and the turntable drive pulley away from the take-up turntable. In the place of these two driving devices, the fast drive pulley is pulled against the capstan flywheel and either the take-up turntable (for fast forward) or the reversing pulley (for fast rewind). The greater diameter of the fast drive pulley will initiate a faster rotation of the selected turntable. See Fig. 12-9.

Referring once more to Fig. 12-6, it will be noted that the tape moves from left to right, or from the supply to the take-up turntable. Thus it moves past the erase head before it passes the record/playback head. The heads are usually mounted with the pinch roller on a slide plate that is linked to the record and play buttons. When one of these modes is selected, the heads and the pinch roller are pushed against the tape and capstan. In addition, when the record mode is selected, a large erasing current (usually at the bias frequency) is fed to the erase head, which erases the previous record from the monaural or stereo tracks that are to be recorded on. For a discussion of the action of erasing, recording, or playback, the reader is referred to Chap. 3. The theory of operation and circuit analysis of direct recording and playback is covered in Chap. 9.

Additional Features and Limitations of the Audio Cassette To prevent accidental erasure of prerecorded tapes, a small push-out tab (lock-out tab) is provided for each track of the tape at the rear of the cassette (Fig. 12-10). If a tab is removed, a small space is left in the cassette body. When the cassette is placed in the recorder, a spring-loaded sensing arm fits into the space, and the record mode is locked out. If it is desired to record on the tape again, a small piece of transparent tape placed over the holes returns the cassette to its original state, and the sensing arm is inhibited.

To provide continuous, nonstop listening, automatic tape changers have been developed. These permit the stacking of a number of cassettes for feeding to the recorder/player. A stack of six C120 cassettes will provide 6 hr of continuous, nonstop listening.

Cassette recorders may have one or more of the following features: level-input meters, automatic level control to reduce distortion caused by overloading the microphone, input for radio and output for external speakers or amplifiers, remote or built-in microphones, digital counters

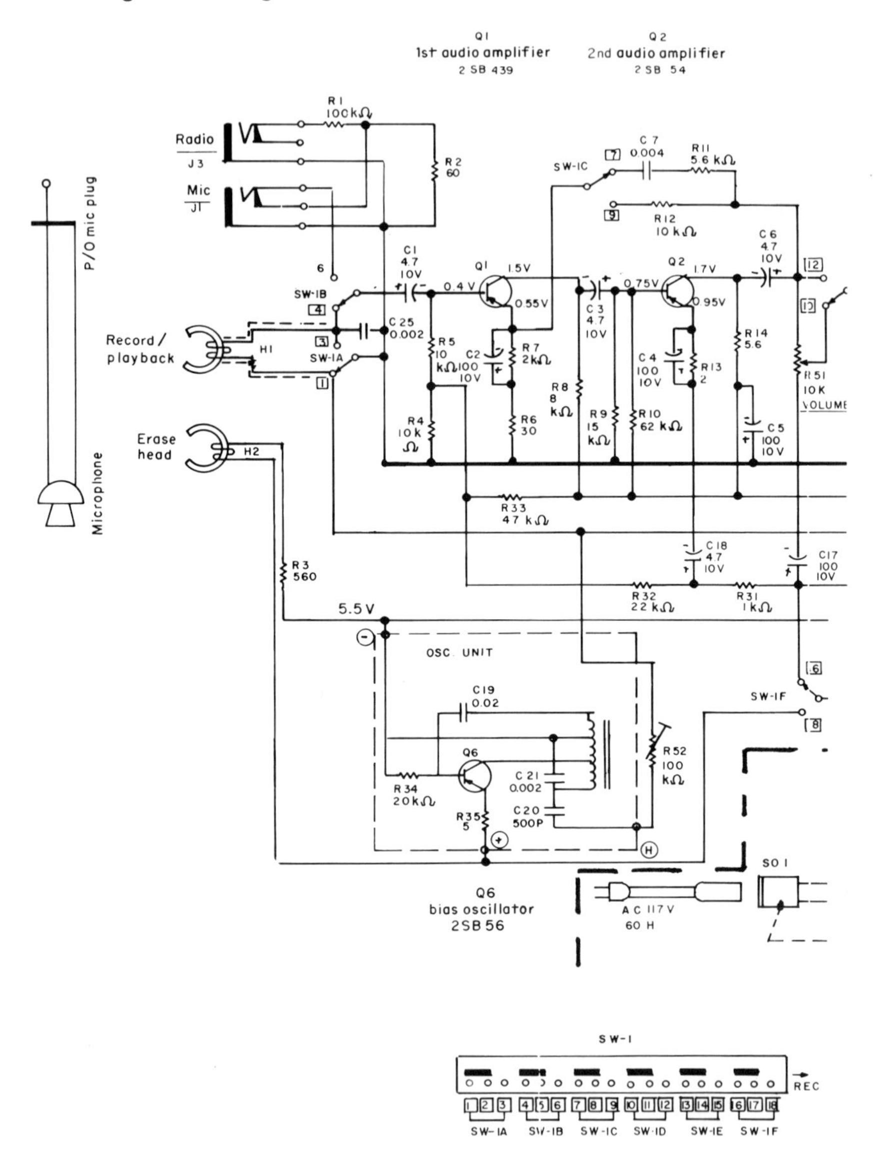

Fig. 12-5 Electrical schematic diagram of cassette recorder/player.

for indexing, and extra speakers for use as a public address system. The recorders may also be found in combination with radios and slide and movie projectors.

Unfortunately, tape editing with cassettes is extremely difficult, due to the thinness of the tape backing material, but by use of quality splic-

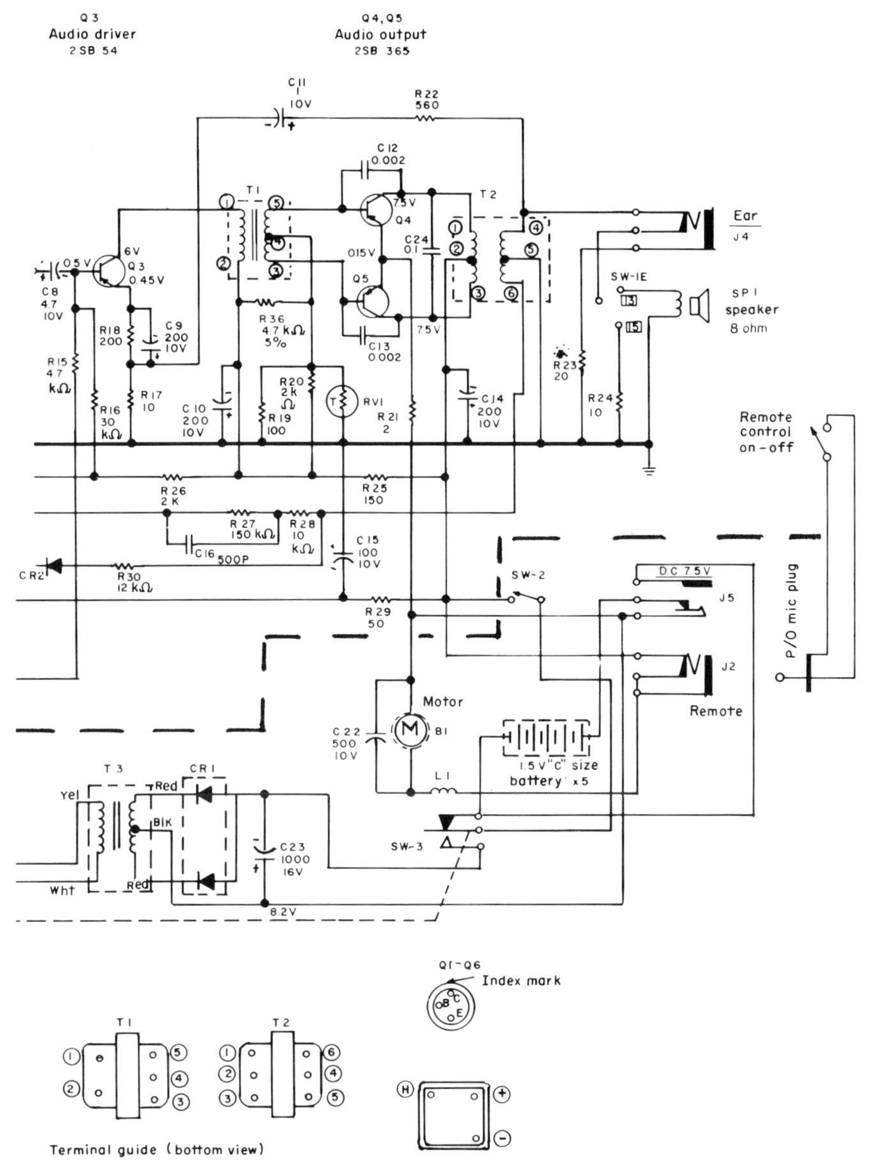

Power circuit is enclosed in heavy dashed lines. (*Ampex Corporation.*)

ing tape and with a great deal of patience, it can be done. Many of the cassettes purchased at discount prices turn out to be a complete disaster. The cassettes are often cheaply made and assembled under unfavorable conditions, using tape not designed for cassettes. They jam easily, and the tape may damage the heads. Cassette users should

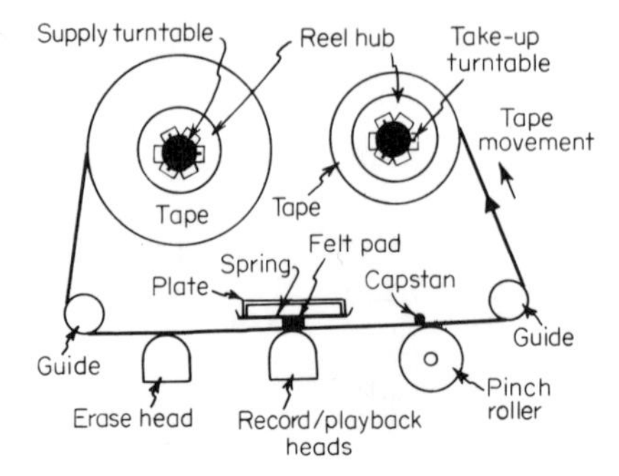

Fig. 12-6 Cassette tape path in recorder/player.

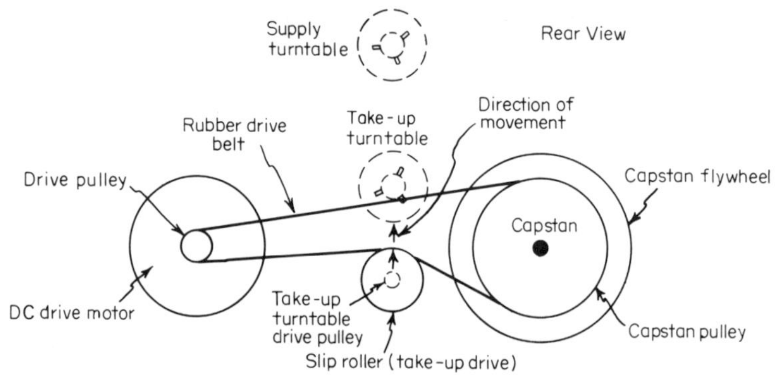

Fig. 12-7 Record and playback operation—capstan drive.

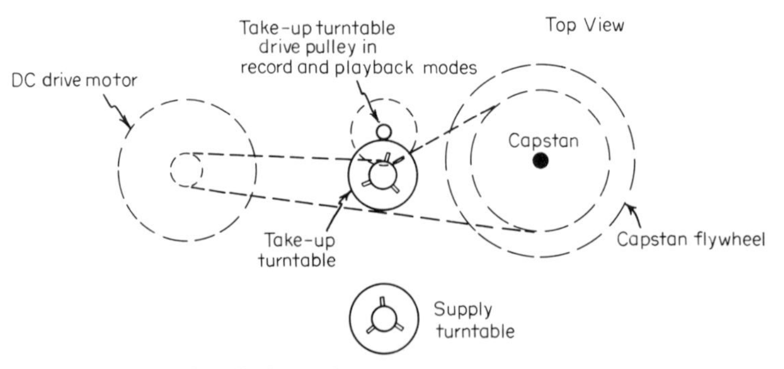

Fig. 12-8 Record and playback operation—turntable drive.

also be aware that most head cleaners are injurious to the plastic parts of the cassette recorder. Alcohol should be used as the cleaning agent.

The Digital Cassette/Cartridge

Digital cassette recorders made their appearance in 1968. They were designed primarily to provide a low-cost solution to the problem of larger-capacity storage requirements for small digital systems (the mini-

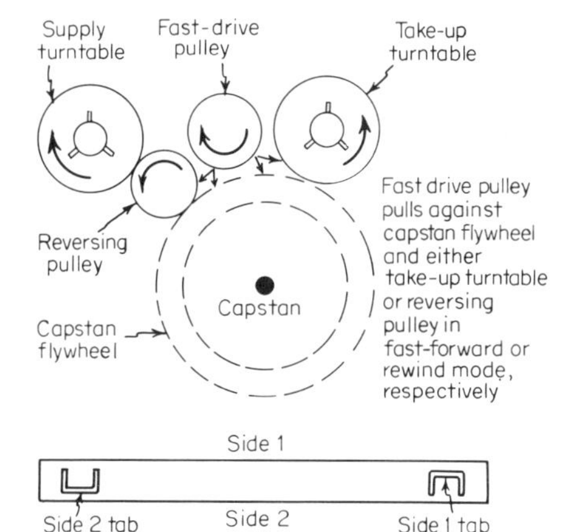

Fig. 12-9 Fast-mode operation —turntable-drive fast modes.

Fig. 12-10 Cassette lockout tabs.

computer). They were modest-performance devices that used Philips computer-tape cassettes (most computer tape is 100 percent tested for dropouts). With the Philips format the tape is 0.15 in. wide and the track width is 0.057 in. Other systems were introduced later that used modified cassettes or cartridges with tape widths of 0.25 and 0.500 in. Of the 23 major manufacturers of digital cassette and cartridge recorders, more than two-thirds use the Philips format.

Like the tape widths, most of the parameters of the cassette/cartridge recorders vary widely, for example, in the number of tracks. One, two, or four tracks are used with 0.15-in. tape; two, four, or eight tracks with 0.25-in. tape; and seven or nine tracks with 0.50-in. tape. This last is used with a cartridge recorder which uses a plastic cartridge that houses a standard 8½-in. reel. It is self-threading. The read/write speeds also vary, from 1.875 to 37.5 ips, speeds of between 2 and 10 ips being the most prevalent. The storage capacities and packing densities also change from system to system, being based on tape length, tape width, tape speed, number of tracks, etc. Storage capacities are quoted anywhere from 1×10^6 to 6.5×10^6 for 300 ft of tape. Bit packing densities are listed as from 250 to 1,600 bpi, 1,000 bpi and below being the most common. Start-stop times also vary over a wide range. They are limited by the maximum safe tape tension and mass of the container. Due to the variety of systems, start-stop times vary from 15 msec start and 10 msec stop to 300 msec start and 150 msec stop. The most commonly quoted time is 20 msec for both.

Cassette Tape Drive Systems

The design of the Philips cassettes (Figs. 12-1 and 12-3) permit the use of a capstan on either side of the heads. Thus, by using dual capstans that rotate in opposite directions and two sets of pinch rollers, the movement of the tape can be made bidirectional. In addition, with the dual capstan, if both capstans are made to rotate in the same direction, then when both pinch rollers are closed, the first capstan provides isolation between the heads and the reel. Using this technique, very accurate tape speed and position can be maintained. The motion of the tape can be continuous or incremental, depending on the type of control circuit used for the capstan drive system. If an optical tachometer is attached to the capstan shaft (similar to those already described in Chap. 8), precise control of the capstan speed can be obtained.

Another system of tape drive is also used. Although it is not as popular as the capstan-drive method, it is worthy of description. It does not use a capstan to move the tape, but rather the tape is moved by controlling the torque of the leading reel motor. During the read mode (playback), a prerecorded clock signal is compared with a reference signal in a phase comparator. Any difference between the two signals in either frequency or phase creates an error signal that is used to control the torque of the take-up reel motor.

A standard capstan-and-reel drive system is illustrated in Fig. 12-11. In the forward-drive mode, heads mounted on a moving arm are closed against the tape, and the capstan pinch roller marked *F* closes on the capstan (also marked *F*). The tape is pulled by capstan *F* from the left and fed to the right. The belt tightener marked *F* tightens against the reel-spindle pulley belt, and the reel spindle is turned (counterclockwise), taking up the tape fed to it by the capstan. The pinch roller and belt tightener marked *R* are deenergized.

In the reverse direction, the opposite pinch roller and belt tightener are energized (those marked *R*), and the ones marked *F* are deenergized. To stop the recorder, small spring-loaded brake pads are applied to both reel spindles, the pinch roller(s) are deenergized, and the heads

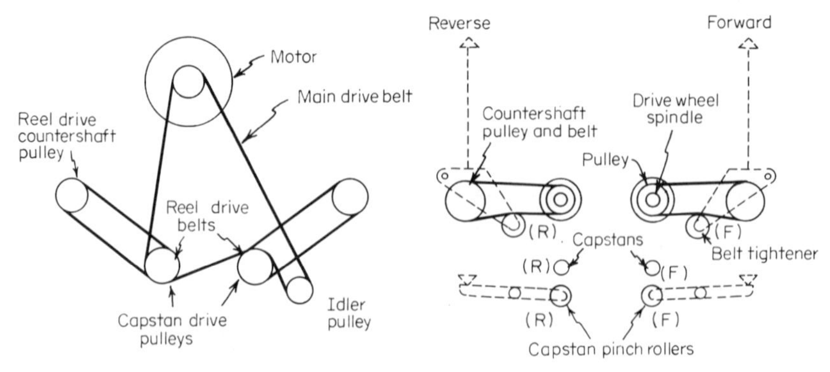

Fig. 12-11 Capstan-and-reel-spindle drive systems.

are retracted. In the fast modes, neither the pinch roller nor the head-closing arm is energized. The tape is moved by the reel-spindle drives only.

A single dc motor is used to provide the drive power for these systems. Heavy-mass flywheels on each capstan shaft reduce the speed variations during operation. Beginning of the tape (BOT) and end of the tape (EOT) are usually sensed via photocells and the clear leader at the beginning and end of each cassette. Many systems use a rigid casting as the tape deck that serves as a chassis for the cassette drive system, the cassette holder, magnetic head assembly, and the electronic control and signal systems. It provides a stable base for the tape-drive components. Most digital cassette recorders can be mounted either horizontally or vertically without affecting their operation.

The control and signal electronics are very complex, and thus are much beyond simplified discussion. The reader is referred to the appropriate manufacturer's handbook for details on these circuits and their interface with the cassette recorder.

The Video Cassette/Cartridge Recorder for Television Broadcasting

The video cassette/cartridge recorders used in the television broadcast field have been engineered to fit the needs of a specific broadcasting concept. This is the programming of short segments of prerecorded 2-in.-wide video tape containing short-event material such as commercials, news, sports features, weather, interviews, promotional material, public service announcements, program open and close sequences, or other short video broadcast material with audio segments. Since such activity must be fitted into existing television broadcasting systems, the tape format used is the same as that set up for quadruplex reel-to-reel video recorders. See Chap. 11 for details of tape format.

There are two major manufacturers of this type of equipment, Ampex and RCA. Both use the same basic concept to achieve the end result, but they have chosen different paths to get there. RCA uses a video tape cartridge as the carrier for each recorded tape segment. They stack up 22 of these cartridges into a carrousel-type device they call a cartridge-changer mechanism. There are two playback stations (special video tape recorders and automatic-threading mechanisms) and appropriate switching and control electronics. When properly programmed, a pair of mechanical arms engage a cartridge on the changer mechanism and place it into a playing position at the selected playback station. A special threading mechanism extracts a loop of tape from the cartridge and places it in the operating path. The threading mechanism then retracts so that it does not interfere with the normal play

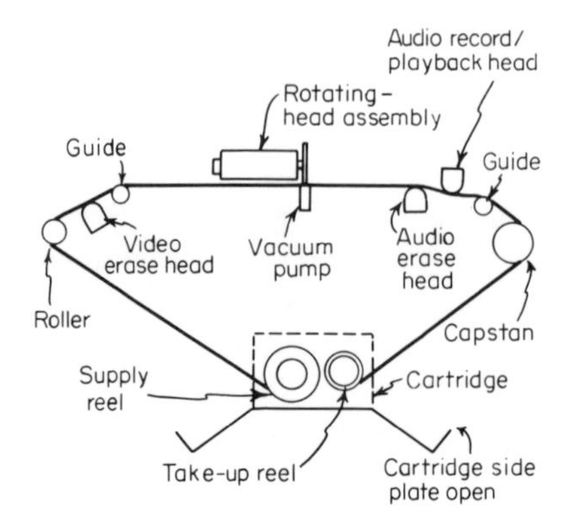

Fig. 12-12 Cartridge automatic-tape-threading path. (*RCA Broadcasting News.*)

modes. The tape is handled from the rear side to prevent damaging the oxide surface. See Fig. 12-12 for the threading path. At the end of play the threading mechanism removes the tape from the tape path, and the tape is drawn into the cartridge. The cartridge is then available for removal, and another one put in its place.

To prevent discontinuity of programming, a second playback station is loaded by the changer mechanism. Just prior to the time that the first tape is ending, an end-of-program cue is used to condition the control circuits, and the second station starts up, so that, as the first program ends, the second starts, and there is no program gap. This is accomplished by the use of automatic cuing. See Fig. 12-13 for the format of recorded cartridge tape.

RCA's video tape cartridge is a molded-plastic container measuring approximately 2½ by 3½ by 5 in. It holds two small spools of 2-in.-wide video tape. The maximum usable tape capacity per cartridge is 236 ft plus leader and tail. The total playing time at standard video tape speed (15 ips) is approximately 3 min. Every cartridge has two permanent mechanical cue marks, one near the beginning and the other near the end of the tape (Fig. 12-13). These are used to prevent the tape from being completely unwound in the recorder. The beginning

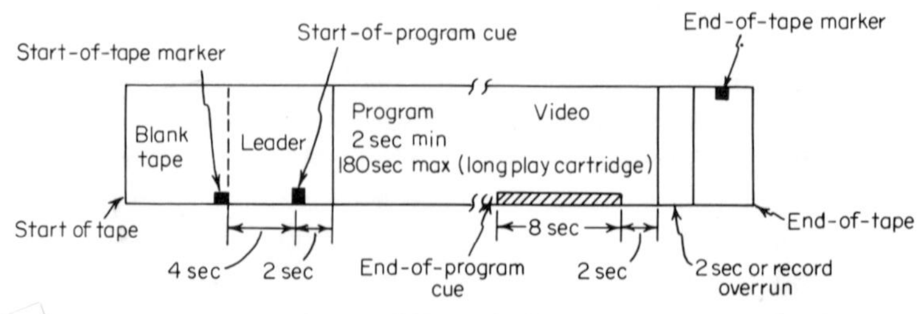

. 12-13 Format of recorded cartridge tape. (*RCA Broadcasting ws.*)

marker (called the start-of-tape marker) is used by the control circuits during the recording to determine where the recording should begin. The start-of-program cue is used in playback to determine the exact point from which the equipment will start when commanded to play back. The playback unit must be given an approximate 2-sec head start so that it can get up to speed and provide a fully synchronized color picture. The tape format also includes an 8-sec tone as an end-of-program cue. The end of this 8-sec tone is used to start off the second-player-station activity so that it will be up to speed by the time the first player completes its tape. This means that one program feeds automatically into the next one without a break. The 2-sec head start is called the preroll time. The changeover is known as the sequential automation.

The sequence of playing the cartridges is determined by the order in which they are placed in the belt changer (Fig. 12-14). Assume cartridge 1 is opposite station *A* to start. On the play-cue command, the cartridge-changer mechanism loads the cartridge into the station *A* playing position. Then the belt moves clockwise until cartridge 2 is opposite station *B*. Cartridge 2 is transferred to the station *B* playing position, and the belt moves again until slot 1 is opposite station *A*, ready to receive cartridge 1 when it has been played. When cartridge 1 is retrieved, the belt moves until cartridge 3 is opposite station *A*. It loads cartridge 3 into the station and moves again until the empty slot 2 is opposite station *B*, ready to receive cartridge 2 when it has been played. This sort of action continues for the length of time for which the system has been programmed.

The Ampex Corporation uses up to 24 video tape cassettes in its ACR 25 automatic video cassette recorder/reproducer (Fig. 12-15). The cassettes are stacked into a carrousel located between two video

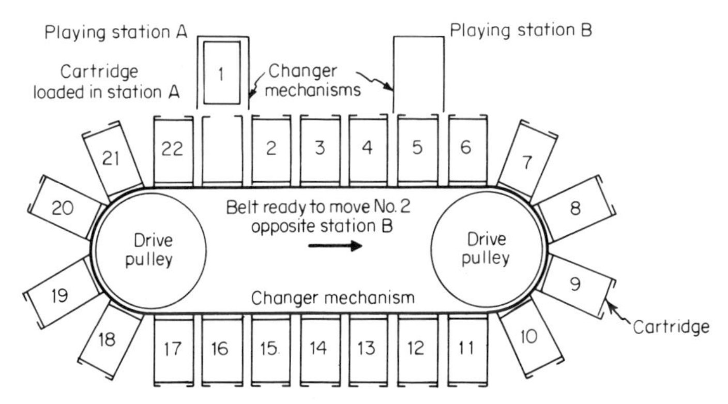

Fig. 12-14 Cartridge-changer system. (*RCA Broadcasting News.*)

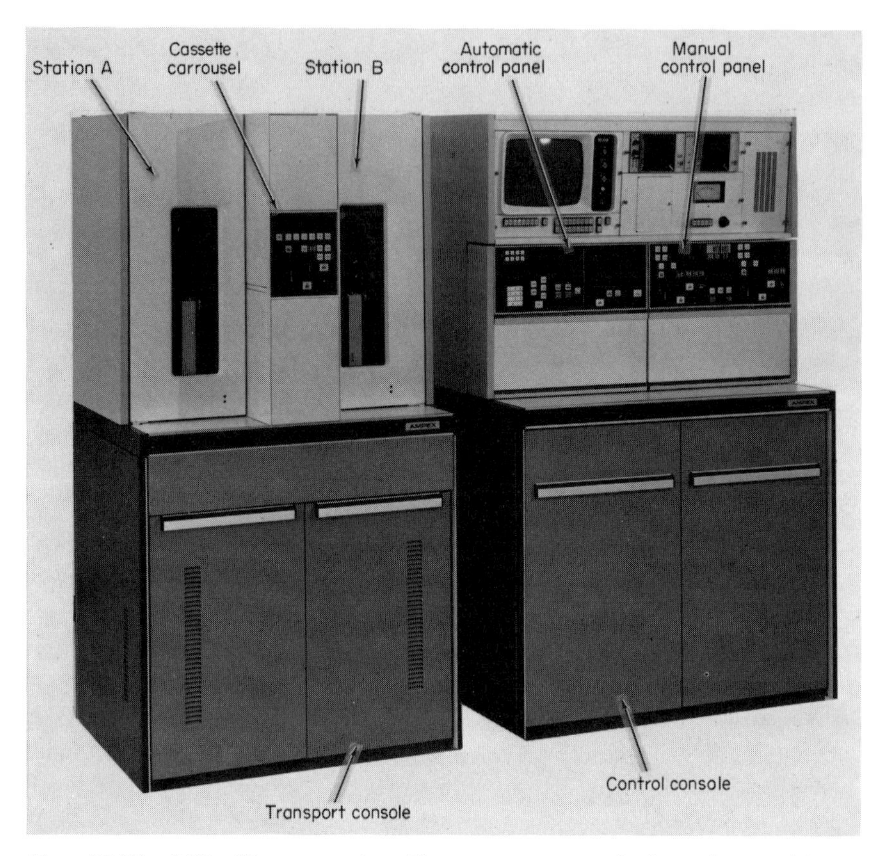

Fig. 12-15 ACR 25, automatic video cassette recorder/reproducer. (*Ampex Corporation.*)

transports. Unlike the RCA TCR 100 cartridge system, these cassettes are available on a random-access basis (selection programmed via the automatic control panel). Any order may be called for; the only requirement is that the correct cassette be in the bin that is being programmed. There are other major differences between the two systems: cassette size and method of tape threading, tensioning, and guiding, among others.

The cassette used is larger than the cartridge, being 2.5 by 3.75 by 6.25 in. as against 2.5 by 3.5 by 5 in. There is, however, 6 min of record/playback time on each cassette, plus leader and trailer. In addition, the tape in the cassettes can be removed for storage or shipping on separate spools. The cassettes can then be loaded with new material and kept active. When the tape is not in use, it is prevented from slipping in the cassette by a spring-loaded brake. During use, the spring is held under tension by a pneumatically operated drive-and-positioning mechanism.

To provide for gentle tape handling, vacuum and air systems are used. Referring to Fig. 12-16, it may be seen that there are six air-lubricated guides, two each at the entrance to the vacuum chambers and one before and after the audio-head assembly. These guides contain openings that are pressurized so that the tape floats across them on an air bearing. Two vacuum chambers are used to provide constant tape tension and isolate the reels from the operating heads and capstan. The vacuum chambers are also used in the threading process.

During the loading sequence, the selected cassette is fitted against the long-vacuum-chamber opening. The capstan, audio-head shield, and female-guide and control-track head assembly are retracted. The brakes are released, and the tape is fed out of the cassettes by velocity-servoed reel motors and tachometer. A large vacuum is applied, and the tape is sucked into the area shown in the dotted lines of Fig. 12-16. When chamber photocells sense that the tape is in the vacuum chambers, vacuum pressure is reduced to a nominal value. The capstan, audio-

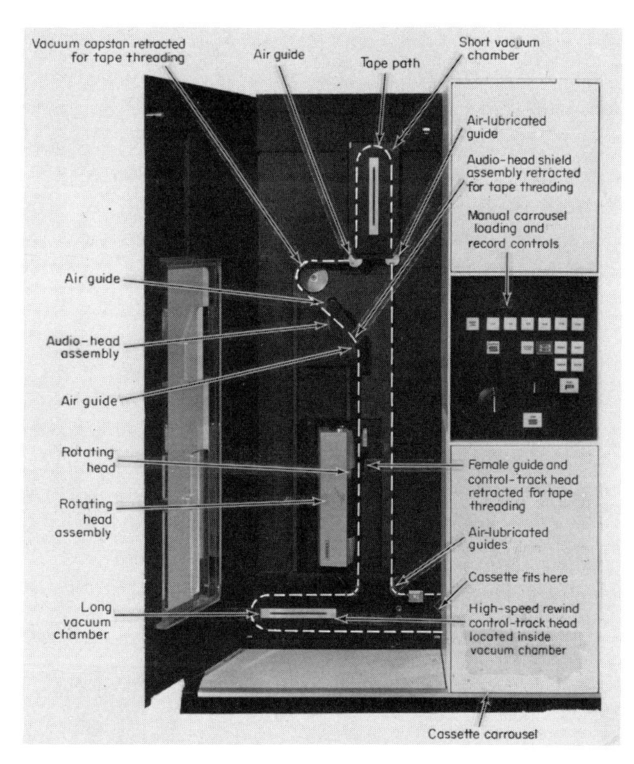

Fig. 12-16 ACR 25 tape-threading path. (*Ampex Corporation.*)

head shield, and female-guide assembly swing up into position, effectively locking the tape into its operating position. A vacuum capstan is used to provide the essential coupling of the tape to the capstan and rule out the need for a pinch roller. Vacuum is also used to hold and form the tape against the female guide (see Chaps. 8 and 11 for details). The tape is now ready for recording or playback. The tape moves clockwise around the transport.

There is a built-in off-line rewind feature in the ACR 25 whereby a cassette can be rewound after the program sequence has been completed. It also uses computer control and has a lockup time of only 200 msec maximum. Thus full back-to-back programming is available. There are no programming restrictions. The ACR 25 is a full VTR with record, edit, etc.

When the cassettes are rewound normally, only the long vacuum chamber is used. The capstan, audio-head shield, and female-guide assembly are retracted. The vacuum in the short vacuum chamber is turned off. The reel motors then take up the slack in the tape, loading the released tape into the cassette. Only the tape in the long vacuum chamber is left out of the cassette at this point. The chamber controls the tension of the tape as it is being drawn into the cassette. In order for the control track to be monitored as the tape is being rewound, a special control-track head is mounted in the long vacuum chamber. When this head senses the disappearance of the control track (tape has been wound past the start of the program), the reel brakes are turned on and the vacuum pressure lowered. The balance of the tape is wound into the cassette during the 0.2-sec deceleration time. The cassette can now be removed.

The Video Cassette/Cartridge Recorder for Home and Office

This group of cassette/cartridge recorders and players may be divided into two classes: the players only, such as Sony's Videocassette,[1] and the recorder/player group, such as the Ampex Instavideo.[2] Normally, they will use the home television set or a separate video monitor for the viewing screen. They are simple to operate, small in size, and relatively inexpensive. They use prerecorded tapes only, or they may record and play back their own tapes as well.

The Videocassette System

The Videocassette system consists of a player, television set, or video monitor and prerecorded cassette tapes. The player is connected di-

[1] Registered trademark of the Sony Corp. of America.

[2] Registered trademark of Ampex Corporation.

rectly to the television set through the antenna terminals by a single-cable hookup. Any standard color or black-and-white television set will do. To operate the system, the Videocassette is inserted into the player, the play button is pushed, and pictures are instantly produced on the television screen (Fig. 12-17).

The Videocassette contains 1,200 ft of $\frac{3}{4}$-in. chromium dioxide tape. The tape backing is polyester, and total tape thickness is 1 mil. The tape is wound on two reels in a flat plane within the Videocassette, which measures $1\frac{1}{4}$ by $5\frac{1}{2}$ by $8\frac{3}{4}$ in. The player uses dual-head, helical-scan principles of operation (these have been discussed in detail in Chap. 8). It produces a video signal of 525 lines, 60 fields. The head-drum speed is 1,800 rpm, video writing speed is approximately 404 ips, and the tape speed is $3\frac{3}{4}$ ips.

Tapes can be backtracked to any point. Threading and unthreading is automatic (Fig. 12-18). When the Videocassette is inserted into the player, the tape is automatically threaded, as indicated in the sequence of pictures. Within 3 sec of pushing the stop button, the tape is returned to the Videocassette. Fast forward and rewind takes place in less than 3 min for a full 60-min color cassette. These functions are performed completely within the cassette without tape being threaded

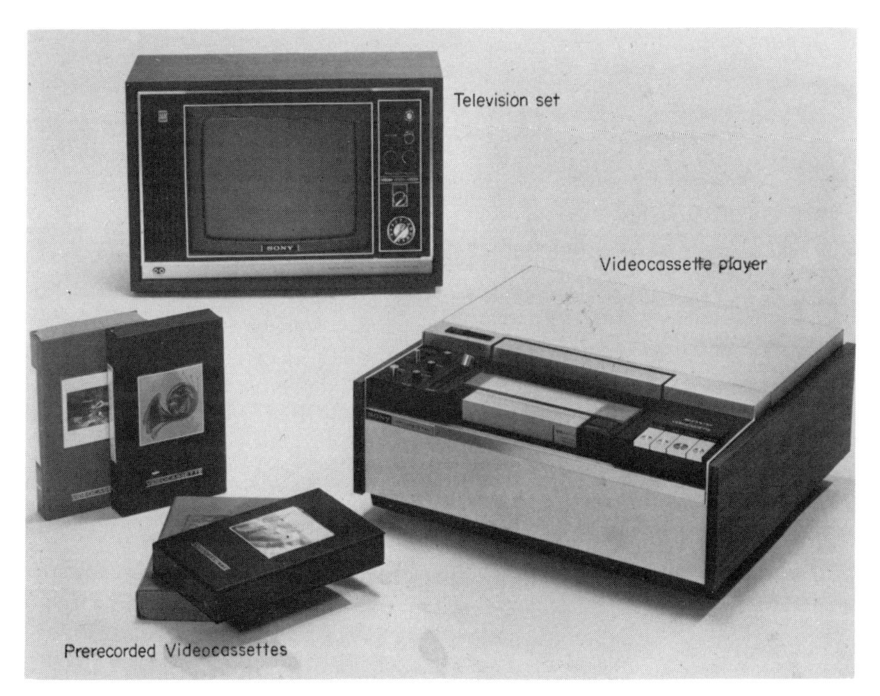

Fig. 12-17 Sony's Videocassette system. (*Sony Corp. of America.*)

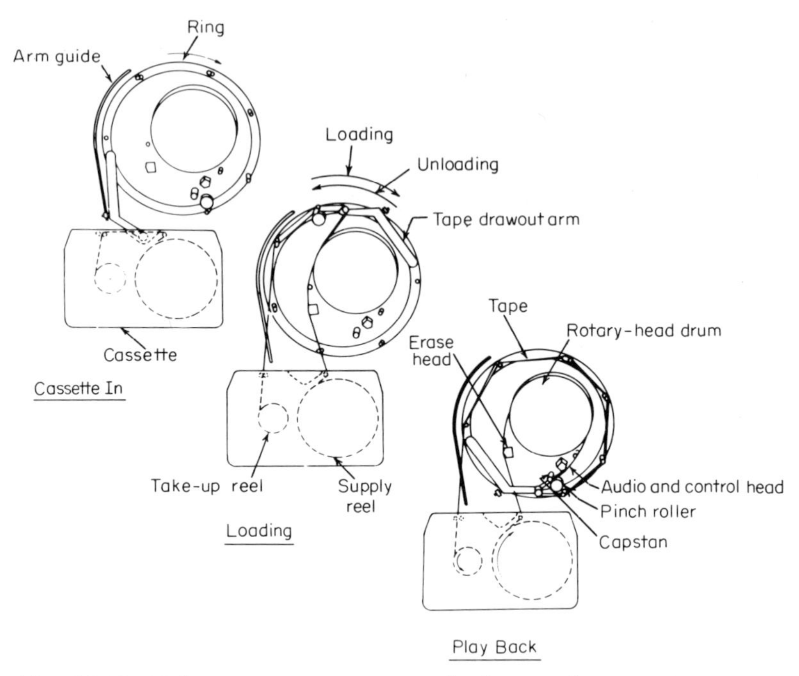

Fig. 12-18 Videocassette automatic tape-loading mechanism. (*Sony Corp. of America.*)

on the drum. Figure 12-19 shows the Videocassette with the cover removed.

The Instavideo System

The recorder/player is 11 by 13 by 4½ in. and weighs less than 16 lb complete with batteries. It uses a 4.6-in.-diameter 0.8-in.-thick cartridge filled with 30 min of ½-in. tape. It uses dual-head, helical-scan recording and playback techniques and conforms to the International video EIAJ Type I VTR Standard (tape speed 7.5 ips and writing speed 437 ips). This means that a tape made on the Instavideo can be played on the Japanese ½-in. helical recorders, and vice versa. It can be used for 525-line 60-field or 625-line 50-field formats. Instavideo records from a camera or off the air through a television set in color or black and white. It plays back instantly from self-recorded or prerecorded tapes. It loads automatically by simply dropping in the cartridge and pushing the play or record button (an automatic threading mechanism takes over). It may be run off standard ac power or be made completely portable by using the built-in rechargeable batteries (Fig. 12-20).

The controls are simple pushbutton type, the same as already used for audio cassettes. It contains an autosearch feature that automatically

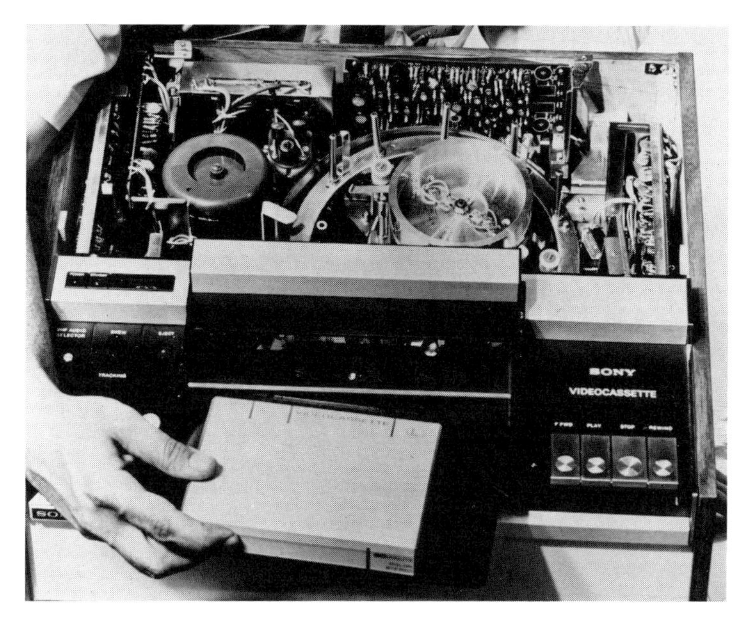

Fig. 12-19 The Videocassette player, top cover removed. (*Sony Corp. of America.*)

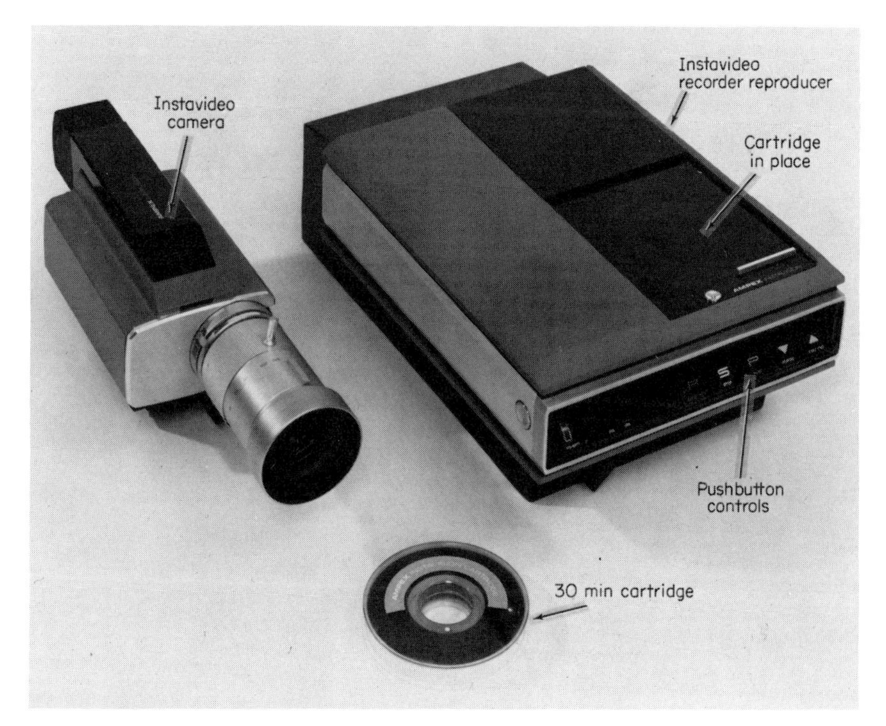

Fig. 12-20 Ampex Instavideo system. (*Ampex Corporation.*)

searches a reinserted cartridge and stops at a point at which it was previously removed. This can be used for indexing by subject matter.

A lightweight camera has been developed for use with Instavideo. It weighs $5\frac{1}{2}$ lb and includes an electronic viewfinder, built-in crystal microphone, and a 6:1 zoom lens. It is hand-held and trigger-operated. When connected to the Instavideo recorder, the camera trigger also operates the recorder. After the recording is complete and the tape is rewound, by going into the playback mode the operator can watch the recorded material on the camera viewfinder (acts as a miniature television screen). With the microphone in the front end, sound can be recorded with the pictures. An external microphone can also be used.

With this sort of simplicity and portability, Instavideo can be used in the home, office, plant, hospital, schoolroom, meeting place, police station and police patrol, sports field, picnic ground, family outing, etc., and by almost anyone. It can be used for entertainment, training, reports and documentation, surveillance, and audio-visual communications. A video tape letter could be sent anywhere in the world.

Magnetic recording has come a long way since Valdemar Poulsen's telegraphone in 1900. Now, with the new family of cassette and cartridge recorders and the day-by-day advances that are being made, it will go much further, much faster. The author hopes that the object of this book, to give the reader some small insight into magnetic recording, has been accomplished.

REFERENCES

1. Educational Product Report: Audio Cassette Tape Recorders, *Educ. Prod. Rep.*, vol. 3, no. 5, February, 1970.
2. Fibush, D. K., and J. P. Sewell: A Cassette Loaded Video-tape Recorder for Automated Station Operation, paper presented at 107th Technical Conference of SMPTE, Apr. 26–May 1, 1970.
3. Hering, J.: Battery-operated and Cassette Tape Recorders, AEG-Telefunken Progress, 1969, no. 4, p. 163.
4. Murphy, W. J.: Cassette-cartridge Transports, *Modern Data,* August, 1970, pp. 64–74.
5. *RCA Broadcasting News:* Fuel for the Imagination-Video Cartridge Tape, December, 1969, p. 30.
6. Wessler, J.: The Case for Cassettes, Standards, and Non-standards, *Datamation.* Sept. 15, 1970, p. 51.

APPENDIX A

Glossary of Audio and Instrumentation Terms

ABRASIVENESS: A relative measure of the roughness of tape and its effect on the wearing of a magnetic head.

AC BIAS: A high-frequency signal, three to five times the highest data frequency, that is linearly mixed with the data signal and fed to the record heads to compensate for the hysteresis effect of the tape. The use of ac bias improves the linearity and sensitivity of the system, as well as providing maximum undistorted output levels. The amount of bias current used is one that represents the best compromise of low distortion, extended high-frequency response, and high output.

AC ERASURE: When a magnetic material is placed in a gradually diminishing cyclic field, a family of hysteresis loops will be formed that get successively smaller until demagnetization is reached (erasure).

ACICULAR: Needle-shaped. Standard magnetic tape is made up of acicular particles of gamma ferric oxide ($Fe_2O_3\gamma$), approximately 20 μin. long and 3 μin. wide.

ADDITIVE: Tape-coating materials other than the magnetic oxide ($Fe_2O_3\gamma$) and the binder (vinyl resins) that lower the coefficient of friction of the binder (lubricants), soften the binder (plasticizers), and retard the growth of fungus (fungicides) and materials that make the coating conductive (antistatic element—carbon black).

AMPLITUDE OF SIGNAL: Excursion strength, displacement from a reference point. In general, expresses the relative strength of a signal.

ANALOG RECORDERS: Devices designed to record continuously variable functions in which the recorded information has a direct relationship to the input information.

ANALOG RECORDING: A method of recording in which the recorded information has a direct relationship to the input information. The direct recording process generally refers to analog recording in that the current presented to the record head is analogous to the frequency and amplitude of the incoming ac signal, the exception being audio recording, where preemphasis is deliberately added to increase the low- and high-frequency responses to conform to the nonlinear characteristics of the human ear.

ANALOG VOLTAGE: A voltage that varies in a continuous fashion in accordance with the magnitude of a measured variable.

ANCHORAGE: The adhesive quality of the magnetic oxide coating to the tape base film. This is as opposed to cohesion qualities of the coating itself, which is the determining factor on relative wear. It may be measured by checking whether the coating can be lifted from the backing with pressure-sensitive adhesive tape, or by the amount of force required to separate the coating from the base film using a specially designed knife.

ANHYSTERESIS: A condition of magnetization which occurs without hysteresis effects. A magnetization process resulting when an ac field, which is varying symmetrically in alternate directions and is gradually decreasing in amplitude, is superimposed on the magnetizing field. It bears a close relationship to direct recording using ac bias, the difference being that the magnetizing field remains constant during the decay of the ac field in the anhysteresis case, whereas both the magnetizing and ac bias decay simultaneously in the direct recording process.

ANTIBLOOM AGENT: An agent that prevents the magnetic coating from shedding excess powdery residue.

ANTIFUNGUS AGENT: An agent that retards the growth of fungus.

ANTISTATIC ELEMENT: Carbon black, used to make the coating conductive.

AZIMUTH ALIGNMENT: Alignment of the recording and reproducing head gaps so that their center lines are exactly parallel with each other. Misalignment causes short-wavelength (high-frequency) losses.

B: Magnetic intensity, measured in gauss.

BACKING THICKNESS: Thickness of the tape backing material: 1.5 or 1 mil is usual for audio, instrumentation, digital, and video: audio cassettes use down to 0.25 mil.

BANDPASS: The complete range of frequencies that a recorder is capable of recording and reproducing. Since it does not contain tolerance factors, this is not a useful specification.

BANDWIDTH: The range of frequencies that a recorder can record and reproduce within a specified amplitude variation in frequency response (± 3 dB) and a meaningful signal-to-noise ratio commensurate with these frequencies.

BASE-FILM BACKING: The plastic backing for the coating. The base film for most instrumentation and digital tapes is polyester. Less expensive base films, often used for audio tapes, are cellulose acetate and polyvinyl chloride.

BIAS: *See AC bias.*

BIAS CROSSTALK: Bias crosstalk is measured bias or erase-frequency-signal level present at the output of an analog magnetic tape reproduce amplifier. It is normally encountered while erasing direct recording on one tape track and reproducing a signal from another tape track, or when monitoring a record process. Under these conditions the bias signal is introduced in the reproduce-head signal circuit, and passed through the reproduce amplifier.

BIAS-INDUCED NOISE: *See Bias crosstalk.*

BIAS OSCILLATOR: An oscillator which is used to furnish the ac bias.

BINDER: A composition of organic resins used to hold the magnetic oxide particles to each other and to the base film.

BLOCKING—LAYER-TO-LAYER ADHESION: The tendency for adjacent layers of tape on a reel to stick together, usually due to long-term storage under high-humidity and -temperature conditions.

B_{max}: The maximum magnetization a given material is capable of supporting. Occurs at saturation.

B_r—RETENTIVITY: Remanence magnetization. The amount of magnetization, or flux density, remaining in a magnetic material after the magnetizing force has been returned to zero. A measure of flux concentration expressed in maxwells per square centimeter or gauss.

BUCKLING: Deformation of the circular form of a roll of magnetic tape, caused, generally, by a combination of adverse storage conditions or improper winding tension.

BULK DEGAUSSER, OR BULK ERASER: A device for erasing a complete roll of magnetic tape at one time. The roll is usually rotated in a 60-Hz ac field that is decreased either by withdrawing the roll from the electromagnet, withdrawing the electromagnet from the roll, or reducing the ac supply to the electromagnet.

BULK-ERASED NOISE: The noise resulting when reproducing bulk-erased tape with the erase and record heads completely deenergized. This noise is controlled by the number of magnetic particles that pass the reproduce head in unit time.

BUS: One or more conductors which are used as a path for transmitting information, voltage, etc., from any of several sources to any of several destinations.

CAPSTAN: An electromechanical system that pulls tape past the heads at a precise rate of speed.

CARRIER FREQUENCY: The frequency of the system at rest or with no input. This frequency is generally deviated in amplitude, frequency, or by interruption to impose recoverable data into the system. The carrier frequency is usually much higher than the modulating frequencies.

CARRIER-FREQUENCY DEVIATION (FM): The swing or change in frequency of the center-carrier frequency f_0 caused by a signal being recorded by use of the FM technique.

CERTIFIED TAPE: Computer tape that has been checked on all tracks throughout the roll and is certified by the manufacturer to have less than a certain total of errors, or more normally, zero errors.

CHANNELS: Separate data-origination sources. This is not necessarily a separate tape track, since numerous channels may be recorded on one track by various multiplexing techniques (frequency- or time-sharing basis).

CINCHING: Longitudinal slippage between the layers of tape in a tape pack when the roll is accelerated or decelerated.

COATING: The magnetic layer that is applied to the base film. It consists of the gamma ferric oxide particles held in a binder.

COATING RESISTANCE: The electrical resistance of the magnetic coating measured between two parallel electrodes spaced a known distance apart along the length of the tape.

COATING THICKNESS: The thickness of the magnetic coating applied to the base film. It ranges from 170 to 650 μin., two basic thickness standards being (1) thin coat, 170–200 μin., and (2) thick coat, 400 to 450 μin. In general, thin coatings have excellent resolution at the expense of reduced output at long wavelengths. Thick coatings give high output at long wavelengths at the expense of resolution.

COEFFICIENT OF HYGROSCOPIC (HUMIDITY) EXPANSION: The relative increase in the linear dimension of a magnetic tape or base material per percent increase in relative humidity, measured in a given humidity range, i.e., polyester film, 11 parts in 10^{-6} in./in. for 1 percent change of relative humidity within the range of 20 to 92 percent RH.

COEFFICIENT OF THERMAL (TEMPERATURE) EXPANSION: The relative increase in the dimension of a magnetic tape or base film per degree rise in temperature (Fahrenheit), measured in a given temperature range, i.e., polyester film, 15 parts in 10^{-6} in./in. for 1° change in temperature within a range of 70 to 120°F.

COERCIVITY (H_{ci}): The magnitude of magnetizing force to reduce the remanence to zero. *See also Intrinsic coercivity.*

CONDUCTIVE COATINGS: Coatings that are treated to reduce the coating resistance, and thus prevent the accumulation of static electrical charge. Carbon black is often used for this purpose. *See also Antistatic element.*

CROSSTALK: Interchannel signal interference in a multitrack magnetic tape recorder is a wavelength effect that is closely related to the spacing of the recorded tracks on tape. Two primary sources of crosstalk are (1) inductive coupling of the leakage flux between the gaps of adjacent heads, and (2) coupling of leakage flux from the signals recorded on one track into the head of the other track.

CUPPING: Curvature of a magnetic tape in a lateral direction. It may occur because of improper curing or drying of the coating or because of differences in thermal or hygroscopic coefficients of expansion of the coating and base film.

DATA: Signals in electrical form that can be recorded or reproduced and are completely flexible as to further use.

DC DRIFT—LONG TERM: Long-term drift vs. time is expressed as a percentage drift, or millivolts change, over a stated period of time (ambient temperature and line voltage constant). Long-term drift vs. ambient temperature is expressed as a coefficient of drift in percent or millivolts change per degree centigrade or Fahrenheit change of ambient temperature (constant line voltage). Long-term drift vs. line voltage is expressed as a coefficient of drift in percent or millivolts per volt or percent change in line voltage (constant ambient temperature). *See DC drift-short term.*

DC DRIFT—RECORD AMPLIFIER: Deviation of the intelligence-bearing component or record-head current from a steady value when the input to the record amplifier is held constant.

DC DRIFT—REPRODUCE AMPLIFIER: Deviation of the output of the reproduce amplifier from a steady-state value when the input to the amplifier from the reproduce head is held constant.

DC DRIFT—SHORT TERM: Falls into two classes: jitter and transient response. Jitter is instability over a period of a few seconds or minutes expressed in the same units as long-term drift. Transient response is defined by describing unusual behavior immediately following a transient or step change in ambient temperature or line voltage.

DC ERASURE: *See Erasure.*

DC NOISE: The noise that results when reproducing a tape which has been nonuniformly magnetized by a record head that was energized with direct current either in the presence or absence of bias. This type of noise can have very high long-wavelength components. At high values of direct current, the dc noise approaches saturation noise, the additional noise connected to the fluctuations in head-to-tape contact.

DECIBEL (dB): Logarithmic expression of a power, voltage, or current ratio.

$$dB = 10 \log \frac{P1}{P2} = 20 \log \frac{E1}{E2} = 20 \log \frac{I1}{I2}$$

Voltage or Current Ratio

Ratio..........	1	1.25	1.41	2	4	10	100	1,000
dB............	0	2	3	6	12	20	40	60

DEFECT: An imperfection in the tape leading to a dropout.

DEGAUSS: The act of reducing all residual magnetization of a given object to zero.

DEVIATION (FM): *See Carrier-frequency deviation FM.*

DEVIATION RATIO (FM): The modulation index for a maximum value of modulating frequency.

DIRECT RECORDING: *See Analog recording.*

DISPERSION: Distribution of oxide particles within the binder.

DISTORTION: Amount of harmonic content permitted in a recorded standard reference frequency (usually measured in percentage).

DISTORTION—HARMONIC: The production of frequencies which have a harmonic relation to the frequency transmitted through the system. It may be measured by attenuating the fundamental (without attenuating harmonics) to some value below the system noise and measuring the remaining signal. The distortion is then expressed as a percentage of the total harmonics to the output amplitude of the frequency being transmitted through the unit. It includes all harmonic-distortion products at the output.

DISTORTION—HETERODYNE: The production of frequencies corresponding to the sums and differences of the fundamentals and harmonics of two or more frequencies which are transmitted through the element or system.

DISTORTION—INTERMODULATION: The production of frequencies corresponding to the sums and differences of the fundamentals and harmonics of two or more frequencies which are transmitted through the element or system.

DOMAIN VOLUME: The smallest useful magnetic volume. By the present state-of-the-art, and using $Fe_2O_3\gamma$ as the magnetic material, it contains 10^{18} molecules and approaches physical size of 20 μin. long by 3 μin. wide. Each domain is spontaneously magnetized to saturation in some direction.

DRIFT: Tape-velocity deviations from nominal velocity, occurring at frequencies dc to 0.7-Hz instrumentation and dc to 0.1-Hz audio. *See also DC drift.*

DROPOUT: Any tape-caused phenomenon that results in temporary or permanent loss of signal for a specified length of time. It is usually expressed in terms of percentage of reduction below average output for a specified length of time. The most prominent cause of dropouts is surface contamination, where a piece of oxide shed or foreign particle adheres to the surface of the tape and lifts the tape from the head. Computer dropout is 50 percent below preset level. Audio, instrumentation, and video take into account both the amplitude and duration of the level reduction. Video dropout is loss of signal (down 12 dB or better) for 10-μsec or longer duration. *See also Dropout—permanent* and *Dropout—temporary*.

DROPOUT—PERMANENT: This type of error results from the physical deformation of the tape coating, or an imperfection of the backing material. No amount of machine cycling will remove or alter this dropout. This is the nodule, or impurity, that is actually coated into the tape during manufacture. Initial testing and all subsequent tests will indicate this as being a dropout.

DROPOUT—TEMPORARY: This is a "here today and gone tomorrow" dropout. It usually is a bit of dust, lint, or foreign matter lightly adhering to the tape coating. Normal machine recycling will completely remove this error.

DROPOUT COUNT: The number of dropouts detected in a given length of tape (normally a full roll for digital recording).

DURABILITY INDEX: Usually expressed as the number of passes that can be made with a tape before significant degradation of output occurs divided by the corresponding number that can be made using a reference tape.

DYNAMIC INTERCHANNEL TIME DISPLACEMENT ERRORS (ITDE): Differential timing errors between two channels in the direction of tape travel. These errors result from the nonperpendicular motion (skew) of magnetic tape as it passes the heads.

DYNAMIC RANGE: The ratio of the maximum signal which can be recorded (at a given level of distortion) to the minimum signal which can be recorded (determined by the inherent noise level of the system) over a narrow frequency range. Over a broader-band spectrum, this is better known as signal-to-noise ratio.

DYNAMIC RANGE—FM RECORD: The maximum output of the FM system is determined by the maximum carrier deviation for which it was designed. Values in excess of this will result in severe distortion. The dynamic range can be found by subtracting the desired minimum signal-to-noise ratio from the signal-to-noise ratio at maximum deviation.

DYNAMIC TAPE SKEW: *See Tape skew.*

ERASE FIELD STRENGTH: The minimum initial amplitude of a decreasing alternating field required to reduce the output of a given recorded signal by a specified amount, i.e., a saturated signal to −60 dB or better. However, to ensure complete erasure under all conditions, the minimum field advisable is increased by a factor of 4.

ERASURE: Clearing magnetic tape of all previous signals and data preparatory to recording new information. Erasure may be accomplished by using (1) ac—the tape is demagnetized by an alternating field which is reduced in amplitude from a high value (reducing both B and H to zero in gradually decreasing amounts), and (2) dc—the tape is saturated by applying a unidirectional field, i.e., passed over a head fed with direct current or over a permanent magnet.

ERASURE FIELD: A recommended field which will reduce a saturated-tape-output signal to better than −60 dB.

ERROR: In digital recording, a dropout or noise pulse (drop-in) that exceeds a particular limit. Normally, it is defined as a 30 percent drop in saturated-tape output or a spurious noise pulse greater than 10 percent of normal saturated-tape-output level.

ERROR AT MAXIMUM PULSE WIDTH: This is the width that cannot be increased without instability being present, and which does not respond to input variations when recorded and reproduced. Error at maximum pulse width is that error present at the threshold point of this condition.

ERROR AT MINIMUM PULSE WIDTH: Minimum pulse width is that duration which, when recorded and reproduced through a system, cannot be decreased further in duration, irrespective of further input reduction. Error at minimum pulse width is that error present at the threshold point where no change in output pulse width exists with reduction of input pulse, and where with increasing input pulse width, a reduction in error is evident.

FAILURE: A detected cessation of ability to perform a specified function or functions within previously established limits. It is a malfunction which is beyond adjustment by the operator by means of controls normally accessible to him. This requires that measurable limits be established to define satisfactory performance of the function.

FAST WIND TIME: The time required to transport a given length of tape in the fast mode from one reel or position to another. Must be specified as the time required to wind a specific length of tape.

FLUTTER: Instantaneous tape-speed error having frequencies of variation between 200 Hz and 10 kHz for instrumentation and 10 Hz and above for audio systems. It is generally the result of friction between the tape and heads or guides.

FREQUENCY MODULATION (FM): A modulation technique where a center-carrier frequency is shifted or deviated by the signal to be recorded in direct proportion to the signal amplitude. The rate of change of deviation is in direct proportion to the frequency of the signal to be recorded (f_m).

FREQUENCY RESPONSE: The variation of sensitivity with signal frequency. *See also Frequency response—direct record* and *Frequency response—FM record.*

FREQUENCY RESPONSE—DIRECT RECORD: This is the amplitude transmission characteristic of a system as a function of frequency. The amplitude response is stated as the actual gain or loss ratio of the particular frequency with respect to the amplitude obtained at a specified reference. The information is presented as the log of the ratio vs. frequency, with the reference taken in the lower midportion of the passband.

FREQUENCY RESPONSE—FM RECORD: Data frequency bandwidth of the system is arbitrarily limited by the output filter which is selected for optimum data bandwidth for a given modulation index at each carrier frequency.

FUNGICIDE: *See Additive.*

GAMMA FERRIC OXIDE ($Fe_2O_3\gamma$): The magnetic oxide used to manufacture a large percentage of present tapes. The gamma distinguishes the ferromagnetic form (acicular particles) from the nonferromagnetic alpha ferric oxide (rhombohedral crystalline structure).

GAP DEPTH: The dimension of the head gap measured in the direction perpendicular to the surface of the head.

GAP LENGTH: The dimension of the gap of a head measured from the edge of one pole piece to the edge of the other. It is the dimension of the gap in the direction of tape travel.

GAP LOSS: The loss in output of a reproduce head attributable to its fixed gap length. The loss increases as wavelength decreases, so that when the wavelength is equal to approximately 87 percent of the head-gap length, the output of the head is zero.

GAP SMEAR: The effect from moderately abrasive magnetic tape which will wear the leading edge of the head gap and set up a cold-flow condition, thereby forcing head material into the gap, creating an effective magnetic short.

GAP WIDTH: The dimension of the head gap measured in a direction parallel to the head surface and right angles to head-tape movement. The record-head gap width establishes the track width.

GAUSS: A unit of flux density. One gauss equals one line of magnetic force (one maxwell) per square centimeter.

H: Magnetization force measured in oersteds.

HARMONIC DISTORTION: Distortion of a signal characterized by the appearance in the output of harmonics other than the fundamental when the input signal is a sine wave. It is expressed as a percentage of the sum of the rms amplitudes of all the harmonics measured against the rms amplitude of the fundamental. Third-order harmonic distortion is used as a measure of distortion in symmetrical systems such as ac-bias recording.

HEAD DEGAUSSER: A device with specially contoured tips used to demagnetize (degauss) a magnetic head.

HEAD-TO-TAPE CONTACT: The amount that the surface of the record or reproduce heads contact the surface of the magnetic tape. Lack of contact will cause losses that follow the formula (separation loss) $= 55d/\lambda$ dB, where $d =$ distance of separation in inches and $\lambda =$ wavelength in inches. Thus it follows that good head-to-tape contact is essential for good high-frequency resolution in a reproduce head.

HYSTERESIS: From the Greek *hysterein,* meaning to fall behind, or lag. The inherent characteristic of a ferromagnetic material for the magnetic intensity (B, in gauss) to lag the magnetizing force (H, in oersteds).

IMPEDANCE—INPUT: Input impedance is the impedance presented by a system to a source. The input-impedance description should include both the resistive and the reactive components. The type of ground reference should be included, i.e., one side grounded, center grounded, balance to ground, dc volts off ground.

IMPEDANCE—OUTPUT: The impedance presented by a transducer to a load. The output impedance includes both the resistive and reactive components.

INHERENT RELIABILITY: The probability of performing without failure of a specified function under a specified test condition for a required period of time.

INPUT—RECORD: Input is an all-inclusive term covering the input requirements of a system. It includes input impedance, voltage range, current range, power range, level range, frequency range, and normal operating level, among others.

INTERCHANNEL DISPLACEMENT ERROR: Defined as the equivalent physical distance, measured in the direction of tape motion, between reproduced pulses that were originally recorded simultaneously on different channels. The displacement is measured in mils, and is related to time and tape speed; $D = st(10^{3})$, where D = displacement in mils, s = tape speed in inches per second (ips), t = time measured between pulses in microseconds.

INTERMODULATION DISTORTION: *See Distortion—intermodulation.*

INTERSTACK SPACING: The lateral distance between the trailing edges of the record-head gaps of the two head stacks or the gap center line to gap center line of the reproduce-head stacks. IRIG Standards set this measurement at 1.5 ± 0.001 in.

INTRINSIC COERCIVITY (H_{ci}): A measurement of the force required in oersteds to reduce the intrinsic induction of the magnetized material to zero, zero being the intrinsic induction of a vacuum.

INTRINSIC FLUX (ϕ_i): The product of intrinsic flux density and the cross-sectional area in a uniformly magnetized sample of magnetic material.

INTRINSIC FLUX DENSITY (B_i): The intrinsic flux density is equal to the excess of normal flux density over the flux density in a vacuum for a given value of magnetizing field strength.

INTRINSIC INDUCTION: *See Intrinsic flux density.*

IRIG: Abbreviation for Inter-Range Instrumentation Group.

IRIG SUBCARRIER CHANNELS: A number of FM proportional- or constant-bandwidth channels employed to frequency-modulate a transmitter or recording device.

JITTER: Momentary movement of playback (reproduce) signal caused by time-base error (TBE) and/or dynamic interchannel time displacement error (ITDE). It is the first order of flutter, usually measured in nanoseconds.

LATERAL DIRECTION: Direction across the width of the tape.

LAYER-TO-LAYER ADHESION: *See Blocking.*

LAYER-TO-LAYER SIGNAL TRANSFER: Also known as print-through. The transfer of a magnetic field from one layer to another within a roll of tape. The magnitude of the induced signal tends to increase with the storage time and temperature and decrease with the unwinding of the tape roll. It is a function of the magnetic instability of the magnetic oxide.

LEVEL—OPERATING: The normal input voltage, current, or power signal required by the equipment.

LINEARITY—AC AND DC: The extent to which the magnitude of the reproduced signal is directly proportional to the magnitude of the input signal. Good linearity and low distortion go hand in hand.

LIVE DATA: *See Data.*

LONGEVITY: Length of life to wearout.

LONGITUDINAL CURVATURE: Any deviation from straightness of a length of tape, i.e., maximum lateral displacement of a given length of tape when laid out on a flat surface and placed under a small amount of tension.

LONGITUDINAL DIRECTION: Direction along the length of the tape.

LUBRICANTS: Substances that minimize the abrasive effects of the oxide particles. *See Additive.*

MAGNETIC COAT: *See Coating.*

MAGNETIC INSTABILITY: A magnetic property that causes variations in the residual flux density of the tape to occur with temperature, time, or mechanical flexing.

MAGNETIC TAPE: Normally consists of a base film coated with magnetic oxide held in a binder. The oxide particles ($Fe_2O_3\gamma$) are acicular in shape and normally of single-domain size.

MAGNETIZING FIELD STRENGTH (H): The instantaneous strength of the magnetic field applied to a magnetic material.

MALFUNCTION: A general term used to denote the occurrence of a failure of a system to give satisfactory performance. It need not constitute a failure if readjustment of operating controls can restore an acceptable operating condition.

MAXIMUM INTERCHANNEL DISPLACEMENT ERROR: The highest value of interchannel displacement error measured, considering all combinations of channels in a given head arrangement. Measurement includes the effects of gap scatter, tape distortion, and variations in electronic characteristics of associated electronics.

MAXWELL: The cgs unit of magnetic flux.

MEAN TIME BETWEEN FAILURES (MTBF): The total measured operating time of a number of equipments divided by the total number of failures of a reparable equipment is defined as the ratio of the total operating time to the total number of failures. The measured operating time of the equipments which did not fail must be included.

MEAN TIME TO FAILURE (MTTF): The measured operating time of a single piece of equipment divided by the total number of failures of the equipment during the measured period of time.

MODULATED-CARRIER RECORDING: The information recorded in the form of a modulated carrier, i.e., amplitude modulation (AM), frequency modulation (FM), pulse-amplitude modulation (PAM), pulse-duration modulation (PDM), pulse-code modulation (PCM), and combinations of these such as PDM/FM or PCM/FM.

MODULATING FREQUENCY—FM: The signal which represents the data being imposed on the carrier frequency (f_m).

MODULATION INDEX—FM: The ratio of carrier-frequency deviation to modulation frequency. (Abbreviated MI.)

MODULATION NOISE: The noise produced when reproducing a tape which has been recorded with a given signal and which is a function of the instantaneous amplitude of the signal. It is related to dc noise and is caused by poor particle dispersion and surface irregularities.

MULTIPLEXING: Time or frequency sharing of a single track on tape by two or more signals, each occupying a different time or frequency domain, i.e., two FM signals recorded together with different center-carrier frequencies.

NOISE: A term for unwanted electrical disturbances, other than crosstalk or distortion components, that occur at the output of the reproduce amplifier. *See also System noise* and *Tape noise*.

NOISE SPIKE: A spurious signal of short duration that occurs during the reproduction of a tape and has a magnitude considerably in excess of the average peak value of the system noise.

NONORTHOGONAL TIME-BASE ERROR (NTE): The worst-case summation of time-base errors (TBE) after tape servo correction plus dynamic interchannel time displacement errors (ITDE). NTE = TBE + ITDE.

OERSTED: The cgs unit of magnetic field strength. One oersted is the intensity of a magnetic field in which a unit pole experiences a force of one dyne.

OPERATION RELIABILITY: The probability that the system will have specified performance for a given period of time.

ORIENTATION DIRECTION: The direction in which magnetic oxide particles are oriented. In tapes designed for rotary-head applications, the orientation direction is transverse (across the tape). For tapes used for longitudinal recording, the orientation direction is longitudinal (length of the tape).

ORIENTATION RATIO: In a magnetic tape coating, the orientation ratio is the ratio of residual flux density in the orientation direction of the particles to the residual flux density perpendicular to the orientation direction. The orientation ratio of conventional tapes is nominally 1.7.

OUTPUT: An all-inclusive description of the output capabilities of the system. Output includes all the following: output impedance, voltage, current, power, level in decibels, frequency response, ground reference, signal-to-noise ratio, rise time, noise level, normal operating level, droop, etc. It is normally specified in terms of maximum output that can be obtained for a given amount of harmonic distortion.

OUTPUT (RELATIVE): An index figure for judging output levels using a standard tape output as a reference at a specific wavelength, i.e., 15 or 7.5 mils.

OUTPUT UNIFORMITY: A maximum deviation in magnetic tape output as measured against a standard tape.

OVERALL INTERCHANNEL DISPLACEMENT ERROR: The apparent distance separating pulses recorded simultaneously on separate channels, and subsequently reproduced on a tape transport other than the one on which the recording was originally made.

OXIDE: *See Gamma ferric oxide.*

OXIDE BUILDUP: The accumulation of oxide or wear products from the tape in the form of deposits on the surface of the heads. Oxide buildup causes extreme loss at short wavelengths and accelerates tape wear.

OXIDE COATING: *See Coating.*

OXIDE LOADING: A measure of the density of the oxide in a coating mix. It is usually specified as weight of oxide per unit volume of coating.

OXIDE SHED: The loosening of particles of oxide from the tape coating during use.

PACKING DENSITY: The amount of digital information (bits) recorded along a length of tape, measured in bits per inch (bpi).

PARITY CHECK: A self-checking code employing binary digits in which the total number of 1s (or 0s) in each code expression is always even or odd.

PARTICLE ORIENTATION: A process during the manufacture of tape where the acicular particles are rotated so that their longest dimensions tend to lie parallel to one another. By increasing the residual flux density, particle orientation increases the output of the tape and its signal-to-noise ratio. *See also Orientation direction.*

PARTICLE SHAPE: The particles of gamma ferric oxide used in the manufacture of conventional magnetic tape are acicular with a dimension ratio of approximately 7:1.

PARTICLE SIZE: The average size of a magnetic particle used for making magnetic tape (conventional) is 20 μin. long by 3 μin. wide. It behaves as one domain volume. *See also Domain volume.*

PERCENTAGE DEVIATION: The ratio of modulating frequency to center-carrier frequency expressed in percent.

PERPENDICULAR DIRECTION: Direction perpendicular to the plane of the tape.

PLASTICIZER: A substance that provides the needed flexibility in the coating of magnetic tape.

PLAYBACK, OR REPRODUCE: The reproduction of the recorded signal. As the tape is transported past the playback heads, the magnetic field extending from the surface of the tape enters the head across the gap and sets up a flow of magnetic flux through the head cores. This flux cuts the windings and produces a voltage proportional to the number of windings and the rate of change of flux.

POLYESTER: An abbreviation for polyethylene glycol terephthalate, a material commonly used as the base film for magnetic tape. It has a higher humidity and temperature stability than most other base-film materials. It also has greater strength and fungus and mildew resistance.

POSTEMPHASIS: Increasing gain at lower frequencies to compensate for inherent system recording losses.

PREEMPHASIS: Peaking or increasing the system gain at high frequencies to compensate for inherent system recording losses. With audio recording the low-frequency end is also peaked. *See also Postemphasis.*

PREVENTIVE MAINTENANCE: A procedure of inspecting, testing, and reconditioning a system at regular intervals, according to specific instructions intended to prevent failures in service or to retard deterioration.

PRINT-THROUGH: *See Layer-to-layer signal transfer.*

PULSE-AMPLITUDE DELTA MODULATION (PAΔM): Like pulse delta modulation, differences in amplitude can be used to modulate the duration or position. Any of the pulse trains can be used to modulate an RF carrier, either amplitude-, frequency-, or phasewise. *See also Pulse delta modulation.*

PULSE-AMPLITUDE MODULATION (PAM): In this type of modulation, the amplitude of each pulse will represent the amplitude of the modulating wave at the time of the pulse. There are two kinds of PAM, unidirectional and bidirectional.

PULSE-CODE MODULATION (PCM): In this type of modulation, amplitudes are represented by groups of pulses. Each group is a code symbol that represents the amplitude of the modulating wave.

PULSE DELTA MODULATION (PΔM): The individual pulse or groups of pulses will represent the difference between the amplitude of the modulating wave at one sampling point and the next. The differences in amplitude can be used to modulate the duration or position.

PULSE-DURATION MODULATION (PDM): The duration, or the width, of the pulse is varied in accordance with the amplitude of the modulating signal.

PULSE-FREQUENCY MODULATION (PFM): The pulse repetition rate is varied.

PULSE-POSITION MODULATION (PPM): The position of each pulse is varied with respect to some timing mark.

RECERTIFICATION AND RENOVATION OF TAPE: Recertification is the act of determining the number of errors on a tape at a given packing density. To be effective, it is necessary to have a cleaning operation incorporated to remove surface accumulation of dirt, redeposits, etc. There is no way to "rejuvenate" tape. Tape can be cleaned and surface deposits removed, but scratches, grooves, edge damage, etc., cannot be corrected.

RECORDING TIME: The time from initiation of signal on the recording medium to cessation of signal, expressed in hours, minutes, seconds, and milliseconds. Tape speed should always be mentioned when recording time is given.

REEL: The flanged hub, made of metal or plastic, on which tape is wound. Audio tape is usually wound on plastic reels. Instrumentation tape is normally wound on NAB metal reels conforming to standards of the National Association of Broadcasters or on precision reels that conform to very stringent specifications.

REEL SIZE: Defined by the diameter of the reel flange and the tape width which the hub of the reel normally carries.

REFERENCE TAPE: A tape used as a reference against which the performance of other tapes is measured.

RELATIVE OUTPUT IN DECIBELS: An index figure for judging tape output levels using a standard tape as zero reference.

RELIABILITY: The probability of performing without failure a specified function under given conditions for a specified period of time.

REMANENT FLUX (ϕ_r): Flux, in maxwells, remaining on the tape after the magnetizing force which produced saturation has been removed. This value represents the signal-producing force that will be picked up by the reproduce head.

REPRODUCING TIME: The time from initiation of signal in the reproduce head to cessation of signal, expressed in hours, minutes, seconds, and milliseconds. Tape speed should always be mentioned.

RESIDUAL FLUX: *See Remanent flux.*

RESIDUAL FLUX DENSITY (B_r): The magnetic flux density at which the magnetizing field strength is zero when a sample of magnetic tape is in a magnetized condition. It is an indication of the output that can be expected at short wavelengths.

RESINS: Substances used to keep the binder pliable, prohibiting cracking and flaking off. *See also Additive.*

RESOLUTION: The degree to which differing amounts of magnetization can be reduced and recorded on tape and still provide a useful and distinguishable separation on reproduction, i.e., the difference between the magnetization representing 0.1 and 0.15 V of signal.

RESPONSE: Tape output expressed in decibel difference compared with a given reference tape, or to the sensitivity of the same piece of tape at various recording wavelengths.

RETENTIVITY: *See Residual flux density.*

SAMPLE—SAMPLING RATE: *See Time-division multiplexing.*

SATURATION: That point in a magnetic material where an increase in the magnetizing force will not cause an increase in magnetic intensity to be exhibited by the sample under test.

SATURATION FLUX (ϕ_m): Maximum flux (in lines per cross-sectional tape area) which the tape will support when magnetized to a state of saturation.

SATURATION NOISE: The noise produced when reproducing a uniformly saturated tape. It is associated with imperfect particle dispersion.

SCRAPE FLUTTER: A part of flutter. The violin-string effect produced when tape under tension moves past points of suspension such as heads and guides.

SELF-DEMAGNETIZATION: The process by which a piece of magnetized tape tends to demagnetize itself by virtue of the opposing fields created within it by its own magnetization. This effect becomes increasingly stronger at short wavelengths.

SENSITIVITY: Tape output at relatively long wavelengths, usually 7.5 to 15 mils. It is expressed as a relative sensitivity relating to some reference, such as another tape.

SEPARATION LOSS: The loss in output that occurs when the surface of the tape is separated from the surface of either the record or the reproduce head. Separation loss may be caused by poor guiding, cupped tape, the presence of projections, dust or wear products on the tape, or the accumulations of dust or wear products on the head. The loss is a function of wavelength and follows the law (separation loss, dB) = $55d/\lambda$ dB, where d = distance of separation in inches and λ = wavelength in inches.

SIGNAL-TO-NOISE RATIO: The ratio of the rms signal to the rms noise appearing across a constant output load. The noise figure is unweighted with a recorded signal.

SIGNAL-TO-NOISE RATIO—DIRECT RECORDING: The signal-to-noise ratio is the rms ratio of signal power to that of the noise in the passband. Noise is generated by several elements in the system. This definition does not include noise components which cannot be measured when the output is connected to a bandpass filter having no more than a 1-dB loss at each edge of the equipment bandpass, and then a loss of 18 dB/octave as the frequency is extended from those points. Such noise as modulation noise, bias noise, etc., are excluded in this measurement. The signal level shall be measured in decibels at normal operating level. The signal source is then removed, the input shorted, and the noise level measured in decibels. The difference between the two values is the signal-to-noise ratio.

SIGNAL-TO-NOISE RATIO—FM RECORDING: The difference in output level between the signal produced by sine-wave modulation of the FM carrier to the full frequency deviation specified and the unmodulated carrier at the center frequency is the signal-to-noise ratio in FM. This produces a voltage ratio expressed in decibels.

SIGNAL-TO-NOISE RATIO—PULSE RECORDING: The ratio of peak-value signal-plus-noise to the noise (with shorted input) as measured at the output of the record amplifier.

SINGLE-DOMAIN PARTICLE: All ferromagnetic materials are composed of permanently magnetized regions in which the magnetic moments of the atoms are ordered. These regions are called domains. The domains have a size determined by energy considerations (10^{18} molecules). When a particle is small enough so that it cannot support more than one of the domains, it is called a single-domain particle.

SKEW: The amount of displacement of the tape edges from a straight line in inches per inch, i.e., 0.112 in. per 96 in. *See also Tape skew.*

SLITTING-WIDTH VARIATION: The amount of variation in tape width that can be tolerated by various recording methods or tape recorders is very small, generally $+0.000$ to -0.004 mil.

SOLVENTS: Substances that promote the mixing of all tape-coating ingredients to assure smooth, even dispersion.

SPOKING: A form of buckling in which the tape pack is deformed into a polygon shape.

SPOOL: *See Reel.*

SQUARENESS FACTOR: The ratio of the remanent flux to saturation flux B_r/B_{max}. It is a measure of the efficiency of the tape. If $B_r/B_{max} = 1$, a square hysteresis loop will be formed, indicating that the tape was 100 percent efficient.

SQUEAL: Audible tape vibrations, primarily in the longitudinal movement, caused by frictional excitation at suspension points (heads and guides, etc.). A form of scrape flutter.

STABLE TAPE-MOTION TIME: The elapsed time between actuating the drive mode and obtaining a flutter reading within the limits specified for the tape transport.

STANDBY TIME: The time during which a system has partial application of power and can be made to function properly almost at once.

START TIME: The elapsed time between actuating the drive mode and obtaining a given speed within the specified flutter of the transport.

STOP TIME: The elapsed time between the actuation of the stop switch and the cessation of tape motion.

STOPPING DISTANCE: The distance the tape travels after actuation of the stop switch.

SURFACE ASPERITIES: Small projecting imperfections on the surface of the tape coating. They limit and cause variations in head-to-tape contact.

SURFACE TREATMENT: Any process by which the surface smoothness of the tape coating is improved after it has been applied to the base film. A smooth coating improves head-to-tape contact and reduces separation loss.

SYSTEM NOISE: The total noise produced by the whole recording system, including tape.

TAPE BASE: *See Base-film backing.*

TAPE LENGTH: The lineal dimension of tape parallel to the direction of the tape motion, expressed in feet, inches, and parts of inches. Length measurements should always be made with the same tape tension as when it passes over the heads. The dimensions of tape are affected by humidity and temperature.

TAPE NOISE: The noise that can be specifically attributed to the tape. It may be broken down into the following general classes: bulk-erase noise, dc noise, modulation noise, saturation noise, and zero-modulation noise. See definitions of each under their own heading.

TAPE PACK: The form taken by the tape as it is wound on a reel. A good tape pack is free of spoking, cinching, and layer-to-layer adhesion.

TAPE SKEW: The deviation of tape from following a linear path when transported across the heads. This causes time displacement error between signals recorded on different tracks and different amplitudes between the outputs from individual tracks owing to variations in azimuth alignment. Static skew indicates steady components, and dynamic skew indicates the fluctuating components of skew.

TAPE SPEED: The speed at which the tape is transported across the heads during normal recording and reproducing modes, expressed in inches per second (ips).

TAPE-SPEED DEVIATION: The long-term deviation of the tape speed from normal recording and reproducing speeds. It may be caused by the stepping action of hysteresis capstan motors, mechanical vibrations such as wow or flutter, or power-line frequency changes (drift). Deviations of less than ±0.5 percent are considered acceptable for low-band recorders and ±0.2 percent for high-band versions.

TAPE-TO-HEAD SPEED: The relative speed of tape and head during normal recording or reproducing. The tape-to-head speed coincides with the tape speed in the conventional longitudinal recorder, but in video and instrumentation rotary or helical-scan systems it is considerably greater than the tape speed; i.e., rotary-head-instrumentation tape speed is 12.5 or 25 ips, whereas tape-to-head speed due to the head rotation is 1,570 ips.

TAPE THICKNESS: The thickness of the base film of the tape, expressed in mils. Most common instrumentation tapes are 1.5- and 1-mil thicknesses. Audio tapes are thinner, starting at 1 mil and going down as far as 0.25 mil for some of the 120-min cassette tapes.

TAPE TRACKING: Tape tracking as pertains to tape recorders may be defined as the characteristic path the tape traverses in its passage through the machine. The quality of tracking will vary between machines. A tape transport may be said to have "good tracking" or "poor tracking." "Perfect tracking" is obtained when all points along any elemental longitudinal fiber of the tape pass through the same series of points in the tape handler. In practice, the tape will wander laterally, transversely, and angularly, the amount being inversely proportional to the quality of tape tracking.

TAPE WIDTH: The lineal dimension of tape perpendicular to tape travel expressed to the nearest ¼ in.

TEAR STRENGTH: The force in grams required to initiate a tear in a specially shaped specimen of tape or base film using a tester such as the Elmendorf tear tester.

TELEMETRY STANDARDS: Refers to the Telemetry Standards of the Telemetry Working Group of the Inter-Range Instrumentation Group of the Range Commanders Council, White Sands Missile Range, New Mexico.

TIME-BASE ERROR (TBE): Timing error that remains in the system after all available correction has been applied, i.e., by servo systems, electronic correction units, etc.

TIME-DIVISION MULTIPLEXING: A recording technique in which time is shared between a number of channels of information. The telemetry standard number of apertures per second is 900. Dividing the number of channels into 900 will give the sampling rate.

TOTAL THICKNESS: The sum of the thicknesses of the base film and the magnetic coating. The total thickness governs the length of tape that can be wound on a given reel and the amount of print-through that will occur.

TRACK: The area of tape surface that coincides with the recorded magnetization produced by one record-head gap.

TRACK SPACING: The distance between the center lines of adjacent tracks. With typical longitudinal-instrumentation tracks 50 mils wide, the track spacing is 70 mils.

TRACK WIDTH: The width of the track corresponds to the width of a record-head gap (*see Track*). Common instrumentation track widths are 50 mils, although some special systems have tracks as narrow as 25 mils. Rotary-head track widths are typically 5 or 10 mils, and audio systems vary over a wide range, depending upon whether they are used for full-track, half-track, four-track, etc. (monaural or stereo).

TRANSDUCER: A device that converts data from its natural form into electrical signal for purposes of recording on magnetic tape.

ULTIMATE TENSILE STRENGTH: The force per cross-sectional area required to break a length of tape or base film.

UNIFORMITY: The extent to which the output remains free from variations.

WAVELENGTH (λ): The distance along the length of a sinusoidally recorded tape corresponding to one cycle. The wavelength is equal to the tape speed divided by the frequency of the recorded signal. Wavelength is usually expressed in mils.

WEAR, OR WEAR LIFE: Generally associated with video, but sometimes used with wideband-instrumentation tape. This term is used to indicate the number of passes that may be expected before there is an appreciable change in tape characteristics. Very specific conditions must be met while making this test, i.e., for tip penetration, tension, etc.

WEAR PRODUCTS: Any material that is freed from the tape during use. This could be oxide particles, portions of the coating, and material loosened from the edges of the tape.

WETTING AGENTS: Agents that "wet" each oxide particle to make it more conducive to receive the other binder ingredients for a uniform mix.

WIND: The way in which tape is wound onto a reel. An A wind is one in which the tape is wound so that the coated surface faces toward the hub. A B wind is one where the coated surface faces away from the hub.

WOW: Instantaneous variations of tape speed at moderately slow rates, overall frequency range being 0.2 to 200 Hz for instrumentation and 0.1 to 10 Hz for audio. It usually results from mechanical movement of transport components such as bearings, flywheels, belts, etc.

YIELD STRENGTH: The amount of force required to distort (stretch) the tape prior to breaking. The point at which elongation occurs. Tensile and yield strength tests are generally made at the same time, using a tensile tester and ¼-in.-tape samples.

ZERO ERRORS AT DENSITY: Also known as dropouts or drop-ins. An error is defined as a 50 percent drop-in saturated-tape output on a one or more bit basis, or as any spurious noise pulse greater than 8 percent of normal saturated-tape output level. This is a digital term.

ZERO-MODULATION NOISE: The noise produced when reproducing an erased tape with the erase and record heads energized but with zero input signal. The noise level is nominally 3 to 4 dB higher than with bulk-erased noise. *See also Tape noise.*

APPENDIX B

Glossary of Television Terms

CREDITS

Much of this glossary of television terms is included in this book with the express permission of the American Telephone and Telegraph Co., Long Lines Department. It is taken from the booklet titled "Television Signal Analysis," as revised by the Network Transmission Committee of the Video Transmission Engineering Advisory Committee (Joint Committee of Television Network Broadcasters and the Bell Telephone System), April 1963, and published by the American Telephone and Telegraph Co., Long Lines Department.

TERMS AND DEFINITIONS

ASPECT RATIO: The numerical ratio of picture width to height.

BACK PORCH: That portion of the composite picture signal which lies between the trailing edge of the horizontal sync pulse and the trailing edge of the corresponding blanking pulse.

BACK-PORCH TILT: The slope of the back porch from its normal horizontal position. Positive or negative refers, respectively, to upward or downward tilt to the right.

BANDWIDTH: The number of cycles per second expressing the difference between the limiting frequencies of a frequency band.

BLACK COMPRESSION: Amplitude compression of the signals corresponding to the black regions of the picture, thus modifying the tonal gradient.

BLACK PEAK: The maximum excursion of the picture signal in the black direction at the time of observation.

BLACKER-THAN-BLACK: The amplitude region of the composite video signal below reference black level in the direction of the synchronizing pulses.

BLANKING (PICTURE): The portion of the composite video signal whose instantaneous amplitude makes the vertical and horizontal retrace invisible.

BLANKING LEVEL: The level of the front and back porches of the composite video signal.

BLEEDING WHITES: An overloading condition in which white areas appear to flow irregularly into the black areas.

BLOOMING: The defocusing of regions of the picture where the brightness is at an excessive level, due to enlargement of spot size and halation of the fluorescent screen of the cathode-ray picture tube.

BOUNCE: An unnatural sudden variation in the brightness of the picture.

BREATHING: Amplitude variations similar to "bounce" but at a slow, regular rate.

BREEZEWAY: In NTSC color, that portion of the back porch between the trailing edge of the sync pulse and the start of the color burst.

BURNED-IN IMAGE: An image which persists in a fixed position in the output signal of a camera tube after the camera has been turned to a different scene.

CAMERA TUBE: *See Pickup tube.*

CATHODE-RAY TUBE: An electron-tube assembly containing an electron gun arranged to direct a beam upon a fluorescent screen. Scanning by the beam can produce light at all points in the scanned raster.

CHROMINANCE SIGNAL: That portion of the NTSC color television signal which contains the color information.

CLAMPER: A device which functions during the horizontal blanking or sync interval to fix the level of the picture signal at some predetermined reference level at the beginning of each scanning line.

CLAMPING: The process that establishes a fixed level for the picture signal at the beginning of each scanning line.

CLIPPING: The shearing off of the peaks of a signal. For a picture signal, this may affect either the positive (white) or negative (black) peaks. For a composite video signal, the sync signal may be affected.

CLOG: A deposit of material which builds up across the head gap. This buildup shorts out the gap, making the head inoperative. It could also wear the tape and lift the tape from the head. This generally results from poor-quality tape that has a bad coating-mix formulation.

COLOR BURST: In NTSC color, normally refers to a burst of approximately nine cycles of 3.6-MHz subcarrier on the back porch of the composite video signal. This serves as a color-synchronizing signal to establish a frequency and phase reference for the chrominance signal.

COLOR SUBCARRIER: In NTSC color, the carrier whose modulation sidebands are added to the monochrome signal to convey color information, i.e., 3.6 MHz (3.579545 MHz).

COLOR TRANSMISSION: The transmission of a signal which represents both the brightness and the color (chrominance) values in a picture.

COMPOSITE VIDEO SIGNAL: The complete video signal. For monochrome, it consists of a picture signal and the blanking and synchronizing signals. For color, additional color-synchronizing signals and color-picture information are added.

COMPRESSION: An undesired decrease in amplitude of a portion of the composite video signal relative to that of another portion. Also, a less-than-proportional change in output of a circuit for a change in input level. For example, compression of the sync pulse means a decrease in the percentage of sync during transmission.

CONTRAST: The range of light and dark values in a picture, or the ratio between the maximum and minimum brightness values. For example, in a high-contrast picture there would be intense blacks and whites, whereas a low-contrast picture would contain only various shades of gray.

CRO: Cathode-ray oscilloscope.

CROSSTALK: An undesired signal interfering with the desired signal.

CUTOFF FREQUENCY: That frequency beyond which no appreciable energy is transmitted. It may refer to either an upper or lower limit of a frequency band.

DAMPED OSCILLATION: Oscillation which, because the driving force has been removed, gradually dies out, each swing being smaller than the preceding, in smooth regular decay.

DEEMPHASIS: *See Restorer.*

DEFINITION: *See Resolution (horizontal and vertical).*

DELAY DISTORTION: Distortion resulting from nonuniform speed of transmission of various frequency components of a signal; i.e., the various frequency components of the signal have different times of travel (delay) between the input and the output of a circuit.

DETAIL: Refers to the most minute elements in a picture which are distinct and recognizable. Similar to definition or resolution.

DIFFERENTIAL GAIN: The amplitude change, usually of the 3.6-MHz color subcarrier, introduced by the overall circuit, measured in decibels or percent, as the picture signal on which it rides is varied from blanking to white level.

DIFFERENTIAL PHASE: The phase change of the 3.6-MHz color subcarrier introduced by the overall circuit, measured in degrees, as the picture signal on which it rides is varied from blanking to white level.

DISPLACEMENT OF PORCHES: Refers to any difference between the level of the front porch and the level of the back porch.

DISTORTION: The departure, during transmission or amplification, of the received signal waveform from that of the original transmitted waveform.

DRIVING SIGNALS: Signals that time the scanning at the pickup device.

DROPOUT: A reduction of signal level by 12 dB or better for a period of 10 μsec or longer. When video tape is tested for dropouts and yields a count of more than 15 per minute, the tape is rejected.

ECHO (OR REFLECTION): A wave which has been reflected at one or more points in the transmission medium, with sufficient magnitude and time difference to be perceived in some manner as a wave distinct from that of the main or primary transmission. Echoes may be either lagging or leading the primary wave and appear in the picture monitor as reflections, or ghosts.

EDGE EFFECT: *See Following whites* or *Leading whites* and *Following blacks* or *Leading blacks.*

EIA: Abbreviation for Electronic Industries Association.

EQUALIZING PULSES: Pulses of one-half the width of the horizontal sync pulses which are transmitted at twice the rate of the horizontal sync pulses during the blanking intervals immediately preceding and following the vertical sync pulses. The action of these pulses causes the vertical deflection to start at the same time in each interval, and also serves to keep the horizontal sweep circuits in step during the vertical blanking intervals immediately preceding and following the vertical sync pulse.

EXPANSION: An undesired increase in amplitude of a portion of the composite video signal relative to that of another portion. Also, a greater than proportional change in the output of a circuit for a change in input level. For example, expansion of the sync pulse means an increase in the percentage of sync during transmission.

FIELD: One-half of a complete picture (or frame) interval, containing all the odd or even scanning lines of the picture.

FIELD FREQUENCY: The rate at which a complete field is scanned, nominally 60 times a second U.S. Standard and 50 times a second European Standard.

FLASH: Momentary interference to the picture of a duration of approximately one field or less, and of sufficient magnitude to totally distort the picture information. In general, this term is used alone when the impairment is of such short duration that the basic impairment cannot be recognized. Sometimes called hit.

FLYBACK: *See Horizontal retrace.*

FOLLOWING (OR TRAILING) BLACKS: A term used to describe a picture condition in which the edge following a white object is overshaded toward black. The object appears to have a trailing black border. Also called trailing reversal.

FOLLOWING (OR TRAILING) WHITES: A term used to describe a picture condition in which the edge following a black or dark-gray object is shaded toward white. The object appears to have a trailing white border. Also called trailing reversal.

FRAME: One complete picture consisting of two fields of interlaced scanning lines.

FRAME FREQUENCY: The rate at which a complete frame is scanned, nominally 30 frames per second.

FRAME ROLL: A momentary roll.

FRONT PORCH: That portion of the composite picture signal which lies between the leading edge of the horizontal blanking pulse, and the leading edge of the corresponding sync pulse.

GAIN-FREQUENCY DISTORTION: Distortion which results when all the frequency components of a signal are not transmitted with the same gain or loss. A departure from "flatness" in the gain-frequency characteristic of a circuit.

GHOST: A shadowy or weak image in the received picture, offset either to the left or right of the primary image, the result of transmission conditions which create secondary signals that are received earlier or later than the main or primary signal. A ghost displaced to the left of the primary image is designated as leading, and one displaced to the right is designated as following (lagging). When the tonal variations of the ghost are the same as the primary image, it is designated as positive, and when it is the reverse, it is designated as negative.

GLITCH: A form of low-frequency interference, appearing as a narrow horizontal bar moving vertically through the picture. This is also observed on an oscilloscope at field or frame rate as an extraneous voltage pip moving along the signal at approximately reference black level.

HALO: Most commonly, a dark area surrounding an unusually bright object, caused by overloading of the camera tube. Reflection of studio lights from a piece of jewelry, for example, might cause this effect. With certain camera-tube operating adjustments, a white area may surround dark objects.

HEAD WEAR: The tip projection of each video head is measured at intervals. If the amount of wear of the heads is determined to be within specifications, the tape tested within that interval is released. Head wear is quoted in microinches of wear per hour, using a particular tip penetration, i.e., 2 μin./hr average. This term is associated with video tape testing.

HEIGHT: The size of the picture in a vertical direction.

HIGH-FREQUENCY DISTORTION: Distortion effects which occur at high frequency. Generally considered as any frequency above the 15.75-kHz line frequency.

HIGH-FREQUENCY INTERFERENCE: Interference effects which occur at high frequency. Generally considered as any frequency above the 15.75-kHz line frequency.

HIGHLIGHTS: The maximum brightness of the picture, which occurs in the regions of highest illumination.

HIT: *See Flash.*

HORIZONTAL (HUM) BARS: Relatively broad horizontal bars, alternately black and white, which extend over the entire picture. They may be stationary or may move up or down. Sometimes referred to as a venetian-blind effect. Caused by approximately 60-Hz interfering frequency, or one of its harmonic frequencies.

HORIZONTAL BLANKING: The blanking signal at the end of each scanning line.

HORIZONTAL DISPLACEMENT: Describes a picture condition in which the scanning lines start at relatively different points during the horizontal scan. *See Serrations* and *Jitter.*

HORIZONTAL RETRACE: The return of the electron beam from the right to the left side of the raster after the scanning of one line.

HUE: Corresponds to "color" in everyday use, i.e., red, blue, etc. Black, white, and gray do not have hue.

ICONOSCOPE: A camera tube in which a high-velocity electron beam scans a photoemissive mosaic which has electrical-storage capability.

IEEE: *See IRE.*

INTERFERENCE: In a signal transmission path, extraneous energy which tends to interfere with the reception of the desired signals.

INTERLACED SCANNING (INTERLACE): A scanning process in which each adjacent line belongs to the alternate field.

ION: A charged atom, usually an atom of residual gas in an electron tube.

ION SPOT: A spot on the fluorescent surface of a cathode-ray tube which is somewhat darker than the surrounding area because of bombardment by negative ions which reduce the sensitivity.

ION TRAP: An arrangement of magnetic fields and apertures which will allow an electron beam to pass through but will obstruct the passage of ions.

IRE: Abbreviation for the Institute of Radio Engineers. This organization combined with the American Institute of Electrical Engineers, effective Jan. 1, 1963, to form the Institute of Electrical and Electronic Engineers (IEEE).

IRE ROLL-OFF: The IRE Standard oscilloscope frequency-response characteristic for measurement of level. This characteristic is such that at 2 MHz the response is approximately 8.5 dB below that in the flat (low-frequency) portion of the spectrum, and cuts off slowly.

IRE SCALE: An oscilloscope scale in keeping with IRE Standard 50, IRE 23.S1, and the recommendations of the Joint Committee of TV Broadcasters and Manufacturers for Coordination of Video Levels.

JITTER: A tendency toward lack of synchronization of the picture. It may refer to individual lines in the picture or to the entire field of view.

KINESCOPE: Frequently used to mean picture tubes in general; however, this name has been trademarked.

KINESCOPE RECORDING: A motion-picture-film recording of the presentation shown by a picture monitor. Also known as television recording (TVR), Vitapix, etc.

LEADING BLACKS: A term used to describe a picture condition in which the edge preceding a white object is overshaded toward black. The object appears to have a preceding, or leading, black border.

LEADING WHITES: A term used to describe a picture condition in which the edge preceding a black object is shaded toward white. The object appears to have a preceding, or leading, white border.

LINE FREQUENCY: The number of horizontal scans per second, nominally 15,750 times per second.

LOW-FREQUENCY DISTORTION: Distortion effects which occur at low frequency. Generally considered as any frequency below the 15.75-kHz line frequency.

LOW-FREQUENCY INTERFERENCE: Interference effects which occur at low frequency. Generally considered as any frequency below the 15.75-kHz line frequency.

LUMINANCE SIGNAL: That portion of the NTSC color television signal which contains the luminance, or brightness, information.

MESHBEAT: *See Moiré.*

MICROPHONICS: In video transmission, refers to the mechanical vibration of the elements of an electron tube resulting in a spurious modulation of the normal signal. This usually shows as erratically spaced horizontal bars in the picture.

MICROSECOND: One-millionth of a second (10^{-6}).

MOIRÉ: A wavy, or satiny, effect produced by convergence of lines. Usually appears as a curving of the lines in the horizontal wedges of the test pattern and is most pronounced near the center where the lines forming the wedges converge. A moiré pattern is a natural optical effect when converging lines in the picture are nearly parallel to the scanning lines. This effect, to a degree, is sometimes due to the characteristics of color-picture tubes and of image orthicon pickup tubes (in the latter termed meshbeat).

MONOCHROME TRANSMISSION (BLACK AND WHITE): The transmission of a signal wave which represents the brightness values but not the color (chrominance) values in the picture.

MULTIPLE BLANKING LINES: The effect evidenced by a thickening of the blanking-line trace or by several distinct blanking lines as viewed on an oscilloscope. May be caused by hum.

NAB: Abbreviation for National Association of Broadcasters.

NANOSECOND: 10^{-9} sec.

NEGATIVE IMAGE: Refers to a picture signal having a polarity which is opposite to normal polarity and which results in a picture in which the white areas appear as black, and vice versa.

NOISE: The word noise is a carryover from audio practice. Refers to random spurts of electrical energy or interference. May produce a salt-and-pepper pattern over the picture. Heavy noise sometimes is called snow.

NTSC: Abbreviation for National Television System Committee.

ORTHICON (CONVENTIONAL): A camera tube in which a low-velocity electron beam scans a photoemissive mosaic on which the image is focused optically and which has electrical-storage capability.

ORTHICON (IMAGE): A camera tube in which the optical image falls on a photoemissive cathode which emits electrons that are focused on a target at high velocity. The target is scanned from the rear by a low-velocity electron beam. Return-beam modulation is amplified by an electron multiplier to form an overall light-sensitive device.

ORTHICON EFFECT: One or more of several image orthicon impairments that have been referred to as orthicon effect, as follows:

1. Edge effect—usually a white outline of well-defined objects
2. Meshbeat, or moiré
3. Ghost—appears in connection with bright images and is not limited in position to leading or lagging the main image
4. Halo
5. Burned-in image

OUTPUT: Each video tape manufacturer uses some form of standard reference tape to set up the measuring equipment to 100 percent. The output of the tape under test is then continuously checked against the reference. If it should drop below 95 percent of the reference level, it is rejected.

OVERSHOOT: An excessive response to a unidirectional signal change. Sharp overshoots are sometimes referred to as spikes.

PAIRING: A partial or complete failure of interlace in which the scanning lines of alternate fields do not fall exactly between one another but tend to fall (in pairs) one on top of the other.

PEAK TO PEAK: The amplitude (voltage) difference between the most positive and the most negative excursions (peaks) of an electrical signal.

PEDESTAL: This term is obsolete.

PEDESTAL LEVEL: This term is obsolete; blanking level is preferred.

PERCENTAGE SYNC: The ratio, expressed as a percentage of the amplitude of the synchronizing signal to the peak-to-peak amplitude of the picture signal between blanking and the reference white level.

PHOTOEMISSIVE: Emitting, or capable of emitting, electrons upon exposure to radiation in and near the visible region of the spectrum.

PICKUP TUBE: An electron-beam tube used in a television camera where an electron current or a charge-density image is formed from an optical image and scanned in a predetermined sequence to provide an electrical signal.

PICTURE MONITOR: This refers to a cathode-ray tube and its associated circuits, arranged to view a television picture.

PICTURE SIGNAL: That portion of the composite video signal which lies above the blanking level and contains the picture information.

PICTURE TUBE: A cathode-ray tube used to produce an image by variation of the intensity of a scanning beam.

PIGEONS: Noise observed on picture monitors as pulses or bursts of short duration, at a slow rate of occurrence—a type of impulse noise.

POLARITY OF PICTURE SIGNAL: Refers to the polarity of the black portion of the picture signal with respect to the white portion. For example, in a black negative picture, the potential corresponding to the black areas of the picture is negative with respect to the potential corresponding to the white areas of the picture. In a black positive picture the potential corresponding to the black areas of the picture is positive. The signal as observed at the broadcasters' master control rooms and telephone-operating centers is black negative.

PREEMPHASIS: A change in level of some frequency components of the signal with respect to the other frequency components at the input to a transmission system. The high-frequency portion of the band is usually transmitted at higher level than the low-frequency portion of the band.

RASTER: The scanned (illuminated) area of the cathode-ray picture tube.

REFERENCE BLACK LEVEL: The level corresponding to the specified maximum excursion of the luminance signal in the black direction.

REFERENCE SIGNALS (VERTICAL INTERVAL): Signals inserted into the vertical interval at the program source which are used to establish black and white levels. Such a signal might consist of 5 μsec of reference black at 7.5 IRE divisions and 5 μsec of reference white at 100 IRE divisions located near the end of lines 18 and 19 of the vertical interval.

REFERENCE WHITE LEVEL: The level corresponding to the specified maximum excursion of the luminance signal in the white direction.

REFLECTIONS, OR ECHOES: In video transmission this may refer either to a signal or to the picture produced.

1. Signal
   a. Waves reflected from structures or other objects
   b. Waves which are the result of impedance or other irregularities in the transmission medium
2. Picture: Echoes observed in the picture produced by the reflected waves

RESOLUTION (HORIZONTAL): The amount of resolvable detail in the horizontal direction in a picture. It is usually expressed as the number of distinct vertical lines, alternately black and white, which can be seen in three-quarters of the width of the picture. This information usually is derived by observation of the vertical wedge of a test pattern. A picture which is sharp and clear and shows small details has good, or high, resolution. If the picture is soft and blurred and small details are indistinct, it has poor, or low, resolution. Horizontal resolution depends upon the high-frequency amplitude and phase response of the pickup equipment, the transmission medium, and the picture monitor, as well as the size of the scanning spots.

RESOLUTION (VERTICAL): The amount of resolvable detail in the vertical direction in a picture. It is usually expressed as the number of distinct horizontal lines, alternately black and white, which can be seen in a test pattern. Vertical resolution is primarily fixed by the number of horizontal scanning lines per frame. Beyond this, vertical resolution depends on the size and shape of the scanning spots of the pickup equipment and picture monitor, and does not depend on the high-frequency response or bandwidth of the transmission medium or picture monitor.

RESTORER: As used by AT&T, a network designed to remove the effects of preemphasis, thereby resulting in an overall normal characteristic.

RETMA: Abbreviation for Radio Electronic Television Manufacturers Association.

RETRACE (RETURN TRACE): *See Horizontal retrace* and *Vertical retrace.*

RF PATTERN: A term sometimes applied to describe a fine herringbone pattern in a picture. May also cause a slight horizontal displacement of scanning lines, resulting in a rough, or ragged, vertical edge of the picture. Caused by high-frequency interference.

RINGING: An oscillatory transient occurring in the output of a system as a result of a sudden change in input. Results in close-spaced multiple reflections, particularly noticeable when observing test patterns, equivalent square waves, sine-squared signal, or any fixed objects whose reproduction requires frequency components approximating the cutoff frequency of the system.

ROLL: A lack of vertical synchronization which causes the picture as observed on the picture monitor to move upward or downward.

ROLL-OFF: A gradual attenuation of gain-frequency response at either end or both ends of the transmission passband.

SATURATION (COLOR): The vividness of a color, described by such terms as pale, deep, pastel, etc. The greater the amplitude of the chrominance signal, the greater the saturation.

SCANNING: The process of breaking down an image into a series of elements or groups of elements representing light values and transmitting this information in time sequence.

SCANNING LINE: A single continuous narrow strip of the picture area containing highlights, shadows, and halftones, determined by the process of scanning.

SCANNING SPOT: Refers to the cross section of an electron beam at the point of incidence in a camera tube or picture tube.

SERRATED PULSES: A series of equally spaced pulses within a pulse signal. For example, the vertical sync pulse is serrated in order to keep the horizontal sweep circuits in step during the vertical-sync-pulse interval.

SERRATIONS: A term used to describe a picture condition in which vertical or nearly vertical lines have a sawtooth appearance. The result of scanning lines starting at relatively different points during the horizontal scan.

SETUP: The separation in level between blanking and reference black levels.

SHED: A powdery residue that occurs around the guides and heads. It is the result of bad slitting, edge damage, or poor coating-mix formulation.

SMEAR: A term used to describe a picture condition in which objects appear to be extended horizontally beyond their normal boundaries in a blurred, or smeared, manner.

SNOW: Heavy random noise.

SPIKE: *See Overshoot.*

STREAKING: A term used to describe a picture condition in which objects appear to be extended horizontally beyond their normal boundaries. This will be more apparent at vertical edges of objects when there is a large transition from black to white or white to black. The change in luminance is carried beyond the transition, and may be either negative or positive. For example, if the tonal degradation is an opposite shade to the original figure (white following black), the streaking is called negative; however, if the shade is the same as the original figure (white following white), the streaking is called positive. Long streaking may extend to the right edge of the picture and, in extreme cases of low-frequency distortion, can extend over a whole line interval.

SYNC: An abbreviation for the words synchronization, synchronizing, etc. Applies to the synchronization signals, or timing pulses, which lock the electron beam of the picture monitors in step, both horizontally and vertically, with the electron beam of the pickup tube. The color sync signal (NTSC) is known as the color burst.

SYNC COMPRESSION: The reduction in the amplitude of the sync signal with respect to the picture signal, occurring between two points of a circuit.

SYNC LEVEL: The level of the tips of the synchronizing pulses.

SYNCHRONIZATION: The maintenance of one operation in step with another.

TAPE WEAR: Just as a video head should not wear, neither should the video tape wear excessively. The end-of-life point of tape is where the output drops 25 percent, recording only once, or it begins to show excessive dropouts, clog, or shed. Like head life, this test is normally done with a particular tip penetration, i.e., better than 250 passes with a tip penetration of 2.5 mils.

TEARING: A term used to describe a picture condition in which groups of horizontal lines are displaced in an irregular manner. Caused by lack of horizontal synchronization.

TELEVISION RECORDING (TVR): *See Kinescope recording.*

TRANSIENTS: Signals which endure for a brief time prior to the attainment of a steady-state condition. These may include overshoots, damped sinusoidal waves, etc., and therefore additional qualifying information is necessary.

VERTICAL BLANKING: Refers to the blanking signals which occur at the end of each field.

VERTICAL RETRACE: The return of the electron beam from the bottom to the top of the raster after completion of each field.

VESTIGIAL SIDEBAND TRANSMISSION: A system of transmission wherein the side-band on one side of the carrier is transmitted only in part.

VIDEO: A term pertaining to the bandwidth and spectrum position of the signal which results from television scanning and which is used to reproduce a picture.

VIDEO BAND: The frequency band utilized to transmit a composite video signal.

VIDEO TAPE RECORDING (VTR): A magnetic tape recording of the composite video signal.

VIDEO-IN-BLACK: A term used to describe a condition as seen on the waveform monitor when the black peaks extend through reference black level.

WAVEFORM MONITOR: Refers to a cathode-ray oscilloscope used to view the form of the composite video signal for waveform analysis. Sometimes called A-scope.

WHITE COMPRESSION: Amplitude compression of the signal corresponding to the white regions of the picture, thus modifying the tonal gradient.

WHITE PEAK: The maximum excursion of the picture signal in the white direction at the time of observation.

WIDTH: The size of the picture in a horizontal direction.

Index

A oxide, 58
Abrasiveness, 50, 55, 243
Ac bias (*see* Bias, ac)
Ac erasure, 68, 69, 243
Acetate-base tape, 52, 53
  table of properties, 53
Acicular particles, 48, 49, 243
Acquisition of data (*see* Magnetic recording systems, applications of)
Adhesion, 50–51
Advantages of magnetic recording, 14, 15
Air-gap loss (*see* Reproduce losses, gap)
Air-lubricated guides, 97, 237
Alfesil, 42
Amplitude equalization (*see* Equalization, amplitude)
Amtec, 211, 212, 220
Analog recording (*see* Direct recording)
Anchorage (*see* Adhesion)
Antistatic agent (*see* Conductive agent)
Aperture corrector, 157–159
Apparent errors, 100
Applications of magnetic recording systems (*see* Magnetic recording systems, applications of)
Armature, printed circuit, 99, 100
Arthur, Dr. Paul, Jr., 48
Atomic structure of iron, 28
  (*See also* Ferromagnetic materials)
Audio- and cue-track circuit, 143, 144
  block diagram, 144
Audio cartridge recorder, 224, 225
Audio cassette recorder (*see* Cassette recorders, audio)
Audio recording (*see* Direct recording)
Audio reproduction (*see* Direct reproducing)
Audio tape parameters:
  consumer, 57
  mastering, 56, 57
  professional, 57
Audio terms, glossary of, 243–262
Audio track, rotary instrumentation, 143, 144
Auto-chroma, 208, 209

Automatic cueing, 233, 234
Automatic equalizer (*see* Auto-chroma)
Automatic gain control (AGC), 156
Automatic-threading mechanism, 233
Autosearch, 240
Auxiliary heads, video recorders, 43
Auxiliary track recording, video, 141–144
Azimuth alignment, 78, 244
Azimuth loss, 78–80

B oxide, 58
B-H curve, 67–69, 71–73
Back porch (*see* Horizontal sync pulse)
Backing material, 19, 52
  table of differences, 53
  (*See also* Magnetic tape)
Backing thickness, 46, 47, 244
Ball mill, 51
Bandwidth, video recorder, 39
Base film (*see* Magnetic tape)
Baseplate assembly, 94
Basic carrier frequency, 162
Basic elements of magnetic recorder (*see* Magnetic recorders, basic elements of)
Basic response in reproduction, 76, 77
Bass boost, 86–89
Belt changer, 235
Berliner, Emile, 10
Beryllium copper (*see* Brake)
Bias:
  ac, 71–74
    action, 71–74
    amount of, adjustment of, 74
    demagnetization losses from, 75
    introduction, 12–14
    vs. recorded level, 74
    used with FM, 161, 162, 172, 176
    uses of, 21, 22
  dc, 71, 72
    with demagnetized medium, 71
    introduction, 12
    with presaturated medium, 72
    uses of, 21
  requirements for, 69, 70
Bias driver, 155
Bias erasure (*see* Record losses)
Bias oscillator, 146, 147, 152, 155, 162
Bias trap, 143
Biasing systems:
  intermediate band, 146, 147
  1.5-MHz wideband, 152
  rotary instrumentation audio, 143, 144
  rotary video audio and cue, 143, 144
Bin-loop adapter, 128, 129
Bin-loop recorder, 128, 129
  tape-tension system, 129
Binder material, 50, 51, 245
  properties of, 50
  special, for video, 61
Bit packing density, 231
Blanking pulse, 204
Bloch, F., 28
Bloch wall, 28, 29
B_{max} (maximum magnetization), 67, 68
B_r (*see* Remanence magnetism)
Brake:
  band, 134
  disc, 132–134
  string, 134
Broken-tape sensing, 137–139
  mechanical, 137
  photoelectric, 138
  pneumatic, 138
  vacuum, 139
Brushes, motor, 99
B_s (*see* Saturation magnetism)
Bulk degaussing (*see* Bulk erasing)
Bulk erasing, 69
Butler oscillator (*see* Bias oscillator)
B_y (emerging flux), 64–66

Canoe area, 107–110, 187
Capstan, 94–98, 107
  bin-loop tape tension, 129, 130
  polymer-coated, 97
  rubber-sleeved, 117
  stuffing, 129
Capstan motor:
  ac split-phase induction, 98
  dc printed-circuit, 99
  delta-wound, 110
  hysteresis-synchronous, 94, 96
Capstan servo, video (*see* Quadruplex video recorders)
Capstan servo system, 100–107
  direct, 18.24-kHz and 17-kHz, 101–103
    block diagram, 102
  FM, 103, 104

Capstan servo system (*Cont.*):
photoetched tachometer, 101, 102, 104
200-kHz, 104–107
block diagram, 105
slotted tachometer, 100
Carpenter, T. W., 12
Carrier:
AM (*see* Record process, direct)
FM (*see* Record process, FM)
Carrier frequency (*see* Record process, FM)
Carrousel, cassette recorder, 235–238
Carson, William L., 12
Cartridge, 224, 225, 234, 235, 240
Cartridge recorders, 224, 225, 230–235, 240–242
audio, 224, 225
changer mechanism for, 233–235
digital, 230–233
Lear-Jet, 224, 225
video: home and office, 238–242
camera, 242
Instavideo, 240–242
television broadcasting, 233–235
automatic cueing, 234, 235
automatic-threading mechanism, 233, 234
cartridge-changer mechanism, 233–235
playback station, 233–235
tape format, 234
Cassette, 223, 224, 235–240
Cassette recorders, 223–240
audio, 226–230
accessories for, 227
automatic cassette changers, 227
schematic diagram, 228, 229
tape drive, 226, 227, 230
bi-directional, 232
cleaning agents, 230
digital, 230–233
capstan-and-reel spindle drive, 232, 233
tape-drive system, 231–233
limitations of, 227–230
lockout tab, 227
tape path, illustrated, 230
video, 233–242
home and office, 238–240
Videocassette, 238–240
television broadcasting, 235–238
tape-threading system, 237
Cassette recorders, video, television broadcasting (*Cont.*):
vacuum capstan, 238
vacuum chamber, 236–238
Cassette recording format, 224
CCIR curves, 88
(*See also* Equalization, amplitude)
Cellulose acetate (*see* Magnetic tape, base film)
Chrominance, 214
Chromium dioxide magnetic tape, 48, 49, 239
Closed-loop drive system (*see* Longitudinal drive systems)
Coating, magnetic: chromium dioxide, 48
iron oxide, 49, 50
Coating process (*see* Magnetic tape, manufacturing processes)
Coating roughness, 55
Coating thickness, 46, 47
Coefficient of humidity, 53
Coefficient of thermal expansion, 53
Coercivity (H_{ci}), 50, 67–69, 246
intrinsic, 50, 67, 252
Cohesion, 50
Color processor, 213–214
Color reproduction, 220, 221, 233–242
Colortec, 212, 213, 220
Computer tape, 59, 60
Conductive agent, 51
Consumer tape (*see* Audio tape parameters)
Continuous-loop recorder, 130
Control track:
instrumentation recorders, 101–107
video recorders, 110, 111, 216, 217, 237, 238
Control-track filters, 104
Control-track generator (CTG) (*see* Capstan servo system)
Control-track head, video recorders, 42, 43, 110, 187, 238
Cosine equalizer, 157–159
Crolyn, 48
Crosstalk, 37, 247, 265
Cue track, 143, 144

D wrap format, 115
Data acquisition (*see* Magnetic recording systems, applications of)

Dc amplifier (*see* Capstan servo systems)
Dc bias (*see* Bias, dc)
Dc drift, 247
Definitions:
of audio terms, 243–262
of instrumentation terms, 243–262
of television terms, 263–274
of video terms, 263–274
Degauss (*see* Demagnetization)
De Kempelein, 9
Demagnetization, 19, 75
losses, 64, 75
recording, 64, 75
Demagnetization curve, 19
Demodulator (*see* FM demodulator)
Development of magnetic recording, 9
Deviation (*see* FM modulator)
Digital cartridge recorder, 230–233
Digital cassette recorder (*see* Cassette recorders, digital)
Digital comparator (*see* Capstan servo system)
Dihedral error, 115
Dilutants, 51
Direct record amplifier, 21, 141–160
Direct record/reproduce system:
advantages of, 159, 160
limitations of, 159, 160
Direct recording, 141–160
intermediate-band, 145–147
block diagram, 146
low-band, 141–144
1.5-MHz wideband, 151–153
block diagram, 153
2.0-MHz wideband, 153–156
block diagram, 154, 156
video auxiliary track, 143, 144
voice-log, 142
Direct reproducing, 144–159
intermediate-band, 147–151
block diagram, 148
low-band, 144, 145
block diagram, 145
1.5-MHz wideband, 151–153
block diagram, 153
2.0-MHz wideband, 153–159
block diagram, 154, 156
equalization curve, 157
Disc construction, illustrated, 121
Disc pack recorder, 121
Disc-recorder systems, 121–128
in-contact head-to-disc, 121–124
instrumentation, 121–126
spaced heads, 121, 126, 127
television (*see* Television disc-recorder systems)
track numbering, 123
Disc-recording pattern, 125
Discriminator (*see* FM demodulator)
Dispersants, 51
Domain theory, 26–30
Domain vector, 28–30, 72
Domains, magnetic, 19, 20, 26–30, 72
change in volume of, 28
magnetization of, 28–30
Drift, 20, 248
dc, 247
Drive pulley, differential, 129
Dropout, 248–249, 266
Dropout correction, video, 209–211
Dropout sensor, 209–211
Dropout-tested tape, 60, 231
Durability index, 249
Dynamic range, 249

E oxide, 58
EBU (European Broadcasters Union), 183, 188
Eddy-current loss (*see* Record losses; Reproduce losses)
Eddy currents, 33
Edison, Thomas Alva, 10
EE mode, 193, 203, 204
EEG (electroencephalograph), 6
EIAJ Type I VTR Standard, 116, 240
EKG (electrocardiograph), 5, 6
Electrons:
orbital motion, 26–28
spin angular momentum, 26–28
uncompensated spin, 27, 28
Elemental volume (*see* Domain theory)
Emerging flux (B_y), 64–66
End-of-tape sensing, 136, 137
mechanical, 136
photoelectric, 136, 137
pneumatic, 136
vacuum, 137
Equalization:
AME (*see* amplitude *below;* audio *below*)

Equalization (*Cont.*):
amplitude, 23, 86–89, 147, 148, 153, 157–159
audio (CCIR, NAB, AME), 86–89
CCIR (*see* amplitude *above;* audi *above*)
FM, 162, 173, 178
instrumentation, 86
intermediate-band, 147–151
amplitude, 147, 148
phase, 148–151
NAB (*see* amplitude *above;* audio *above*)
need for, 23, 85
1.5-MHz wideband: amplitude, 153
phase, 153
phase, 23, 89, 90, 148–151
playback, quadruplex video recorder, 206, 207
record: audio, 86–89
quadruplex video recorder, 202, 203
2.0-MHz wideband, cosine, 157–159
Equalization curves, 23, 87, 89
Erase head, 19, 43, 187, 188, 227
cassette recorders, 227
video, 43, 44, 187, 188
audio and cue, 187, 188
Erase process, 68, 69
ERMA (Electronic Method of Accounting), 7, 8
European Broadcasters Union (EBU), 183, 188
European equalization standard, CCIR curve, 88
Exclusion principle, 27
Extrusion coating (*see* Magnetic coating, processes)

F and *Q* circuit, quadruplex head resonance, 205, 206
Faraday's law, 18, 76, 77, 85
Female guide, 40, 41, 107–109, 184, 186, 197, 237, 238
height relationship to head drum, 186, 187
Ferric oxide:
alpha ($Fe_2O_3 \cdot H_2O$), 49
gamma ($Fe_2O_3{}^{\gamma}$), 49
particle size, 39, 49–50
Ferrite head, 122
Ferromagnetic materials, 26, 27
atomic structure, 27, 28
magnetic behavior, 27–30
Ferromagnetism, 26–28
Field, 266
Flutter, 20, 95–97, 250
Flywheel, viscous damped, 95
FM carrier deviation, 162, 166
FM demodulator, 24, 166–168, 172–174, 177–180
intermediate-band, 166–168
block diagram, 164
wideband group I, 172–174
block diagram, 173
wideband group II, 177–180
block diagram, 177
FM/FM systems, 4
FM modulator, 21, 162–166, 168–172, 174–177
intermediate-band, 162–166
block diagram, 164
wideband group I, 168–172
block diagram, 170
wideband group II, 174–177
block diagram, 176
FM record amplifier (*see* FM modulator)
FM record parameters:
choice of center carrier, 21
single carrier, 162
wideband, 162
FM record/reproduce:
advantages of, 180
limitations of, 180
major applications of, 180
FM recording (*see* FM modulator)
FM reproducing (*see* FM demodulator)
FM servo systems (*see* Capstan servo systems)
Follower arm (*see* End-of-tape sensing)
Forward-backward counter (*see* Capstan servo systems)
Frame, 267
Frame pulse, 30-Hz, 196, 219, 220
Frequency-division multiplexing, 3, 4
Frequency standard, crystal, 103, 104, 111
Front porch (*see* Horizontal sync pulse)

Gap azimuth, 36, 37
Gap depth, 33, 251

Gap loss (*see* Reproduce losses)
Gap scatter, 36, 37
Gap separator, 32, 33
  material, 33
Gap smear, 36, 251
Gap tilt (*see* Azimuth loss)
Gap-to-gap spacing, 35
Gap width, 33–35, 251
  record, 33–35
  reproduce, 33–36
Glossary of terms:
  audio, 243–262
  instrumentation, 243–262
  television, 263–274
  video, 263–274
Goudsmit, Samuel A., 27
Gramophone, 11
Gravure coating (*see* Magnetic tape coating, processes)
Guard bands for quadruplex video tape tracks, 185, 186
Guides:
  air-lubricated, 97, 237
  female (*see* Female guide)
  fixed, 95, 109
  rotating, 95, 107
    floating, 130
  vacuum (*see* Female guide)

H (magnetic force), 17, 28–30, 64–70
  (*See also* Coercivity)
Head, flying, 121
Head bounce, 81
Head bump, 83, 84
Head cores, 31–33
Head current, 63, 64
Head degausser (*see* Demagnetization)
Head-driver electronics, 152, 153
Head drum, 107–109, 183, 184
Head-drum servo, 111, 112, 190–195
  (*See also* Quadruplex video recorders, servo systems)
Head-drum tachometer, 110
Head-drum timing, 189–197
Head-numbering systems, 37–38
  illustrated, 38
Head-stack parameters, 36, 37
Head tilt (*see* Azimuth loss)
Head tip, 108, 109
Head-tip penetration, 41, 42, 185, 197–199
Head-to-tape contact speed, 39, 40, 116, 251
Head wear vs. gap smear, illustrated, 59
Head writing speed, 116, 117
Heads, magnetic (*see* Magnetic heads)
Helical-scan drive systems, 112–118
Helical-scan recorders, compatibility between, 115, 116
Helical-scan servo systems, 118–120
Helical-scan track patterns, 116
Helical-scan transports, 112–120
  capstan drive systems, 117, 118
  capstan servo systems, 120
    block diagram, 120
  drum servo system, 118–120
  heads, 113–117
  wrap format, 113–115
Helix angle, 115–117
Horizontal sync pulse, 119, 120, 197, 207
Hysteresis loop, 67–70, 72
  effect during recording, 64
  losses, 64

Instavideo, 240–242
  camera for use with, 242
Instrumentation disc recorder, 125–128
  servo control, 125
    pilot signal, 125
  soft switching, 125
Instrumentation head-numbering systems, 37, 38
Instrumentation tape, 57, 58
Instrumentation terms, 243–262
Interleaving, magnetic heads, 36, 37
Intermediate-band direct (*see* Record process, direct; Reproduce process, direct)
Intermediate-band FM (*see* Record process, FM; Reproduce process, FM)
Inter-Range Instrumentation Group (IRIG):
  head standards, 37, 38
  standards, 3, 4, 37, 103, 162, 172
  subcarrier channels, 3, 4, 103
Intersync, 190, 214–221
Intersync modes, 214–221

Intrinsic coercivity (H_{ci}), 50, 67, 252
IRIG (*see* Inter-Range Instrumentation Group)
Iron (*see* Ferromagnetic materials)

Jitter, 253, 269

Knife-blade coating, 53, 54

Lear-Jet cartridge, 224, 225
Lockout tab, 227, 231
Long-wavelength losses, 82–85
Long-wear tape, instrumentation, 58
Longitudinal drive systems, 94–98
  closed-loop, 96
  open-loop, 94, 95
  optimum-loop, 96, 97
  zero-loop, 97, 98
Losses, recording (*see* Record losses)
Low-band FM (*see* Record process, FM)
Low-band magnetic heads, 33, 34
Low-band recording (*see* Record process, direct)
Low-band reproduce (*see* Reproduce process, direct)
Low-frequency loss, 82–85
Lubricants (*see* Magnetic coating, materials)
Luminance, 214

Magnetic coating, 46–54
  materials, 50, 51
Magnetic disc (*see* Disc-recorder system)
Magnetic domains (*see* Domains, magnetic)
Magnetic flux (ϕ), 17, 18, 63–69
Magnetic flux pattern in tape:
  illustrated, 64, 66
  record process, 63–65
  reproduce process, 65–67
Magnetic force (H), 17, 28–30, 64–70
  effect on domain vectors, 29
  (*See also* Coercivity)
Magnetic heads, 17–19, 31–44
  action of record/reproduce, illustrated, 38
Magnetic heads (*Cont.*):
  analog, 31–38
  audio, 31–38
  cassette, 227
  cores, 32, 33
  erase, 19, 69, 187, 188, 227
  ferrite, 121
  gap separator, 32, 33
  gap size, table, 35
  instrumentation numbering systems, illustrated, 38
  interleaving, 36, 37
  longitudinal, 31–38
    table of characteristics, 35
  point of recording, 35, 38
  point of reproduction, 35, 38
  record, 17, 18
    current, 17
  reproduce, 18
    output, 33, 65, 66
  spring-loaded, 97
  wideband rotary, 39–44
Magnetic induction vs. magnetizing force, illustrated, 68
Magnetic materials (*see* Ferromagnetic materials)
Magnetic oxide particle size, 48–50
  effect on rotary-head design, 39
Magnetic phenomena, 26–30
Magnetic recorders:
  bandwidth, 39
  basic elements of, 16–25
    illustrated, 16
    magnetic heads, 17–19
    magnetic tape, 19, 20
    record amplifier, 21, 22
    reproduce amplifier, 23–25
    tape transport, 20
Magnetic recording:
  advantages of, 14, 15
  development of, 9–14
  first patent for, 12
Magnetic recording systems, applications of, 1–8
    accounting, 7, 8
    aerospace, 2
    filing, 7
    FM/FM, 4
    frequency-division multiplexing, 3, 4
    science and medicine, 4–7

Magnetic recording systems, applications of (*Cont.*):
time-base contraction, 3
time-base expansion, 3
underwater seismology, 7
Magnetic saturation, 29, 30, 67–69
Magnetic tape, 19, 45–61, 253
base film, 19, 52, 53
cellulose acetate, 52, 53
polyester, 52, 53
polyvinyl chloride, 52
coating process, 52–54
computer, 59, 60
design considerations, 45–49
drying process, 55
helical-scan video, 60, 61
instrumentation, 57, 58
longitudinal, stationary-head video, 60, 61
manufacturing processes, 49–56
particle orientation, 47, 54
polishing, 55, 56
rotating-head instrumentation, 60, 61
rotating-head video, 60, 61
signal-to-noise ratio, 47
slitting, 56
testing, 56
transverse video, 60, 61
Magnetic tape coating:
binder material, 50, 51
materials used: chromium dioxide, 48
ferric oxide gamma, 19, 49
particle size, 39, 48–50
processes, 52–54
thickness, 46, 47
Magnetic tape recorder, first commercial, 11
Magnetic vectors, 19, 20, 29, 30, 64
Magnetism, theory of, 26–30
Magnetization curve, 28–30, 67–70
Magnetizing field strength (*see* Magnetic force)
Master equilizer (*see* Equalization, playback, quadruplex video recorder)
Mastering tape (*see* Audio tape parameters)
MDA (motor drive amplifier), 101–103, 190–192
Mechanical splices (*see* Quadruplex video recorder)
Medical electronics, 4–7
Microphone preamplifier, 143, 144
Microwave link, 3
Mix preparation (*see* Magnetic tape, manufacturing processes)
Modulating frequency (f_m), 167, 254
Monitor record head, 44
Monitor reproduce head, 43
"Morning-glory" horn, 11
Motors:
ac split-phase induction, 96, 98
capstan (*see* Capstan motor)
dc printed-circuit, 96, 99, 118
dc stepping, 121, 122
delta-wound, 110
head carriages, 121
head-drum, 108–112
hysteresis-synchronous, 94, 96, 101, 117, 119
induction, 118
pancake (*see* dc printed-circuit *above*)
reel, 131–134
three-phase hysteresis-synchronous, 194

National Association of Broadcasters (NAB) Standards, 88
Negative image, 270
Nickel (*see* Ferromagnetic materials)
Noise, 254, 270
Nonorthogonal time-base error (NTE), 254

Old Standard head-numbering system, 37, 38
Omega wrap format, 113, 114
"Once-around tach" (*see* Television disc-recorder systems)
Open-loop drive system (*see* Longitudinal drive systems)
Operational amplifier, 169, 170
block diagram, 170
Optimizing (*see* Quadruplex heads)
Optimum loop-drive system (*see* Longitudinal drive systems)
Orbiting electrons, 26–28
Overbiasing, 74–75
Oxide categories, 58
Oxide conversion (*see* Magnetic tape, manufacturing processes)

Particle orientation, 47, 54, 255
Pauli, Wolfgang, 27
PCM (pulse-code modulation), 24, 25
Pebble mill, 51

Phase comparator (*see* Capstan servo system)
Phase equalization, 23, 89, 90, 148–151
Phase relationship between record and reproduce processes, 66
ϕ_r (remanent flux), 17, 18, 64–66
Philips cassette, 223, 224
Philips Corporation of Holland, 223
Philips format:
  audio, 224, 225
  digital, 230–233
Phonautograph, 9
Phonograph, 10
Photoelectric sensing, 131
Pilot signal (*see* Instrumentation disc recorder, servo control; Rotary-head transport, head-drum servo)
Pinch-roller assembly, 94–96, 107, 129
  bin-loop and loop transports, 129, 130
  helical-scan transports, 117–118
Plasticizers (*see* Magnetic coating, materials)
Playback equalization, 206–207
Playback station, 233–235
Plenum, 132–134
Polyester (*see* Magnetic tape, base film)
Polyvinyl chloride (*see* Magnetic tape, base film)
Poulsen, Valdemar, 11
Precision plate (*see* Baseplate assembly)
Preroll time, 235
Print-through, 46
Printed Motors Inc., 99, 100
Proc amp (*see* Processor)
Processor (proc amp), 204
  color, 213–214
Professional tape (*see* Audio tape, parameters)
Pulse-code modulation (PCM), 24, 25, 256
  reproduce sequence, illustrated, 24
Pulse-repetition rate (*see* FM modulator)

Quadrature error, 199, 200
  illustrated, 199
Quadruplex head resonance, *F* and *Q* circuit, 205, 206
Quadruplex heads, 183, 184
  optimizing, 203
  simplified drawing, 184
Quadruplex heads (*Cont.*):
  vernier tracking control, 200
Quadruplex video recorder, 183–221
  capstan servo, 195, 196, 214–221
  control-track signal, 195, 196
  dropout accessory, 208–211
  head-drum servo, 191–195, 214–221
  intersync, 214–221
  introduction, 182, 183
  mechanical splices, 196
  modulators, 200–202
  playback process, 204–214
    block diagram, 205
  quadrature error, 199, 200
  record process, 200–204
    block diagram, 201, 202
    EE signal, 203, 204
  record timing, 192, 193, 197
  servo systems, 190–197
    capstan, 190, 195, 196
    head-drum, 190–195
    tape-guide, 197, 198
  tape-speed control, 188–190
  time-base error correction, 211–213
  velocity errors, 198, 199
Quadruplex video tape tracks:
  audio, 185, 188
  cue, 185–188
  guard bands for, 185, 186
  illustrated, 185
Quantum theory, 27, 28

Radioactive tracers, 6
Real and apparent errors, 100
Recirculating charge dispenser, 179
  simplified diagram, 179
Record amplifier, 21, 22, 141–147, 151–156, 161–177
  direct, 21, 141–147, 151–156
  FM, 21, 161–177
  pulse types, 22
Record current, 64, 65
Record equalization, 86–89, 143, 202
Record head (*see* Magnetic heads)
Record losses:
  bias-erasure, 75, 76
  demagnetization, 75
  eddy-current, 76
Record player, 11
Record process, 63–76, 141–147, 151–156 161–177

Record process (*Cont.*):
with ac bias, 71–74
without bias, 70
with dc bias, 71, 72
direct, 63–65, 141–147, 151–156
FM, 161–177
mathematics of, 64, 65
(*See also* Direct recording; FM modulator)
Reel or tape storage systems, 128–136
Reel servo and tensioning systems, 128–136
bin loop and bin-loop adapter, 128–130
continuous loop, 130
mechanical control, 134–136
plenum, 132–134
vacuum and photoelectric sensing, 131
vacuum chambers and vacuum sensing, 131, 132
Remanence magnetism (B_r), 29, 30, 50, 64–71
Remanent flux (ϕ_r), 17, 18, 64–66, 257
vs. magnetizing force, illustrated, 70
Remanent induction (*see* Remanence magnetism)
Reproduce amplifier, 23–25, 141–159, 166–180
direct, 23, 141–151, 157–159
FM, 24, 166–180
pulse types, 24, 25
Reproduce gap width, 33–36, 39
Reproduce head, 18, 19, 65, 66, 76–80
azimuth error, illustrated, 80
gap to wavelength relationship, illustrated, 78
response curve, 77
(*See also* Magnetic heads)
Reproduce losses, 77–85, 159, 160
eddy-current, 81
Faraday's law, 85
gap, 77, 78
head-azimuth, 78–80
head-bump, 83
head-contact-area, 82
long-wavelength, 82–85
low-frequency, 82–85
separation, 82
surface, 82
tape-coating-thickness, 83, 84
Reproduce process, 65, 66, 76–90, 141–151, 157–159, 166–180
direct, 141–151, 157–159
Reproduce process (*Cont.*):
FM, 166–180
(*See also* Direct reproducing; FM demodulator)
Retentivity (*see* Remanence magnetism)
Reverse roll (*see* Magnetic tape coating, processes)
Rotary-head recorder:
instrumentation, 39–44, 107–112
television, 183–221
Rotary-head transports:
capstan drive servo systems, 107–111
block diagram, 111
head-drum servo, 107–112
tape drive systems, 107–112
tracking error, 110
Rotary transformer, 40, 118, 203

Saturation magnetism (B_s), 29, 30, 67–70
Scalloping, 198, 199
Scott, Leon, 9
Sensing slot, 131, 132
Separation loss (*see* Reproduce losses)
Servo systems:
direct, 18.24-kHz and 17-kHz, 101–103
FM, 103
quadruplex video recorders, 190–197
rotary-head transports, 107–112
Single-carrier FM record parameters, 162
Single-sideband FM, 205
6 dB/octave curve, 18, 33–35, 77
Skew, 259
Skewing (*see* Venetian-blind effect)
Slitting (*see* Magnetic tape)
Slow-motion pictures, 124
Smith, Oberlin, 11
Soft switching, 125, 126
Spaced-head disc systems, 126–128
Spacing loss (*see* Reproduce losses)
Spray coating (*see* Magnetic tape coating, processes)
Squareness factor, 57–60, 259
Stabilizers (*see* Magnetic tape coating)
Standard alignment tape, 198, 199
Start of program cue, 234, 235
Start of tape marker, 234, 235
Stationary heads, video systems, 43, 44, 187, 188
Still pictures, 124
Surface loss (*see* Reproduce losses)

Sync generator, 193, 194, 204
Sync separator, 194
Synchronizer (*see* Capstan servo system)

Tach mode (*see* Capstan servo system, 200-kHz)
Tachometer:
  head-drum, 110–112, 189–195
  slotted, 100
Tachometer disc:
  photoetched, 101–105
  quadruplex video recorder, 189–193
Tape backing (*see* Magnetic tape, base film)
Tape flap, 46, 61
  (*See also* Flutter)
Tape-speed errors (*see* Capstan servo system)
Tape-surface polishing, 55, 56
Telegraphone, 12, 13
Television disc-recorder systems, 124, 125
  recording pattern, 125
  servos, 124
  slow-motion and still pictures, 124
Television field, 184
Television frame, 124, 184
Television recorder (*see* Quadruplex video recorder)
Television standards, 182–185, 197
Television terms, glossary of, 263–274
Tension arm (*see* Reel servo and tensioning systems)
Thickness loss, 83, 84
Time-base expansion and contraction, 3, 103, 104
Timing rings (*see* Head-drum timing)
Tip penetration, 41, 42, 185, 197–199
Tip width, 184
Track spacing, 35, 36, 261
Track width, 33–36, 261
Tracking control, 217
Tracking error (*see* Rotary-head transports)
Transducers, 2, 7, 131, 262
  radioactive tracers and, 6, 7
  vacuum sensing, 131, 132
Transport, 20, 92–139
  baseplate, 94
  ideal, 93
  selection of, 92–94
Treble boost, 87–89
Triggering time, 167, 171
Turnaround idler, 96

Uhlenbeck, George E., 27
Uncompensated spinning electrons, 27, 28
Underwater seismology, 7
U.S. Television Standards, 182–185, 197

Vacuum capstan, 96, 97, 238
Vacuum chamber, 97, 131, 236–238
Vacuum sensing, 131
Velocity compensation, 198, 199
Venetian-blind effect, 197, 198
Vertical field, 124
Vertical sync pulse, 124, 184, 191–193
Video cartridge recorder (*see* Cartridge recorder)
Video cassette recorder (*see* Cassette recorder)
Video dropout correction, 209, 210
Video tape cartridge, 233–235
Video tape cassette, 235–238
Video tape recorder, 182–221
  (*See also* Quadruplex video recorder)
Video time-base error (*see* Quadruplex video recorder, time-base error correction)
Videocassette system, 238, 239
Voice-log circuit, 142–144

Wavelength, 18, 78, 262
Weiss, Pierre, 26
Wideband FM record parameters, 162
Wideband group I FM (*see* Record process, FM)
Wideband group II FM (*see* Record process, FM)
Wideband head, 33
Wideband rotary heads, 39–44
WIL band FM (*see* Record process, FM)
Wow, 20, 262

Zero-bias recording, 70
Zero-crossing detector, 173, 178, 179
Zero-loop drive system (*see* Longitudinal drive systems)